Freshman Chemical Engineering

Freshman CHEMICAL ENGINEERING

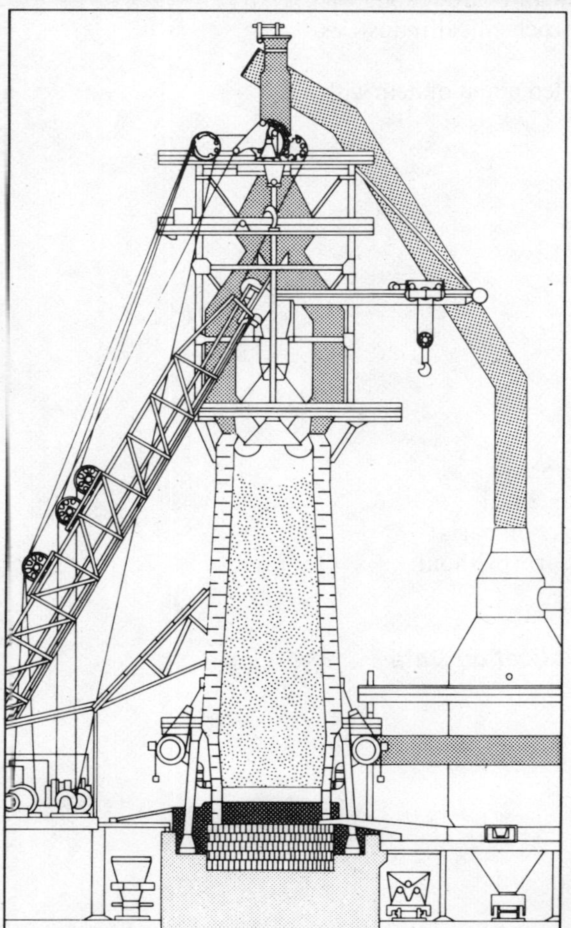

Michael E. Leesley

Department of Chemical Engineering
The University of Texas at Austin

Gulf Publishing Company
Book Division
Houston, London, Paris, Tokyo

for Barbara

Acknowledgments

This is the best place for me to thank two people whose help with this book was invaluable.

The first, my wife Barbara, typed the first draft and the final manuscript. The second, Mr. T.M. Newsom, drew upon his wide experience of the chemical and petrochemical industries to help with the tables in Appendix 1.

To these two friends I offer my acknowledgment of help with sincere gratitude.

Freshman Chemical Engineering

Library of Congress Cataloging in Publication Data

Leesley, Michael E 1942-
Freshman chemical engineering.

Includes bibliographies and index.
1. Chemical engineering. I. Title.
TP155.L37 660.2 79-18418
ISBN 0-87201-106-2

Contents

Author's Introduction

If you have bought this book, you are probably a college freshman interested in chemical engineering. I am writing specifically for the high school graduate who wishes to know more about this one topic.

By way of introduction, I was brought up in Sheffield, England, the South Yorkshire town famous for its steel and cutlery. As a child, I was fascinated by the nighttime flares over the steelworks. I liked to see the red-hot ingots being rolled out from the huge furnace shops and taken into cavernous mills, whose presence was indicated by a distant rhythmic thundering that was distinctly comforting on wintry nights. I will never forget the strong, slender girders or unending coils of strip or wire, and the power that I felt must have been employed to mold such shapes, especially when starting with a large and uncooperative-looking steel ingot.

The big drawback to visiting the steelworks was the depressing effect of the terrible filth that seemed to go hand-in-hand with the industry. Most level surfaces had a brown coating on them; cars rusted quickly and were eyesores after only a year of corrosion in the city's atmosphere unless polished zealously and weekly. A rank sulfurous odor was everywhere and the sky was dark with smoke and steam.

Later, as an undergraduate in the Fuel Technology and Chemical Engineering Department at Sheffield University, I learned that the battle to clean up our environment was on and that industry would never make those mistakes again. Indeed, Sheffield is now the cleanest industrial city in Europe, to the amazement of everyone save those who know how it was done. There is a chapter in this book on the technology that makes steel manufacture possible because it is an industry which employs chemical engineers.

For each industry described in this text, the unit operations which are used are discussed, and the basic chemical engineering principles are referred to in simple terms. For now, it is not necessary to understand the mathematics which describe these principles. Most scientific phenomena can be explained either by words or equations. In both cases, the levels of complexity can range from high to low. I have used words to explain the concepts, principles, and applications of chemical engineering. I have included simple diagrams with descriptive equations in some cases, but they merely supplement the verbal descriptions rather than form an essential part of them.

I have written deliberately for the introductory-level reader. If you get bored, it means one of us is not trying. So, if you get bored, go back and decide whether you are not concentrating hard enough or whether I have obscured the point. If the latter, then write a letter to me and suggest a better way of expressing it. The exercise will help us both, and you will certainly benefit by exercising your powers of articulation.

If you have trouble understanding something, try again. Persevere until it is clear. There is nothing that's beyond you. If you have particular difficulty with a new concept, see your professor and ask about it. Your professor can explain it again in several ways if necessary until you are sure you grasp it. By the way, when you have understood it, go back and find out exactly why *my* words were not clear enough. Again, I invite you to write to me and suggest a better way, for you will have thoroughly learned the concept if you can explain it in your own words.

At this point in your education you should write as much as possible to prepare for the shock of discovering that, except for a few privileged engineers, the profession mainly involves using *writing* instruments, *not* drawing instruments.

By the end of this text, you will know a lot more about the various industries in which you will practice your profession. You will also know where basic chemical engineering principles are employed, and you will have a feel for how they are related and how they affect your endeavors. This will smooth the transition into those classes where you will learn how the principles work chemically and mathematically and how they can be used to provide you with scientific information which your experience will turn into practical know-how. You'll realize that engineering is essentially the application, with the object of making money, of the laws of science and the laws of man. The former are definitions of nature's ways and the latter are design manuals, the engineering codes of practice and regulations imposed upon industry by government agencies.

There are essay assignments at the end of the book. Some are about the various industries described in the text and some are about the basic chemical engineering principles mentioned in each chapter. If you read carefully, you will understand what is meant by each term in a wide variety of situations. You will not be assigned the latter type of essays until you have read the entire text. I allow my students four weeks to read the entire text, and I use this time to introduce them to computers.

You can hardly practice chemical engineering today without a working knowledge of FORTRAN. I have included sufficient FORTRAN to enable you to solve the FORTRAN assignments. In class you will also be provided with some introduction to the assignments. There are more advanced FORTRAN facilities which I have omitted, but if you grasp the FORTRAN chapter and become confident in using FORTRAN, you will easily learn any further FORTRAN refinements as you need them. In any case, advanced FORTRAN techniques will be needed only if you will be writing innovative software, and by then you will easily pick them up.

Start reading right away. I hope you will find chemical engineering a fascinating story. Your professors will probably speak in greater depth on some of the basic chemical engineering principles upon which the essay questions are based. This way, you get two versions—mine and your professor's. If there's something you don't understand, your professor won't mind talking to you. If you go and see your professor, he or she will learn the specific topics that students find most difficult and will develop new methods of explaining them.

Write the essays carefully. They will be graded on how much you have learned about engineering and how well you can express the information in your own words. Writing is very important in chemical engineering, and if you intend to take a pride in engineering, it should be in your writing every bit as much as in a drawing, a computer program, or a worked problem.

Don't be impatient at learning to write well. Improvement will occur automatically if you read the work of others critically and get plenty of practice yourself. Furthermore, you will begin to develop a crisp, personal style and find a joy in self-expression as you realize how essential it is that your written work be concise and convincing. Ambiguity could lead to costly delays or disasters. Misspellings, grammatical errors, awkward sentence construction, or the use of vague, weak words and cliches are sloppy. Such faults indicate a careless writer who has little interest or pride in turning out good work—a poor advertisement for your abilities! But worse, your reader's attention may wander. Your reader may lose concentration and become irritated with you. Obviously, if you have taken the trouble to write something, you want it to be taken seriously. One day, you will be writing proposals for project funding, or explaining how your product works, or negotiating your salary. In none of these cases do you want your audience working against you. You will have to learn from your own mistakes and those you notice in the written work of others.

There are lots of ways to learn to write well, but there is one way in which you are your own best teacher. As you write, carefully plan at each level, from individual words and sentences up to paragraphs.

Choose the exact word you need. English has the richest vocabulary of any language on earth (though I expect that statement would be denied in Paris). There is hardly a word without dozens of synonyms. For example, look in *Roget's Thesaurus* under "plan" and see how various shades of meaning are represented in the spectrum of similar words which fringe it. Incidentally, never be afraid to use a thesaurus to increase your word power. You will also improve your word power the more often you read. Shortly, you will read many technical books, and have an opportunity to examine how certain words have been used to describe complex ideas. Sometimes you will think of other words that would more crisply convey the shade of meaning intended.

At the sentence level, if you have any doubt about its length, remember you will sometimes get the point over more clearly if you write sentences which are short and to the point. Sharp, terse sentences are appreciated if nicely balanced with long, multiclause sentences. They give your reader a rest. You will develop your own style of sentence construction as your writing improves and, again, reading will help. Good writing is more likely to be desired and more easily achieved if it has been recognized and admired in the works of others.

The largest building blocks in reports, essays, and chapters are their paragraphs. Use the first paragraph to introduce your ideas. An old motto advises how to transfer information from teacher to learner: "Tell 'em what you're going to tell 'em, tell 'em, and tell 'em what you told 'em." In other words, begin your essay with an introduction which tells your reader what is about to be discovered and brings him up-to-date on knowledge or background information needed to understand your message.

When the scene is set, you can then "tell 'em" by progressing to a second paragraph which details some important aspect of the topic. When you are certain you have said enough to cover this particular aspect, then draw the reader's attention, in the third paragraph, to another important aspect of the topic. This pattern of developing each aspect of a theme in separate paragraphs both helps you to organize your essay and makes it easier for your reader to build up a structured set of ideas.

When you have finished your essay, tell your reader the importance of all these ideas when they are strung together and explain and emphasize this significance. Use a concluding paragraph to do this, and at the same time, stress again the most important points of your message.

Your final paragraph, then, should summarize, for your readers' benefit, the information you have set down. In other words, "tell 'em what you told 'em." Don't be repetitive, though. It is no use boring readers by repeating parts of earlier paragraphs word for word. Rather, it should be the kind of summary your readers would give if they were asked to summarize the material for someone else.

How you write your essays will be decided by you and will depend on your personal style. Remember, however, that unless you are a recluse or a hermit writing stories which no one else will read, it is to your advantage to write so that your reader's interest is maintained until the end of your message.

As you write more and more, your individual style will develop, if you let it. When it comes to writing an interesting message, your only limitations are self-created. There is little point in writing a technical report to your superiors which makes difficult and tedious reading for them. They may never even bother to read it! If you are proud of the work as you describe it, then be proud of the means by which you describe it. After all, the words you use to describe your creation are just as much a product of your own inventiveness as the creation itself. Just as an expert takes pride in applying his skill to carry out a craft, so should that professional take similar pride in writing about that craft.

Finally, here are some hints on developing a good writing style. Make sure you thoroughly understand the topic before you begin writing. Then you will experience a pride in being able to tell about your newfound knowledge and its significance. Before actually beginning to write, fit the information you have into the context of your previous experience, marshall your facts, and then, with perhaps a few sketchy notes, begin the first draft. Begin with a clear purpose in mind. Perhaps this will be to convince your reader of the importance of the information. Or it may be to convince a professor that you have thoroughly understood the topic. Or a boss that you deserve a raise.

Learn *now* how to do it well. In the future your purpose may be to explain the workings of a complex process. An ambiguity could be disastrous. How many times have you found it difficult to understand a series of instructions for even simple, domestic appliances? Next time that happens, after you have understood the instructions, take a moment to see how you could have written them more clearly. If you do this, you will notice that changing even one word can often make a dramatic reduction in the time needed to understand a sentence.

Even if you approach your essay with your information clearly structured and your objectives clear, you may still run into problems. If you are having trouble with a sentence or paragraph, try reading your sentence aloud as though you were making a speech. If it sounds clumsy, rewrite it. If it sounds pompous, patronizing, sarcastic, rhetorical, or has any other emotional appeal, make sure you can truly justify such a stance. Each sentence has a job to do. Build it carefully to do the job as briefly and clearly as possible.

You can sometimes find improvements by dividing the pieces of the sentence into two categories—those you like and those you don't. Usually, you'll be trying to change parts you don't like in order to achieve an improvement. When this strategy leads to a deadlock, try changing the parts you like. It may be impossible to fit in that beautifully lucid phrase or evocative clause you liked without forcing the rest of the sentence to be clumsy. In that case you must rewrite it. Try to remain flexible.

Sometimes, you will think of a particularly good metaphor or simile which seems so graphical and apt that you are certain it will instantly clear up an obscure point or add emphasis to some vital message by illuminating it. However, you may be reluctant to use the phrase for fear of being accused of using flowery language inappropriate to a technical document. My advice is to leave it in. It would be a pedantic reader indeed that would ridicule your use of an elegantly phrased sentence aimed at facilitating understanding. Use your own judgment. Try to imagine what you would think if you came across the phrase in someone else's work. You'll spot overindulgence and remove it immediately, e.g., describe a flare stack using phrases like those Keats used in his odes and you'll lose your reader in an avalanche of laughter.

Writing well about engineering is as much an art as engineering itself. Indeed, the task of any engineering design, be it for a heat exchanger or a refinery, is akin to constructing an essay. There are an infinite number of totally acceptable solutions to a design problem and an infinite number which vary from almost acceptable to totally unacceptable. The real difference between a professional and an amateur is the length of time needed to make the necessary decisions leading to a successful product, whether it be a profitable refinery or an elegantly precise sentence.

There is another reason why writing is like engineering. Sometimes, designers develop a solution which appears less economic than it should be or which just "feels" clumsy. A possible tactic in the design office is the one described above for use in untangling an awkward sentence by dividing it into favored and nonfavored pieces.

The difficulty may be in a favorite and elegant part of the design. When discarded, however reluctantly, and replaced, many difficulties may suddenly be solved.

There have been many breakthroughs made using this trick. For example, a famous British automobile engineer was given the task of designing a car which would be economical, light, highly maneuverable, easy to park, and yet remain comfortable and attractive in appearance. He drew many different body shapes on tracing paper and placed them over the silhouettes of the company's various standard engine and transmission systems. Several possibilities suggested themselves, but he felt he could do better. Finally, he threw away the engine and transmission silhouettes and worked backwards. Unconstrained, he developed a body shape to fit around four human shapes. Then he demanded a completely new engine and transmission system. The engine designers came up with a transverse-mounted engine with a built-in front-wheel drive. The resulting car (the "Mini") was a huge success and millions were sold. The innovation eventually spawned a whole series of models with ascending engine size and increased comfort and luxury.

Although all breakthroughs are innovative, few have as much impact. As we go through the great industries dependent upon chemical engineering, I will point out some of the more remarkable breakthroughs and describe the situation that led to their occurrence. There will be millions of other breakthroughs in the future. I expect you will be responsible for some of them.

Further Reading

Barrass, R. *Scientists Must Write.* London: Chapman and Hall, 1978.

Beakley, G.E., H.W. Leach, and J.K. Hedrick. *Engineering: An Introduction to a Creative Profession.* New York: Macmillan, 1977.

Gajda, W.J., and W.E. Biles. *Engineering: Modeling and Computation.* Boston: Houghton Mifflin, 1978.

Guth, H.P. *Concise English Handbook.* Belmont, CA: Wadsworth, 1977.

Herman, W. *The Portable English Handbook.* New York: Holt, Rinehart and Winston, 1978.

Hodges, J.C., and M.E. Whitten. *Harbrace College Handbook.* New York: Harcourt Brace Jovanovich, 1978.

Morriss, G.E. *Engineering: A Decision-Making Process.* Boston: Houghton Mifflin, 1977.

Strunk, W. Jr., and E.B. White. *Elements of Style.* New York: Macmillan, 1972.

Watkins, F.C., W.B. Dillingham, and E.T. Martin. *Practical English Handbook.* Boston: Houghton Mifflin, 1978.

Chapter 1

The Chemical Engineering Industry and the Structure of this Text

This text is not intended to give an exhaustive description of every industry which employs chemical engineers. Rather, it is intended to select sufficient industries to demonstrate the application of basic chemical engineering principles to industrial practice. There are many books available which describe the chemical industries in detail, and two of them are named at the end of this chapter. I have selected representative industries and described them in a particular order so that I can introduce, one by one, the basic operations of chemical engineering. I felt it was unsatisfactory for you to be told of an operation that wouldn't be explained until later in the text. Obviously, this did have to occur occasionally, but I have minimized it.

Below are some general headings of the chemical engineering industries categorized for the sake of this text.

Metallic Industries

The manufacture of iron, steel, aluminum, copper, tin, lead, and titanium is similar in many ways, but there are differences. Since the iron and steel industries contain many of the metal-working operations and because they produce most of the construction materials used in chemical plants, I have chosen it to be the subject of Chapter 3.

Power and Steam Generation

The principles are similar whether considering a utility boiler operation producing electricity for a city, a marine boiler, or a boiler plant in a refinery. Chapter 4 is concerned with power and steam generation because these two utilities are the driving forces of the modern chemical plant. The descriptions in Chapter 4 set the scene for your later introduction to thermodynamics.

The Chemical Industries

The oil refining industry has been chosen for Chapter 5 because it produces such a high proportion of feed stocks for other industries. The fuel and lubrication oils industries were chosen for Chapter 6 because of their importance both to the consumer (gasoline) and the chemical engineering designer who needs fuel for a plant. For Chapter 7, I needed a chemical-producing industry, and I chose the petrochemical industry rather than the fertilizer, dye stuffs, food additives, sulfuric acid, or other industries simply because we had already seen in Chapter 5 how feed stocks for the petrochemical industry are prepared. Also, the petrochemical industry is the largest and most variable of the chemical industries, and examples are more easily chosen.

Miscellaneous Industries

Included here are air liquefaction, glass manufacture, brewing, sugar refining, grain distillation, food processing, paints, explosives, dye stuffs, gases, inorganic salts, and many other industries. Of these, sugar manufacture is a convenient way to introduce the unit operation known as "evaporation," and the brewing industry was chosen to demonstrate that biological processes can be made to conform to chemical engineering principles and methods.

Plant Construction and Environmental Protection

There are two industries which are important to all others: waste treatment (or pollution control) and plant construction. These industries employ a high proportion of chemical engineering graduates, and it is appropriate that you have a glimpse of the role these industries perform in our profession. In the chapter on the construction industry, I have included a description of the design methods employed and indicated how new computer-aided design (or CAD) techniques are being used with astonishing success.

Assignments

To tie the book together I have presented a tabular summary in Appendix I which will help you with your essays. If you are writing an essay on distillation, for instance, you could draw examples from this text, and in addition, using Appendix I, you could do some library research and find uses of this operation in other industries.

Finally, a list of essay assignments which your professor may use for homework and testing purposes appears at the end of the text. However, it is likely your professor will lecture on topics not covered by this book but which are important in *your* geographical location, and accordingly, your professor will almost certainly create other assignments.

Incidentally, in compiling the information you will need to develop the essays into a form that will give you a passing grade, you will be exposed sufficiently to chemical engineering to be able to decide whether or not it is the career you want to pursue. In the unlikely event that you decide that chemical engineering is not for you, your professor will help you decide on a major more suited to your talents and temperament.

Further Reading

Cook, G.A. *Survey of Modern Industrial Chemistry*. Ann Arbor: Ann Arbor Science, 1975.

Shreve, R.N. *Chemical Process Industries*, 4th Edition. New York: McGraw-Hill, 1977.

Chapter 2

FORTRAN

In this chapter you will learn how to communicate with computers so you can instruct those arithmetical robots to carry out your calculations. Remember that the computer cannot think intelligently. It merely has the ability to obey simple instructions. Such an instruction could be:

Add 371.2 and 493.8

However, you could do that mentally or with your pocket calculator with little trouble. A computer must be able to do more. It does offer more—lots more. A computer can store thousands of instructions that describe the manipulation of data values to give the desired answer.

To understand the basic philosophy of giving a program of instructions or programming a computer, it may help to compare the computer to an imaginary robot which can be programmed to carry out household duties. Suppose we are able to build into the robot a tape recorder which can tape a sequence of verbal commands. Further suppose, as the tape runs, the robot will obey each instruction as it passes across the tape head. Thus, when programming the robot to walk, we could have a series of instructions as follows:

1. Transfer weight to right foot
2. Move left foot forward
3. Transfer weight to left foot
4. Move right foot forward
5. Transfer weight to right foot
6. Move left foot forward
7. Etc.

We would also have to include other instructions between steps in order to avoid accidents. For example:

1. Transfer weight to right foot
2. Move left foot forward
3. Transfer weight to left foot
4. Move right foot forward
5. Transfer weight to right foot
6. If not within four feet of an obstacle rewind this tape to Instruction #2
7. Etc.

You can see that if the robot never comes up against an obstacle, these six instructions alone would cause the robot to keep walking until it ran out of power or failed mechanically. However, you can also see how we could add other instructions. Since, in our example, the tape recorder in the robot will never run past Instruction #6 unless the robot *does* come within four feet of an obstruction, we can assume that if the tape recorder ever reaches Instruction #7, then the robot is standing in front of an obstruction. We could continue with our program thus:

7. If obstruction is not a door, run this tape forward to Instruction #10
8. Decide if door is open or closed
9. If door is closed, open it and rewind to Instruction #2
10. Decide which way to avoid obstacle
11. Swivel feet in direction decided
12. Rewind to Instruction #2
13. Etc.

Notice how we are aided by the fact that the computer reads these instructions sequentially. For instance, Instruction #6 says that if the obstruction is not a door, go to Instruction #10. If it *is* a door, however, the tape recorder will *not* be wound to Instruction #10 but will go on to Instruction #8 which tells how to cope with the door. The importance of this sequential effect upon your program of instructions to a real computer will soon become apparent.

I am sure you can see by now that with the ability to write a program of instructions to a robot, we can make it do any operation which we can break down into simple commands such as those above. The ability to move up and down the list of instructions, by rewinding and winding the tape, is also vital to us because it saves time when writing instructions. This advantage is made clear in the first six instructions above. Also, the ability to give conditional orders (such as Instruction #9 above) is extremely useful.

Before we leave this simple household robot analogy, let us ponder the problem of programming it to do simple tasks. We would soon find it convenient to develop new instructions to save our time when programming—in other words, a kind of shorthand. The first one would probably be a "repeater" command. For instance, if we wished to program the robot to polish shoes, the instruction list would look something like this:

106. Hold shoe in left hand
107. Hold shoe brush in right hand
108. Place left hand side of shoe brush on front of shoe
109. Do all instructions down to #111 fifty times
110. Move brush to left
111. Move brush to right
112. Swivel shoe 10 degrees to the left
113. Unless brush is touching the shoe's sole, go to #109
114. Etc.

The robot would not begin to obey Instruction #112 until it had done Instructions #110 and #111 for the prescribed 50 times. Furthermore, having done #110 and #111 for the fiftieth time, it will obey #112 and #113 when it will return to #109 for another 50 strokes with the shoe in a new position. Because of the conditional instruction in #113, it will keep returning to #109 to give 50 strokes with the brush until the shoe is turned so far over that the sole is exposed. Then, and only then, will it go to Instruction #114.

Another useful element would be the ability to have a second tape recorder with a special library of operations which the robot must perform frequently. Then, whenever we wish some particular activity to be carried out, we would just tell the robot to find a certain piece of subprogram, or subroutine, stored in the library in the second tape recorder and obey the instructions in it. The last instruction in the subroutine would be one returning the robot's attention to the main program so that it could continue with its chores. For instance, this would enable us to issue the commands:

136. Load dishwasher using Library Subroutine #273
137. Switch dishwasher on
138. Etc.

You can see how the robot would reach Instruction #136, and then, switching on the second tape recorder, it would run through the tape to find the dishwasher-loading routine called #273 and obey the instructions therein before returning to Instruction #137 in the main program on its first tape recorder.

FORTRAN—A Language of Numbers

We no longer need to work with our analogy. The few simple concepts conjectured for the housemaid robot also form the basis of the many instruction codes which are available for programming computers.

Because the computer is essentially an electronic device, the instructions will have to be given in some kind of numerical code. Early programmers, faced with having to memorize hundreds of numbered commands and four-digit storage addresses, soon learned to use the power of the computer to translate simple English commands and formulae into the numerical computer language. Translators, or *compilers*, were developed. Improvements were suggested and incorporated until a standard computer language (FORTRAN) was developed. FORTRAN allows us to use terse English commands to instruct a computer to carry out manipulations of numbers at a very complex level.

Remember, though, that the computer has no mind of its own. It cannot translate a word that you misspell, and it cannot guess at the meaning of an unrecognized command. Such mistakes are fatal to the chances of the program being satisfactorily translated or compiled into the computer's own language. For this reason, such an error is termed a *fatal error*. To guard against fatal errors, the programmers who write compilers also write a manual which sets down a way of doing things which the computer is guaranteed to recognize. Use the manual for your computer as a guide. It is a miniature book of grammar rules which your computer will stick to most rigidly.

Let us now look at some of the things we can tell our computer to do. In the following examples, the language will be FORTRAN IV in all cases. There may be some slight differences between the FORTRAN used below and the FORTRAN which is available on your computer because there is no standard FORTRAN IV observed by all computer manufacturers. Thus, "dialects" are possible and occur frequently. Also, your computer may have some simplifications or added features not used below. These will have been added by helpful programmers at your computer center. However, my advice is to understand thoroughly the FORTRAN used in this book because I have used a common dialect of FORTRAN which is available on most machines. You can later check any small differences which may occur. I assure you that you will not have to relearn FORTRAN, but you may need to make some slight modifications to your knowledge.

Because our computer is going to work with numbers, we need to provide a way of storing numbers and referring to them later. The simplest way is to refer to each number by a name, as in algebra, except that we can use up to six characters instead of one letter. It is perhaps easiest to visualize the storage of numbers by imagining a giant electronic set of pigeon holes inside the computer. Each pigeon hole is called a *storage register* and is labeled by you whenever you first name the value you will store in it. For example:

```
1      X=10.0

2      Y=23.9
```

When the computer obeys Instruction #1, it creates a label for one of its storage registers and stores in it the value 10.0.

Let us continue with a simple program to add two numbers together.

```
1      X=10.0

2      Y=23.9

3      C=X+Y
```

That is all there is to it. The answer to the addition will be stored in a register which will be labeled "C" for this purpose. It may help you not to read the equal sign as "is equal to" but to read it as "becomes." The value in C becomes the value in X added to the value in Y. Note that we can also include on the right-hand side of the equal sign the old value of C. Thus,

```
4      C=C/3.72
```

In this case, the computer looks in the register called C, finds the value, divides that value by 3.72, and then places the new value into C. The old value of C is lost forever.

There is another way of assigning values into named registers. It is called the "READ" statement. This is useful when the values we wish to manipulate in our program are not known in advance (which is usually the case).

I should mention before going further that you will write the program on paper, transfer it to punched cards, and then feed the cards to your computer together with the data values, also on cards, that you know the program will need. How you do this depends on your computer system, and each one is different.

Consider these changes to the program:

```
1      READ 2,X,Y

2      FORMAT(2F10.3)

3      C=X+Y

4      PRINT 5,C
```

There are several new concepts here, so ignore the "FORMAT" instruction for a moment. Remember that the program is to be stored and run later. The READ instruction (#1) tells the computer in shorthand that when this program is run, it will be provided with two values which it will store in registers called X and Y. You will learn later how to do the providing. The addition instruction (#3) is unchanged. The instruction "PRINT 5, C" causes the computer to give the value of C back to you. You will learn how later. The FORMAT instruction used above tells the computer that the two values X and Y will be read from the same punched card and will be made up of 10 characters, three of which will be to the right of the decimal point. Table 2-1 contains four examples of numbers in this format (F10.3) and examples of other formats also.

It would appear that only the first of the examples of F10.3 in Table 2-1 actually meets the specifications outlined. However, you can see that the other three numbers are the exact same format if you count the suppressed spaces which the computer will assume to be present. The incorrect examples in parentheses with NOT before them could *not* be read incorrectly with a format of F10.3. Can you see why not? (Don't forget to count the decimal point in working out the total length.) You will need to remember the significance of these FORMAT instructions when learning how to provide data to your programs. When the computer prints out the value of C, it is programmed to use the specified format, but it will suppress unnecessary spaces or zeros.

The simple program developed above is for carrying out a task better done by hand. Suppose, however, that we have to add 10,000 values together! This would be a lengthy boring task with a high risk of error. Imagine trying to tell the computer to read all those values. Whatever would we call them all? To write a computer program to do the job, a more efficient way of naming and manipulating that many registers is needed. Simply choosing names like X, Y, and Z arbitrarily would mean a longer time typing the program than it would take to add up the numbers. FORTRAN provides a facility for "subscripting" variables which makes it easier to give names to all those values. Here it is used to feed in 10,000 numbers and to find their sum.

```
1      SUM=0.0

2      DO 5 I=1,10000,1

3      READ 4,A(I)

4      FORMAT(F10.3)

5      SUM=SUM+A(I)

6      PRINT 7,SUM

7      FORMAT(F12.3)

8      END
```

Table 2-1
Examples of FORMAT Instructions

FORMAT	Examples	Use
F10.3	654321.123 10.641 1.1 47.82 (Not 7654321.1) (Not 0.1234)	To define the format for READ or PRINT instructions when dealing with REAL values (numbers with a point). The first character, F, indicates that a REAL value is involved. The next integer is the total number of characters in the value (including the decimal point) and the second integer indicates the number of characters which follow the decimal point. Note that zeros are suppressed.
F14.6	1.1 1234567.666666 0.000001 (Not 12345678.123456) (Not 0.0000004)	
F4.1	92.0 1.2 77.7 10.8 (Not 123.1) (Not 0.0007)	
F5.0	9991. 4. 1234. 271. (Not 123.1) (Not 44444.)	
I2	13 99 4 27 (Not 101) (Not 10042)	To define the format for READ or PRINT instructions when dealing with a whole number or INTEGER value. The first character, I, indicates an INTEGER value. The second tells the number of digits in the integer.

(Table 2-1 continued on next page)

Table 2-1 (continued)
Examples of FORMAT Instructions

FORMAT	Examples	Use
I1	4 3 8 0 (Not 17) (Not 114)	
I6	123456 7 666666 9871 (Not 1234567) (Not 88888888)	
E8.4	1.1234E2 7.771E10 1.1E-3 (Not 1.1234E-11) (Not 1.11111E2)	To define the format of a REAL number written in scientific notation. The first character, E, signifies that the value will be written in scientific notation.
E10.4	1.1E06 1.1E6 3.3333E-13 1.0E10 (Not 1.678678E-12) (Not 4936142.9123E4)	

Now, one by one, let us examine the new ideas introduced here. First, the "DO" instruction is exactly the same as the repeat device used for the domestic robot. This one says to the computer, "Carry out every instruction from here down to *and including* Instruction #5 for values of I = 1 up to 10000 in steps of 1." In other words, do the instructions with I set to 1, then 2, then 3, and so on, up to I = 10000. You will see the significance of the I in a moment. After carrying out Instructions #3, #4, and #5 ten thousand times, the computer will then, and only then, obey Instruction #6—the PRINT instruction. Note that I is always an integer in the example.

It is worthwhile at this point to explain the difference between INTEGER and REAL, or noninteger, values (examples of each category are given in Table 2-2). The electronic register system described previously is a mechanism for storing integer values. This is ideal for integers or whole numbers, but the size of the register limits the magnitude of values that can be stored. For instance, some computers store decimal values between plus and minus 8388608. If you would like to know why, ask your professor. In order to use such a register to store an integer, we assign to it a name which implies that the value to be stored in it later on will always be an integer, and no matter how many times the actual value is changed by the program, its status will always be INTEGER. In order that the FORTRAN compiler will recognize this intention, there is a simple rule. Making the first letter of the register name belong to the group of letters I to N inclusive indicates that the value the register will hold shall be an INTEGER (see Table 2-2).

To store REAL values, i.e., values where the position of a decimal point is not in the extreme right position and also to enable us to use much higher values than 8388608, the FORTRAN compiler uses two registers. The first contains an integer. The second contains another integer which is a scientific notation exponent. Thus, the REAL value stored is the integer in the first register times 10 raised to the power of the value stored in the second register. Fortunately, all this organization of numbers, registers, and values is done automatically by the computer if we remember to indicate REAL and INTEGER by their names. Because of their "whole number" quality, INTEGERs are used for DO LOOP counters (which is the purpose that the INTEGER "I" served in the last example), subscripts, indices, and any other variables which only have whole number values, such as dates, times, and reference numbers. REAL names are given to variables which may have values which are not whole numbers, i.e., which have a decimal part. This description includes all scientific data and most engineering data. Because many FORTRAN compilers prohibit the use of arithmetic instructions which include both types (except where INTEGERs are used as subscripts or indices), we will avoid "mixed mode arithmetic" in the examples in this chapter.

Now let us go back to the example. You can see how the first move is to label a variable with the name SUM. This is quite acceptable for a REAL value, and in fact, we can use names up to six characters long. There are some examples of variable names in Table 2-3. SUM is a real variable (for it does not begin with a letter in the range I–N), and it is assigned the value zero (written 0.0). The decimal point must be written even though it is apparently unneeded. This is because a REAL variable must always have its decimal point. Next, Instructions #2, #3, #4, and #5 are carried out 10,000 times. The first time, I has the value 1. Thus, the computer "reads" the first value in the deck of data cards and assigns it to a register called A(1), which is the first element in an ARRAY or list of numbers called A. You will learn more about ARRAYs later. This one is 10,000 elements long and consists of A(1), A(2), A(3), and so on, to A(10000). The program calculates a *new* value for SUM by adding its old value (0.0) to the value of A(1). The old value of SUM is lost forever. Upon reaching Instruction #5, control is automatically transferred back to Instruction #2, and the value of I is increased to 2. Again, the program reads in a value and assigns it to the variable called A(2), and again, a new value is calculated for SUM by adding its last and current value (A(1)) to the value of A(2). Again, a jump back to Instruction #2 raises the value of I by 1 to 3. You can see

Table 2-2
Examples of Variable Types

Type	Example	Notes
REAL	1.234567 103.783 0.001 0.0 6.0 99999.9999 111.0 6.	If the REAL number is a whole number but you wish it to be classified as a REAL, give it a decimal point. The zero is not important, but the decimal point is.
INTEGER	109 47 0 10 400000	Absence of a decimal point defines an INTEGER.

Table 2-3
Examples of Variable Names

Variable Type	Typical Names	Notes
REAL	X Y Z XYZ XX DIAM TEMP VALUE A XTWO XONE X1 X2 CP3O R2D2 (Not LENGTH unless the over-ride rule is used--see later)	The appearance of number in these names is acceptable but the number has no numerical significance used in this way. It is merely used as a character as part of a name.
INTEGER	I J K L M N II LONG IAGE NUMBER NO NYEARS NTRAYS NTRAY NCOUNT KOUNT (Not COUNT unless the over-ride rule is used--see later)	The above comment also applies.

Table 2-3 (continued)
Examples of Variable Names

Variable Type	Typical Names	Notes
ARRAY (which implies a REAL ARRAY)	A(10) X(1) DIAM(12) TEMP(100,7) PO2(3,10,4) TRAY(N) BOOK(KOUNT) X(I) (Not A(X)) (Not X(1.7))	Each dimension of the array is separated by a comma. If the subscript is unknown, its representative symbol must be INTEGER.
INTEGER ARRAY	I(10) NAME(4,4) NODDY(NOO) MEL(10) LBOW(I) NKEL(J) NUMBER(10000) (Not X(X) Not Y(10) unless the override rule is used--see later)	

that after 10,000 times, the value of SUM will be A(1) + A(2) + A(3) + A(4) + . . . A(10000).

When the value of I finally reaches 10,000, control is transferred back to #2, and I is increased to 10001. The computer then knows that it has overshot, and it will then go to the next instruction after the DO LOOP, which in our case is Instruction #6. It causes the answer to be printed out. The last new feature in the example was Instruction #8—"END." Its purpose is to tell the computer that this is the end of the program.

The intention of using the previous program was to teach you the rule about subscripted variables and the DO LOOP instruction. You will find more examples of both of these facilities in Table 2-4 and some ARRAY elements in Table 2-3. There is a more efficient way of writing a program to add 10,000 values together. We have no need to *store* the 10,000 values; we merely want to read in one value at a time and add it into the sum as the following example shows:

```
1     SUM=0.0

2     DO 5 I=1,10000

3     READ 4,A

4     FORMAT(F10.3)

5     SUM=SUM+A

6     PRINT 7,SUM

7     FORMAT(F12.3)

8     END
```

You can see that the third piece of information from the DO instruction is missing. When the FORTRAN compiler sees that the last comma is missing from its normal place,

Tabe 2-4
Examples of DO Instructions

FORTRAN Code	Notes

`DO 7 I=1,10,1`
`AREA(I)=4.0*PI*R(I)*R(I)`
`7 VOL(I)=(4.0*PI*R(I)**3)/3.0`
	The last "1" in the DO instruction is unnecessary but it can be left in.

`DO 213 KOUNT=1,1000`
`READ 6, A(KOUNT)`
`6 FORMAT(F10.5)`
`213 CONTINUE`
	The "1" has been omitted here.

`DO 7191 J=2,10,2` `7191 A(J)=Y(J-1)+4.0*Y(J)+Y(J+1)`	Here is an example of complex subscripts which have to be evaluated before use.

`DO 9 I=1,10`
`DO 9 J=1,25`
`9 TEMP(I,J)=0.0`
	Two "nested" DO LOOPS

it assumes the step is 1. If we need to have integer jumps of two or more, then we must say so. The examples in Table 2-4 will make this clear. The other improvement made in the program is to read each value into a temporary variable called A. After adding A to SUM, a new value replaces the first, and the process is repeated 10,000 times. Of course, this new improvement is only practical if there is no need to store all the 10,000 values for other manipulations later on in the program.

Next, imagine that you do not know, at the time of writing the program, just how many data items will have to be added together. Such a difficulty can be resolved because you will have the opportunity to tell the computer the number of values in your list at the time you actually run the program (i.e., when at last you know). In the following example, the number of values to be read in is N:

```
      READ 1,N
1     FORMAT(I6)
      M=1
      SUM=0.0
2     READ 3,A(M)
3     FORMAT(F10.3)
      SUM=SUM+A(M)
      M=M+1
      IF (M.LE.N) GOTO 2
      PRINT 4,SUM
4     FORMAT(F12.3)
      END
```

Several new points here. First, we have now adopted the more common FORTRAN practice of labeling only those lines which will be referred to later. Although the above example has labels with numbers in ascending numerical order, this is not necessary. Second, notice the difference in the way we must format INTEGER and REAL variables. This FORMAT statement is I6, which tells the computer that the value to be read in is an INTEGER with up to six digits. Also, notice that when assigning a value to M, you must write M = 1 with no decimal point. There are examples of INTEGER formats in Table 2-1 and values in Table 2-2. The third new facility above is the "LOGICAL IF" instruction. Notice its construction. The word IF is followed by a logical expression in parentheses and then by an instruction. If the logical condition in parentheses is true, then the computer will obey the instruction at the end of the line. If the condition is not true, the computer will ignore this line altogether and go on to the next instruction.

In the example, the IF instruction with "GOTO 2" will cause control to be returned to label 2 as long as M is less than N, i.e., until all N numbers have been read in. Then, when M exceeds N, the "GOTO 2" instruction will not be obeyed, and control will pass to the next instruction: "PRINT 4, SUM." The GOTO instruction is always used to transfer control.

Notice also that the statement of the logical condition uses shorthand symbols surrounded by periods. The complete list of possible symbols is as follows:

```
.LT.     Less than
.GT.     Greater than
.LE.     Less than or equal to
.GE.     Greater than or equal to
.EQ.     Equal to
.NE.     Not equal to
.AND.    And
.OR.     Or
```

The use of the first six shorthand notations is obvious. The last two are used when it is necessary to impose more than one condition inside the parentheses as demonstrated by the examples in Table 2-5. This instruction is called the "LOGICAL IF" instruction to distinguish it from the "ARITHMETIC IF" instruction, the use of which we can see in the next example. The following program will read in an undefined number of values and sum them. It also counts the number of values which are positive (NPOZ), the number which are negative (NNEG), and the number which are zero (NZERO).

```
      NPOZ=0
      NNEG=0
      NZERO=0
      SUM=0.0
      READ 1,N
1     FORMAT(I6)
      DO 6 I=1,N
C     READ IN AND CHECK VALUE
      READ 2,A
2     FORMAT(F10.3)
      SUM=SUM+A
      IF (A) 3,4,5
3     NNEG=NNEG+1
      GOTO 6
4     NZERO=NZERO+1
      GOTO 6
5     NPOZ=NPOZ+1
6     CONTINUE
C     COMMUNICATE ANSWERS
      PRINT 7,N,NNEG,NZERO,NPOZ,SUM
7     FORMAT(4I6,F12.3)
      END
```

You can see how the ARITHMETIC IF instruction was used. It consists of the letters IF followed by parentheses containing a value which in our case is simply a single variable but could be an arithmetic expression as you will see from the examples in Table 2-6. After the closing parenthesis there are three labels. The computer understands this instruction to mean: Evaluate the expression inside the parentheses; if it has a negative value, go to the first label in the list; if zero, go to the second label; and if positive, go to the third label.

There are two other new points in this example. First, notice the lines which are labeled with a C in the part of the line that we normally reserve for labels. This C stands for COMMENT and means to the computer that the rest of the line is to be totally ignored and no attempt made to compile it. It is merely there as an aid to memory when you check back through your programs perhaps looking for mistakes or wishing to make modifications for other purposes. It is a device for the programmer's convenience only and serves no other purpose. You may write whatever you wish on a COMMENT line, and as you can imagine, it has many uses. It can be used to describe the function of a piece of FORTRAN code between one COMMENT line and the next. It is often used to indicate the date and author of new lines of code added to an

Table 2-5
Examples of Logical IF Instructions

FORTRAN Code	Longhand Meaning
IF(X .EQ. 0.0)GOTO 17	If X Equals 0.0 go to instruction 17
IF(DIAM .GT. 6.89)THICK=2.0	If DIAM is Greater Than 6.89 then REAL variable THICK becomes equal to 2.0
IF(B*B .LT. 4.0*A*C)GOTO 2	If B squared is Less Than 4 times A times C, go to instruction 2
IF(A .NE. B)PRINT 7,A	If A is Not Equal to B print out the value of A
IF(X .LT. 1.0 .AND. I .EQ. 4)GOTO 6	If X is less than 1.0 and I is equal to 4 go to instruction 6 (if either test fails go to the instruction following this one)
IF(X .LE. 1.0 .AND. Y .GE. 1.0)J=L	If X is Less than or Equal to 1.0 and Y is Greater than or equal to 1.0 then J=L
IF(I .EQ. 1 .OR. I .EQ. 10)N=N+1	Increase N by one if I equals 1 or 10
IF(I .EQ. 0 .AND. J .EQ. 7 .OR. I .EQ. 1)K=K+1	Increase K by one if I=0 and J=7 or if I=1

existing program. You can use it to indicate the units in which the variables must be fed into the program.

Professional programmers agree that if about one-third of a program is made up of COMMENT lines, then it will be easy to work with during future development stages and perhaps even during production use of the program. While you are learning FORTRAN, form the habit of liberally interspersing instructions with COMMENT lines indicating their function. Do it at the time of writing, even though it means extra typing. Do not make the mistake of

Table 2-6
Examples of Arithmetic IF Instructions

FORTRAN Code	Longhand Meaning
IF(A)1,2,3	If the value of A is negative go to label 1. If zero, go to label 2. And if positive go to label 3.
IF(DIAM-6.0)4,18,18	If (DIAM-6.0) is negative go to label 4. If it is zero or positive go to label 18. Note that this also means if DIAM is less than 6.0 go to 4 and if DIAM is equal to, or greater than 6.0, go to 18.
IF(B*B-4.0*A*C)6,103,112	If the value of B squared minus 4AC is negative go to label 6. If zero, go to 103 and if positive go to 112.

a junior programmer who was writing a complex distillation program. I asked him about his progress and he said he had finished, except that he had to go back and put in the COMMENT statements. I saw him later the same day, and he told me he was having difficulty remembering the exact function of each piece of code.

The second new feature is the line labeled 6, which simply instructs the computer to "CONTINUE." This is a "dummy" instruction and serves no other purpose than to provide a line to carry a label. It is often used as the last labeled instruction in a DO LOOP, especially when, as in the above example, it would be awkward to use any of the other instructions as the end one in the DO instruction. It has been unnecessary in the earlier examples of DO LOOPs in this chapter. Some programmers always use a CONTINUE instruction to terminate a DO LOOP. They say it helps them find their way around their program. However, it is not necessary in most cases.

FORTRAN Subroutines

I have been neglecting the manipulation of data values in the examples in order to indicate more administrative instructions. Let us now look at some examples of how numbers can be manipulated by a FORTRAN program. First, we can do simple arithmetic. For example:

```
PROFIT=SALES-(COSTS+TAX)
```

This tells the computer to add together the contents of variables called COSTS and TAX, subtract the total from SALES, and place the result into a register called PROFIT. I am sure you can work out the full meaning of the following arithmetic statement:

```
X=4.0*A+(SUM-17.2)/100.0
```

Note that * is used by the computer as a multiplication symbol. In some cases, there may be some ambiguity. For instance, consider the following two FORTRAN instructions:

```
5      W=Z/2.0*C

6      W=Z/(2.0*C)
```

Here, there is a vital difference between the two instructions. The computer evaluates divisions first, then multiplications followed by additions and subtractions. Instruction #5 means divide Z by 2 and multiply the result by C. Instruction #6 means divide Z by 2 times C. When writing mathematical statements, check carefully to see whether there is any ambiguity of the type shown in this last example. If there is, clarify the ambiguity using parentheses. You need not worry about overdoing it; the computer ignores superfluous parentheses. There are many examples of calculation lines in Table 2-7. The mathematical formula which you wish to put into the program may, of course, be more complex. It may have mathematical functions in it such as logarithmic or trigonometric functions, and we may deal with these in the following way.

Each FORTRAN compiler has access to library subprograms (in this case called FUNCTION SUBROUTINEs) which carry out these functions. They are similar to the ones put in the domestic robot for carrying out routine tasks, i.e., the loading of the dishwasher. You must refer to your own computer system for a full list of available functions, but the most common ones are shown in Table 2-8, which also has an explanation of each function written to the right.

When you use SQRT, the computer looks in its own "private" library for the FUNCTION SUBROUTINE which carries out a square-root calculation and uses it to find the square root of whatever value is contained within the parentheses.

Table 2-7
Examples of Calculation Lines

FORTRAN Code	Notes
`A = B+(C/D)`	$A = B + \frac{C}{D}$
`A = B+C/D`	$A = B + \frac{C}{D}$
`A = (B+C)/D`	$A = \frac{(B+C)}{D}$
`A = (B+2.0*C+D)/D`	$A = \frac{B+2C+D}{D}$
`A = B**2*C**2*D**3`	$A = B^2C^2D^3$
`A = 1/B**2+1/C**2`	$A = \frac{1}{B^2}+\frac{1}{C^2}$
`A = 1/B**(2+N)+N`	$A = \frac{1}{B^{2+N}}+N$
`A = EXP(-B*X**2)`	$A = e^{-BX^2}$

The FUNCTION SUBROUTINE is such a useful device that most FORTRAN compilers allow the user to write special FUNCTION SUBROUTINEs of his own. The layout or syntax of the homemade FUNCTION SUBROUTINE is simple, and to illustrate it I propose to go back to the earlier example of finding the sum of an unknown list of values. Instead of writing a main program to do the job, this time it will be carried out inside the special FUNCTION SUBROUTINE which follows. You will see how to use it in a moment.

```
      FUNCTION SUBROUTINE SUM(N,A)
      DIMENSION A(1000)
      SUM=0.0
      DO 10 I=1,N
10    SUM=SUM+A(I)
      RETURN
      END
```

You can see that you must tell the computer that the special piece of code (which is placed at the end of the main program) is to be a FUNCTION SUBROUTINE and identified by the name you used in your main program when you wished to call it. Note that the argument list is in parentheses as before. The argument list must be in the same order in both the call and the FUNCTION SUBROUTINE definition. In all uses (in the main program) of the FUNCTION SUBROUTINE SUM, the argument list will be two in length, the first argument being an INTEGER (the number of elements in the ARRAY) and the second being the ARRAY name. The FUNCTION SUBROUTINE will assume that the ARRAY is subscripted from 1 to the value of the INTEGER. Thus, in the main program, if you had an array called TEMP which had 24 values and you wished to find the average, you could say:

```
AVG=SUM(24,TEMP)/24.0
```

Another example would be if you had an ARRAY D, containing N values, and you wished to compute an area based upon the average diameter, you could write:

```
AREA=PI*(SUM(N,D)/FLOAT(N))**2)/4.0
```

You can see how the FUNCTION SUBROUTINE call is incorporated into a formula using arguments with names *which may not necessarily be the same* as the

Table 2-8
Some Built-In FUNCTION SUBROUTINES

FORTRAN Code	Longhand Meaning
`Y = ABS(X)`	Set Y to the absolute value of X
`Y = AINT(X)`	Set Y to the integral part of X converted to a REAL (unlike INT below)
`Y = ALOG(X)`	Set Y to the natural logarithm of X
`Y = ALOG10(X)`	Set Y to the logarithm (base 10) of X
`Y = COS(X)`	Set Y to the cosine of angle X (in radians)
`Y = EXP(X)`	Set Y to the exponential of X (e^X)
`Y = FLOAT(I)`	Set Y to the integral value of I converted to a REAL
`J = IABS(I)`	Set J to the absolute value of I
`J = INT(X)`	Set J to the integral part of X and leave it in form INTEGER
`Y = SIN(X)`	Set Y to the sine of angle X (in radians)
`Y = SQRT(X)`	Set Y to the square root of X
`Y = TAN(X)`	Set Y to the tangent of angle X (in radians)

argument names used when defining the FUNCTION SUBROUTINE. The FORTRAN compiler will exchange these on a one-to-one basis. Just make sure that the FUNCTION SUBROUTINE call includes the argument list in the correct order.

There were some other new features in this example. First, the DIMENSION A(1000) instruction simply reminds the computer that the ARRAY A has a maximum number of values of 1,000. This means that N can never exceed 1,000, and so this particular FUNCTION SUBROUTINE is limited to adding up ARRAYs with no more than 1,000 values in them. Next, the double asterisk (**) means "raised to the power of." It will be followed by an INTEGER value. Note that ** is always evaluated before division, multiplication, addition, and subtraction. There are examples of ** in Table 2-7. Finally, note the significance of the RETURN instruction at the end of each FUNCTION SUBROUTINE. This tells the computer to go back into the main program at the position immediately after the FUNCTION SUBROUTINE call. Thus, in the last example, the computer returns to the division sign which precedes the word FLOAT. Incidentally, notice how using mixed mode arithmetic was avoided by converting the INTEGER N into a REAL by using the FUNCTION SUBROUTINE FLOAT provided by the FORTRAN compiler.

Below is another example of the definition and use of a homemade FUNCTION SUBROUTINE. This one is called MAX. It will sort through an ARRAY with up to 8,000 values and return to the main program with the ARRAY subscript of the largest value in the ARRAY.

```
FUNCTION SUBROUTINE MAX(N,X)

DIMENSION X(8000)

MAX=1

DO 1 K=2,N
```

```
      IF (X(K).GT.X(MAX)) MAX=K
1     CONTINUE
      RETURN
      END
```

With the FUNCTION SUBROUTINE provided to the computer by placing it after the END of the main program, it can be called as often as required from the main program. The argument list need not necessarily have the same names as used within the definition of MAX. Here is how you would use MAX to find the largest value in an ARRAY of heights called HEIGHT which contains NUMB values.

```
      READ 1,NUMB
1     FORMAT(I6)
      DO 3 K=1,NUMB
      READ 2,HEIGHT(K)
2     FORMAT(F5.2)
3     CONTINUE
      MAXNO=MAX(NUMB,HEIGHT)
      PRINT 4,HEIGHT(MAXNO)
4     FORMAT(F5.2)
      END
```

In this example, the program will read in the NUMB heights and store them in the ARRAY called HEIGHT, i.e., in HEIGHT (1), HEIGHT (2), HEIGHT (3), etc. Then MAX is called and NUMB and HEIGHT are substituted for N and X in the argument list for the FUNCTION SUBROUTINE MAX. Upon returning from MAX, the subscript of the maximum value is put into the INTEGER variable called MAXNO. Finally, the program prints out HEIGHT (MAXNO), which is the name of the register which holds the maximum value of height in the ARRAY.

Notice again how you need to pass vital information in and out of FUNCTION SUBROUTINEs in a specific order. Also notice that, like all lists in FORTRAN, items in the argument list are separated by commas. Since FUNCTION SUBROUTINEs constitute such a valuable facility, most FORTRAN compilers allow users, or programmers, to write their own. The FORTRAN compiler will accept the names of unknown FUNCTION SUBROUTINEs as it compiles your program, knowing that if you have done your job correctly, the FORTRAN definition of the FUNCTION SUBROUTINE will be at the end of your program. Thus, schematically a block model of your FORTRAN program will be as follows:

Introductory administration lines (see later)

Main (or Master) program with references (or calls) to FORTRAN or user-generated SUBROUTINES

User's own SUBROUTINES and FUNCTION SUBROUTINES in any order

In addition to the FUNCTION SUBROUTINE, there is another kind of subroutine, called simply "SUBROUTINE," which works in a slightly different way. SUBROUTINEs as described in the following examples are contained in any order at the end of the program along with the FUNCTION SUBROUTINEs, if any are present. The purpose of the SUBROUTINE facility is to allow you to write a subprogram and call it by name from the main program. The following is an example of the SUBROUTINE which takes an ARRAY called A with N values and finds the subscripts of the maximum and minimum values as well as the average of all values:

```
      SUBROUTINE SORT(N,A,MAX,MIN,AVGE)
      DIMENSION A(1000)
      SUM=A(1)
      MAX=1
      MIN=1
      DO 21 I=2,N
      SUM=SUM+A(I)
      IF (A(I).GT.A(MAX)) MAX=I
      IF (A(I).LT.A(MIN)) MIN=I
21    CONTINUE
      AVGE=SUM/FLOAT(N)
      RETURN
      END
```

The first two arguments in the SUBROUTINE argument list are names of data *required* by the SUBROUTINE, i.e., the number of values in the ARRAY and its name, but the next three arguments are the names of the answers yet to be generated by the SUBROUTINE. You can regard the argument list of the SUBROUTINE as an in-and-out channel from the main program. This may become clearer when you consider the following piece of FORTRAN code which uses SUBROUTINE SORT to compute the average height and the subscript number of the tallest and shortest person from an ARRAY of 870 heights:

```
      DO 13 I=1,870
      READ 10,HEIGHT(I)
10    FORMAT(F10.3)
13    CONTINUE
      CALL SORT(870,HEIGHT,NTALL,NSMALL,HTBAR)
      PRINT 14,HEIGHT(NTALL)
      PRINT 15,HEIGHT(NSMALL)
      PRINT 16,HTBAR
14    FORMAT(11HTALLEST IS ,F4.1,7H INCHES)
15    FORMAT(12HSHORTEST IS ,F4.1,7H INCHES)
16    FORMAT(11HAVERAGE IS ,F4.1,7H INCHES)
      END
```

Note that the PRINT instructions have FORMAT instructions which contain text to be output along with the calculated values. The H symbol tells the computer that the text which follows is to be printed. The number of characters in the message (including all spaces which *must* be counted) is written immediately before the H as shown. Notice that there is no space after the H. The above output instructions will give (in a typical case):

```
TALLEST IS  75.3 INCHES
SHORTEST IS 62.6 INCHES
AVERAGE IS  68.9 INCHES
```

There are more examples of PRINT and FORMAT in Table 2-9 and more examples of FUNCTION SUBROUTINEs and SUBROUTINEs in Tables 2-10 and 2-11, respectively.

You need to know just one more instruction—a powerful variation of the GOTO instruction. Here is one example:

```
GOTO(17,3,93,104,17,16)N
```

It is shorthand for "Go to the label which is Nth in the list in parentheses." Of course, you must ensure that N never is greater than the number of labels inside the parentheses—6 in this example.

Notice that it is quite acceptable to have the labels in any order and labels may be repeated. There are more examples of GOTO instructions in Table 2-12.

Other Instructions

By now, you know sufficient FORTRAN to write quite sophisticated programs and certainly enough to solve all the problems at the end of this book, except for one or two small extras which are needed for the administrative purposes of the FORTRAN compiler. For instance, when you intend to use an ARRAY, then you need to reserve space in the register store. In addition, although there are rules about how to decide upon the name of a variable depending upon whether it is an INTEGER or a REAL value, we can override those rules if we first make our intentions known to the FORTRAN compiler. For instance, you may wish to have an INTEGER counter called COUNT, or you may find it desirable to use KVALUE to represent a REAL variable. The following statements carry these messages and another new feature:

```
REAL KVALUE,MOLE,MF
INTEGER COUNT,PLATE
DIMENSION TEMP(120),MF(9,9),PLATE(100)
```

These commands tell the computer that the variable KVALUE will be REAL and so will MOLE and MF; that COUNT and PLATE will be INTEGER values; that PLATE is also to be an INTEGER ARRAY of 100 subscripted variables; and that MF is to be a REAL ARRAY with two subscripts representing a two-dimensional matrix as follows:

```
MF(1,1)  MF(1,2)  . . . . .  MF(1,9)
MF(2,1)  MF(2,2)  . . . . .  MF(2,9)
   .        .     . . . . .     .
   .        .     . . . . .     .
   .        .     . . . . .     .
MF(9,1)  MF(9,2)  . . . . .  MF(9,9)
```

There are more examples of these administrative instructions in Table 2-13.

The two final types of administrative instructions to

Table 2-9
Examples of PRINT and FORMAT*

FORTRAN Code	Resulting Printout
`PRINT 610,TEMP` `610 FORMAT(15HTEMPERATURE IS ,F6.1)`	`(for TEMP = 191.3)` `TEMPERATURE IS  191.3`
`PRINT 4,AGEONE,AGETWO` `4 FORMAT(12HTHE AGES ARE,F3.0,4H AND,F3.0)`	`(for AGEONE = 19 and AGETWO = 21)` `THE AGES ARE 19 AND 21`
`PRINT 41,N,X(N)` `41 FORMAT(12HENTRY NUMBER,I3,3H is,F8.3)`	`for N = 8 and X(8) = 273.18` `ENTRY NUMBER  8 IS  273.18`

`DO 7 I=2,10,2`
`PRINT 6,I,DIAM(I)`
`6 FORMAT(9HDIAMETER ,I2,4H IS ,F6.4`
	`DIAMETER  2 IS 6.1291`
	`DIAMETER  4 IS 7.2411`
	`DIAMETER  6 IS 7.9183`
	`DIAMETER  8 IS 8.0014`
	`DIAMETER 10 IS 8.7055`

*Important: Don't forget to count the spaces in an "H" total.

Table 2-10
Some FUNCTION SUBROUTINES

Definition at End of Program	Some Ways to Call It

<pre>   FUNCTION SUBROUTINE AMEAN(N,A)
      DIMENSION A(1000)
      SUM=0.0
      DO 12 I=1,N
12   SUM = SUM+A(I)
      AMEAN=SUM/FLOAT(N)
      RETURN
      END</pre>
	AVERAGE=AMEAN(NHTS,HTS)
	PRINT 7, AMEAN(N,XX)
	A=F+4.0/AMEAN(K,TEMP)

<pre>   FUNCTION SUBROUTINE POZFIND(N,A)
      DIMENSION A(1000)
      SUM=0.0
      DO 3 I=1,N
      IF(A(I))3,3,2
2   SUM=SUM+A(I)
3   CONTINUE
      POZFIND=SUM
      RETURN
      END</pre>
	TOTALP=POZFIND(J,DIFF)
	AVPOZ=POZFIND(N,TOT)/FLOAT(N)

learn are DATA and COMMON. The DATA instruction is a simple way of assigning data values to variables. You would want to use it for data you are sure of at the time you write the program. You should put it in the program immediately after any REAL, INTEGER, or DIMENSION instructions and in the following form:

```
DATA X,Y,Z/7.2,81.7,0.0231/
```

which means "assign the following data values."

```
X=7.2

Y=81.7

Z=0.0231
```

The DATA statement needs one punched card; the assignations would need three. So, DATA saves space, time, and the cost of unnecessary punched cards. There are examples of DATA instructions in Table 2-14. The COMMON statement is a means of shortening the argument lists of SUBROUTINEs in cases where a large amount of data is common to all. It works by being placed up with the other administrative instructions and listing all those data values which are common. Then the COMMON statement is used (again, up with the administrative instructions) to rename the same values within the subroutines themselves. For example:

```
INTEGER RECORD

DIMENSION ARR(100),B(4,4)

READ M1T2

DATA X,Y,RECORD/1.0,1.0,7/

COMMON X,Y,M1T2,RECORD,N,ARR(100),B(4,4)

__________

__________        Main Program

__________

__________

CALL SUBROUTINE ONE(TEMP1,TEMP2)

__________
```

```
___________        Main Program
___________

___________
CALL SUBROUTINE TWO(J,P,TEMP3,REX)

___________
___________        Remainder of Main Program
___________

___________
END

SUBROUTINE ONE(TIN,TOUT)

COMMON A,B,C,K,N,ARR(100),DOT(4,4)

___________
___________        Rest of Subroutine ONE

___________
RETURN

END

SUBROUTINE TWO(M,PRESS,TMEAN,REX)

COMMON P,Q,R,IN,IM,GER(100),HER(4.4)

___________
___________        Rest of Subroutine TWO

___________
RETURN

END
```

Notice how the COMMON variables may have the same *or* different names in each subroutine. Thus, the 100-element ARRAY (called ARR when referred to in the main program) will be called ARR in SUBROUTINE ONE but GER in SUBROUTINE TWO. However, all the 100 values in the ARRAY will be the same in all parts of the program whatever its current name happens to be. Similarly, X in the main program with a value of 1.0 will be called A in SUBROUTINE ONE (but still with a value of 1.0) and P in SUBROUTINE TWO (again with a value of 1.0). There are more examples of COMMON instructions in Table 2-15.

If you have struggled through so far *and* have scrutinized each example to understand the objectives to be achieved by each piece of code, then you are ready to start writing your first FORTRAN program. You have not learned *all* the FORTRAN facilities that exist, but the remainder are so advanced that it is unlikely you will ever need them. If you now concentrate on developing programs and gain confidence in your ability to use FORTRAN commands to carry out the most complex tasks, you will be able to learn more advanced features quickly should you ever need them. And if you ever write programs involving more advanced features, you will almost certainly be in a software development team, in which case there will be at least one professional programmer in the team to help you learn more advanced techniques.

Table 2-11
Examples of SUBROUTINES*

```
      SUM=0.0
      CALL AVERAGE (N,A,AMEAN)
      DO 19 I=1,N
      B(I)=(A(I)-AMEAN)**2
19    SUM=SUM+B(I)
      CALL DIV(N,SUM,X)
      CALL PRINTER(N,A,SUM,X)
      END

      SUBROUTINE PRINTER(N,X,T,P)
      DIMENSION X(1000)
      DO 2 I=1,N
      PRINT 2, X(I)
2     FORMAT(F10.5)
      PRINT 3, N,T,P
3     FORMAT(I4,2F10.5)
      RETURN
      END

      SUBROUTINE DIV(N,SUM,X)
      X=SQRT(SUM/FLOAT(N))
      RETURN
      END

      SUBROUTINE AVERAGE (N,X,AMEAN)
      DIMENSION X(1000)
      SUM=0.0
      DO 1 I=1,1000
1     SUM=SUM+X(I)
      AMEAN=SUM/FLOAT(N)
      RETURN
      END
```

*These examples are deliberately simple in order to demonstrate the structure of a SUBROUTINE. Normally SUBROUTINES are considerably more complex.

Table 2-12
Examples of GOTO Instructions

FORTRAN Code	Longhand Meaning
GOTO 17	Transfer control to the instruction labeled 17
IF(X.GT.0.0)GOTO 4	Transfer control to label 4 if and only if X is greater than zero
GOTO(100,2,61,14,9)MOP	If MOP=1 go to label 100 If MOP=2 go to label 2 If MOP=3 go to label 61 If MOP=4 go to label 14 If MOP=5 go to label 9 If MOP>5 tell user he has a fatal error

Table 2-13
Examples of DIMENSION, REAL and INTEGER Instructions

FORTRAN Code	Longhand Meaning
REAL INK, MOLLY, NOTE	The variables INK, MOLLY and NOTE are to be REAL variables in this program.
INTEGER AGE, RECORD, DATE, TIME	The variables AGE, RECORD, DATE and TIME are to be INTEGER variables in this program.
DIMENSION AGE(10), INK(3,3)	The INTEGER AGE is the name of a 1-dimensional INTEGER ARRAY and INK is the same of a 2-dimensional REAL ARRAY.

Table 2-14
Examples of DATA Instructions

FORTRAN Code	Comments
DATA X,Y/4.971,6.33E7/	Note the use of scientific notation for the value of Y.
DATA TEMP,PRES/-273,0.01/	Here is how you write a negative value.
DATA NTUBE,U,AREA,N/740,2.119,2500.,8/	You can mix modes freely in a DATA instruction.
DATA SUM,SUMSQ,SUMDIF/0.,0.,0./	You can omit the unnecessary zero after the decimal point.

Table 2-15
Examples of COMMON Instructions

FORTRAN Code	Comments
MAIN COMMON A,N,B(1000) S/R COMMON X,I,Z(1000) S/R COMMON X,J,A(1000) S/R COMMON A,JOE,ARR(1000) S/R COMMON SEVEN,LUV,ZZ(1000) and so on	Notice there will be no conflict about using A for the name of a REAL variable in the main program and the name of an ARRAY in the first SUBROUTINE.
MAIN COMMON TEMP(100,100) S/R COMMON T(100,100) S/R COMMON T(100,100) S/R COMMON TEM(100,100)	A good way of making a large ARRAY common throughout the program.

You know adequate FORTRAN already to be able to develop the technical parts of the finished program. Leave the administrative and more complex details to the professionals, or learn how to use them like a professional. If you ever should have need to progress to these advanced refinements, then you know enough FORTRAN to be able to do your learning from a reference text rather than a teaching text such as this one. A list of suitable texts is included at the end of this chapter.

Now, in order to bring together all the features you have learned, I will finish this chapter with a comprehensive example. The example has been deliberately contrived to include almost all the features you have learned. It is a program to solve up to 50 quadratic equations. It checks to see whether the roots are real or imaginary and whether the solution is a "perfect" square. If neither is the case, the program uses the general equation for quadratics:

$$\text{if} \quad Ax^2 + Bx + C = 0,$$

$$\text{then } x = \frac{-B \pm \sqrt{B^2 - 4AC}}{2A}$$

Go through this example carefully. Read the notes in the comment lines. The only new feature in this program is the continuation line necessary to allow the long FORMAT instruction. You will have already noticed that all FORTRAN lines begin six spaces in from the left margin. These spaces can be used for labels, the COMMENT signal, and the continuation line signals. Here are the rules. Labels and the C for COMMENT must fit anywhere in the first five spaces. The sixth space *must* be left blank because *any* character in that position indicates that the line is a continuation of the previous line. Table 2-16 has examples of the use of the six spaces in each line.

Table 2-16
Some Examples of the Correct Use of the First Six Spaces in a FORTRAN Line

```
17    GOTO 6
8     A=B*LOG(X)
C     NOW BEGINS THE OUTPUT
 23   DO 3 I=1,8
C     NOTE LABELS DO NOT
C     NEED TO BE VERTICALLY
C     ABOVE EACH OTHER
   C  NOR DO THE COMMENT SIGNS
      H(I)=Z(I)*A*B
      PRINT 4, H(I),A,X
4     FORMAT(33H THE VALUE OF THE ARRAY
     1MEMBER IS ,F10.5,19HAND THE
     2CONSTANT IS,F10.5,17HT
     3THE VALUE OF X IS ,F10.5)
C     IN THIS EXAMPLE, THE CONTINUATION
C     LINES ARE NUMBERED IN THE
C     SIXTH COLUMN.  THE PRESENCE OF
C     ANY CHARACTER IN THE SIXTH COLUMN
C     INDICATES A CONTINUATION LINE
C
C     NOTE THAT C CAN BE USED TO
C     LEAVE A BLANK LINE TOO
C
3     CONTINUE
```

A Comprehensive Example

Note: This example contains most features except FUNCTION SUBROUTINEs and SUBROUTINEs. Examples of these are to be found in Tables 2-10 and 2-11. This program will solve up to 50 quadratic equations.

```
C     THIS PROGRAM READS IN 50 SETS
C     OF VALUES OF A,B,C AND CALCULATES
C     THE ROOTS OF THE QUADRATIC AX**2+BX+C=0.
C     ALSO IT CALCULATES THE AREA OF A
C     CIRCLE HAVING THE LARGER ROOT AS
C     ITS DIAMETER IF THE ROOT IS POSITIVE.
C
      INTEGER TOTAL
      DIMENSION ROOTS(50,2) ,IMAG(50)
      DATA TOTAL, PI/50,3.1417/
C
C     INTEGER ARRAY IMAG IS TO STORE THE
C     NUMBERS OF EQUATIONS WITH IMAGINARY ROOTS
C
C     BELOW SET A COUNTER OF EQUATIONS WITH REAL ROOTS AND
C     SET ALL ARRAY IMAG TO ZERO
C
      KOUNT=0
      DO 1 I=1,TOTAL
      IMAG(I)=0
C
C     START A DO LOOP TO READ IN, CALC AND PRINT
C
      DO 11 I=1,TOTAL
C
C     READ IN 50 SETS OF A,B,C THREE TO A CARD.
C
      READ 12, A(I),B(I),C(I)
12    FORMAT(3F10.5)
C
C     TEST FOR PERFECT SQUARE OR IMAG ROOTS
C
      IF(B(I)*B(I)-4.0*A(I)*C(I))21,22,23
C
C     CASE OF IMAG ROOTS, FIRST RECORD EQUATION NO
C
      IMAG(I)=1
      PRINT 24,I
24    FORMAT(12HEQUATION NO ,I2,21H  ROOTS ARE IMAGINARY)
      GOTO 11
C
C     PERFECT SQUARE NEXT
C
      ROOTS(I,1)=B(I)/2.0
      ROOTS(I,2)=B(I)/2.0
      GOTO 25
C
C     GENERAL CASE NEXT
C
      Q=AQRT(B(I)*B(I)-4.0*A(I)*C(I))
      ROOTS(I,1)=(-B(I)+Q)/(2.0*A(I))
      ROOTS(I,2)=(-B(I)-Q)/(2.0*A(I))
C
C     PRINTOUT OF REAL ROOTS
C
      PRINT 26, I, ROOTS(I,1),ROOTS(I,2)
26    FORMAT(12HEQUATION NO ,I2,11HROOTS ARE ,2F10.5)
      KOUNT=KOUNT+1
C
C     END OF DO LOOP, KOUNT HOLDS NUMBER WITH REAL ROOTS
C
11    CONTINUE
```

```
C
C     PRINTOUT NO OF EQUATIONS WITH REAL ROOTS
C
      PRINT 27,KOUNT
27    FORMAT(9HTHERE ARE ,I2,25HEQUATIONS WITH REAL ROOTS)
C
C     NOW TO CALCULATE AREAS OF CIRCLES WITH
C     DIAMETER EQUAL TO LARGER ROOT IF POSITIVE
C
      DO 101 I=1 ,TOTAL
C
C     FIRST CHECK IF ROOTS ARE REAL
C
      IF(IMAG(I) .EQ. 1) GOTO 101
C
C     NEXT FIND IF THERE IS A POSITIVE ROOT
C
      IF(ROOTS(I,1) .LE. 0.0 .AND. ROOTS(I,2) .LE. 0.0) GOTO 101
C
C     AND NOW FIND THE LARGER
C
      IF(ROOTS(I,1)-ROOTS(I,2))102,102,103
103   DIAM=ROOTS(I,1)
      GOTO 104
102   DIAM=ROOTS(I,2)
C
C      FINALLY CLACULATE AREA AND PRINT IT OUT
C
       AREA=PI*DIAM*DIAM/4.0
       PRINT 105,I,DIAM,AREA
105    FORMAT(12HEQUATION NO ,I2,24HLARGER POSITI
     1VE ROOT IS ,F10.5,12HAND  AREA is ,F10.5)
C
C      END OF DO LOOP
C
101    CONTINUE
C
C      END OF PROGRAM
C
       END
```

Further Reading

Friedman, F.L., and E.B. Koffman. *Problem Solving and Structured Programming in FORTRAN.* Philippines: Addison-Wesley, 1977.

Gear, W.C. *FORTRAN and WATFIV.* Chicago: Science Research Associates, 1978.

Haskell, R.E. *FORTRAN Programming Using Structured Flowcharts.* Chicago: Science Research Associates, 1978.

Jaffe, R.M. *A Clear Introduction to FORTRAN IV.* Encino, CA: Dickenson, 1972.

Kennedy, M., and M.B. Solomon. *Ten Statement FORTRAN.* Englewood Cliffs, NJ: Prentice-Hall, 1975.

Lipschutz, S., and A. Poe. *Programming with FORTRAN.* New York: McGraw-Hill, 1978.

McCracken, D. *A Guide to FORTRAN IV Programming.* New York: Wiley, 1972.

Chapter 3

The Steel Industry

Steel making is a 3,000-year-old industry which has had its most significant developments in this century. It employs every chemical engineering skill. In this chapter you will learn how coke is made from coal and how it is used to smelt iron from its ore. I shall describe the magnificent blast furnace, which for 200 years has been almost the only means of obtaining iron needed for the industrial revolution. You will then see how the iron is converted to steel in the basic oxygen furnace or the open hearth furnace and how scrap steel is recycled using the electric arc furnace.

Steels vary in composition according to the job they will do. The automobile industry buys steel plate which is ductile enough to be pressed, preferably at low temperature, into shapes for car bodies. They also buy or make iron which, although quite brittle, is extremely strong and can be poured into molds to freeze into complex shapes. The construction industry buys girders and reinforcing bars for concrete. Both products are made of steel with a composition chosen to give high tensile strength.

Steel is used throughout the petrochemical and chemical industries. It is used in (1) reaction vessels, in which reacting chemicals are brought together, often at high pressures and temperatures; (2) distillation towers, which soar over 20 stories high, with internal trays causing a cascading action of the substances being separated or fractionated; (3) storage tanks up to 200 feet in diameter and 50 to 60 feet high; and (4) all the other equipment on a site including pipes, valves, pumps, compressors, structural girders, flooring, flanges, and heat exchangers. The uses for steel in a chemical plant are almost endless. Nevertheless, after the plant is designed, every component, from the largest distillation tower to the last nut and bolt, must be completely described, specified, and ordered for delivery at the time it will be needed during construction. Steel can be bought from some 200 steel manufacturers operating plants in 37 states.

Coming together in an iron and steel plant (see Figure 3-1) are five basic raw materials. Mining companies send coal in its raw state to the steel companies, who keep a 60-day stockpile. They stockpile for two reasons: First, mining production is subject to fluctuations caused by strikes, bad weather, breakdowns in transportation, disasters, and other reasons. Second, as the coal arrives, a mixing operation must take place—the coal is dumped in a flat stockyard according to a carefully calculated pattern and then scooped up in a different pattern. This levels out any minor variation in any characteristic of the coal. Consider what would happen if the sulfur content of the coal doubled as a new seam developed in the local mine. Without a stockpile, the steel companies would have but a few days notice in which to plan and implement modifications to their own plant to deal with the change. With the stockpile, they have a two-month notice and only a gradual change is necessary because of the careful storage and mixing procedures.

The Coke Oven

The coal is placed in coke ovens (Figure 3-2) to concentrate its carbon content by driving off its volatile matter. Cold raw coal is a black, solid rock-like substance which actually contains no volatile matter. However, when coal is heated, some of the carbon atoms join up with atoms of hydrogen, oxygen, sulfur, and nitrogen and leave the coal as hydrocarbons such as methane, benzene, toluene, and many other valuable by-products. A battery of coke ovens is usually made up of 100 individual ovens, each of which is about 20 feet high, 40 feet long, and 20 inches wide. They are stacked in a row with a 6-inch-wide fuel-gas combustion chamber between each one. The coal is dropped in through four manholes which also act as gas seals when they are replaced.

Whenever you are considering a new process in chemical engineering, think yourself *into* the situation. In this

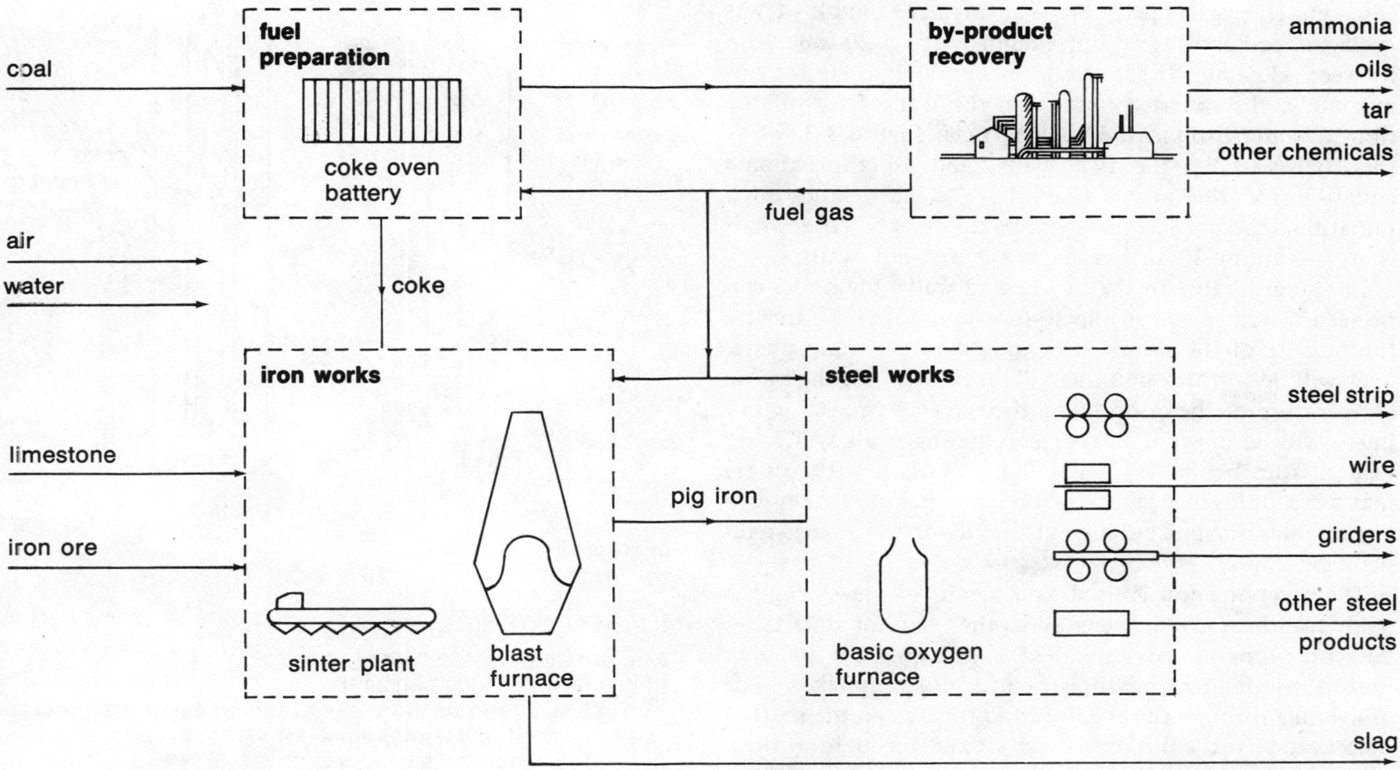

Figure 3-1. *Materials flow in an iron and steel plant.*

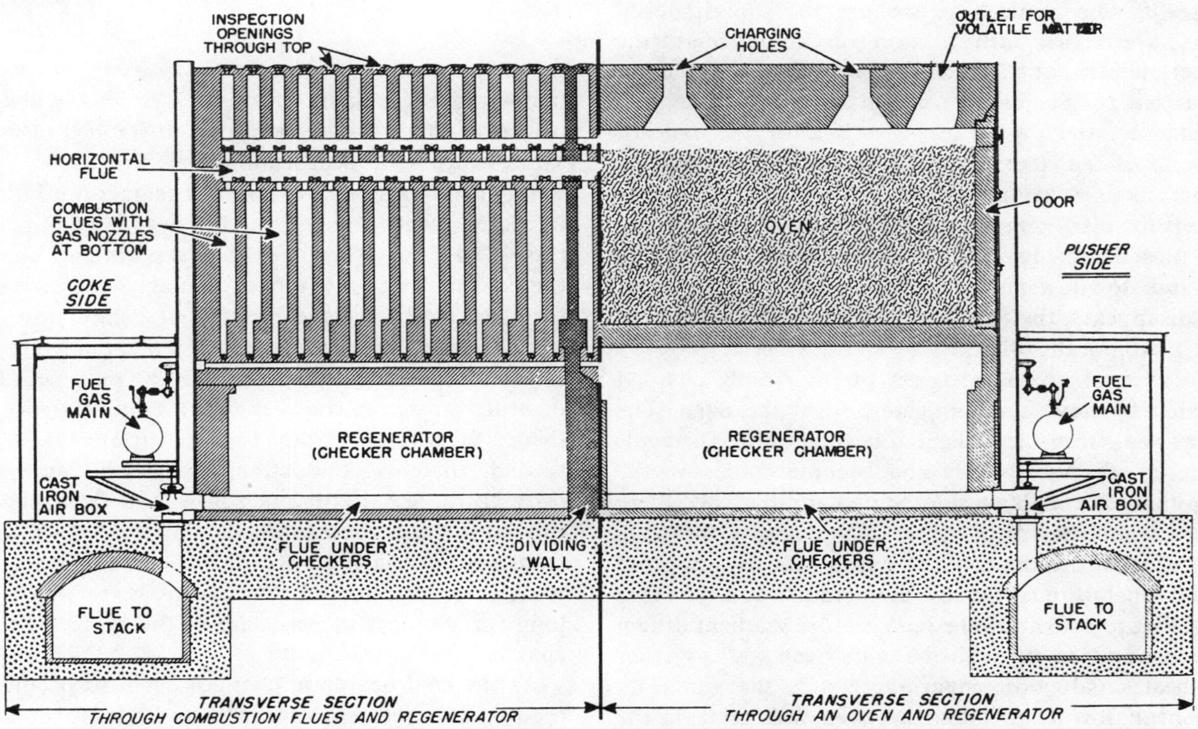

Figure 3-2. *Transverse sections of Koppers regenerative single-divided by-product coke oven battery. The section at left is through the combustion chambers (flues), that at right is through the oven chamber. (© 1971 by U.S. Steel Corp. Used with permission.)*

case, imagine a small (quarter of an inch) chunk of wet coal dropped into a red-hot chamber. It's a 20-foot drop between glowing, bright-red walls (1,500° F). Almost immediately, the radiated heat from the walls dries off the film of water around the coal particle. Figure 3-3 shows the coal particle, the two walls, and an approximate equation for the heat gained by the particle through radiation. The particle itself warms up quickly. After all, it is a maximum 10 inches from the red-hot walls.

Looking parallel to the two heated walls, the doors can be seen at either end of the 40-foot-long furnace. Because the outside of the doors are exposed to the atmosphere, heat will be lost through them. This means that the inside temperature of the doors, even though they are refractory lined, will be lower than the surrounding walls which are heated from behind by a gas flame. Although the doors may be a dull red because of their temperature, they will appear black against the brighter red of the higher-temperature walls.

This phenomenon is used as a means of measuring the temperature of coke ovens and other red-hot objects. A telescopic-looking device called a *disappearing filament optical pyrometer* measures temperatures quickly. The user looks through the eyepiece and focuses on the red-hot object—say, the coke oven wall viewed through an open manhole. Inside the barrel of the pyrometer is a tiny "w"-shaped filament connected through a rheostat to a battery. Because the coke oven wall is emitting radiation, heat, and light, the filament appears as a black "w" superimposed over it. The user then turns a knob (on the outside of the pyrometer) which controls the rheostat. Gradually, more current is fed to the filament and it, too, begins to glow red. Suddenly, the "w" disappears against the red-hot background of the coke oven wall. A dial on the outside of the pyrometer is connected to the control knob and is calibrated to read temperature directly. If the control knob is turned too high, the "w" glows white against the background; too low and it appears black. At the exact point it disappears, the temperature of the coke oven wall may be read on the dial.

Let us go back to the progress of the freshly charged coal, which by now has completely filled the oven. The coal takes heat from the walls and becomes hot. The walls are made of silica refractory and become cooler on the coal side than on the flame side. Stated another way, there exists a temperature drop, or *gradient*, across the wall. This is sufficient driving force to cause heat to flow from the high temperature to the low temperature and thus into the coal. Figure 3-4 shows the temperature gradient drawn over a cross section through the coke oven wall together with a heat conduction equation. Notice the similarity between heat flowing through the oven wall and, on the other hand, water flowing downhill. Water falls down a height gradient, and heat falls down a temperature gradient. Notice from the equation in Figure 3-4 that, as expected

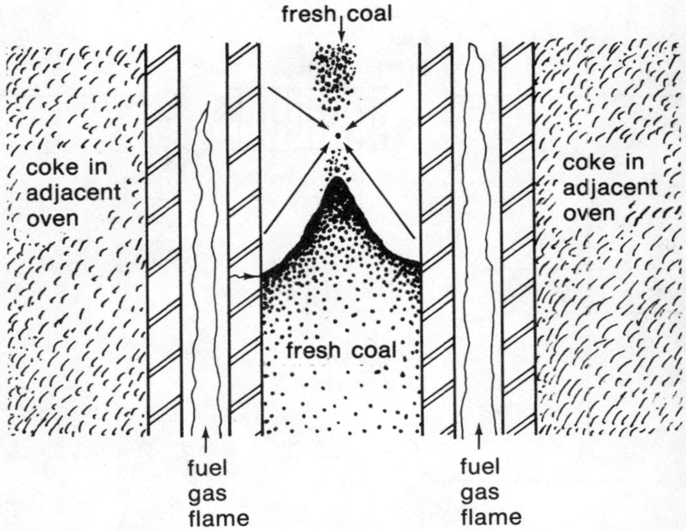

heat received
by descending coal particle

$$q = \sigma F (T_w^4 - T_p^4)$$

where q is heat received by particle
σ is Stefan's constant
F is a ratio of the areas of the wall and the particle
T_w is the wall temperature
T_p is the particle temperature

Figure 3-3. *Charging a coke oven.*

from the analogy with flowing water, the steeper the temperature gradient, the faster the heat will flow. The heat is conducted into the center of the bed of coal by the same process of heat conduction.

Figure 3-4 also shows how the temperature distribution in the coal will change over the next few hours. The heat travels so slowly through the packed bed of coal because the coal particles make poor contact with each other and form an inefficient means of heat conduction. The heat transfer rate depends more on how quickly the heat can flow through the gaps in between the particles. The mode of heat transfer in these gaps is a complicated one: First, there is radiation from coal particle to coal particle; second, there is conduction of the heat across the gap through the gases contained in it; and third, there will be tiny convection currents in the gap, which will carry heat from hot coal particles to cooler ones. These are slow heat transfer processes, and you can now see why it takes so long for the heat to percolate to the center of the bed of coal.

As the coal heats up, changes begin to occur. First, all the remaining water evaporates and rises through the coal and leaves through the pipe shown in Figure 3-5. Then, at about 450° F, the coal softens and becomes sticky, causing some agglomeration of the particles which, incidentally,

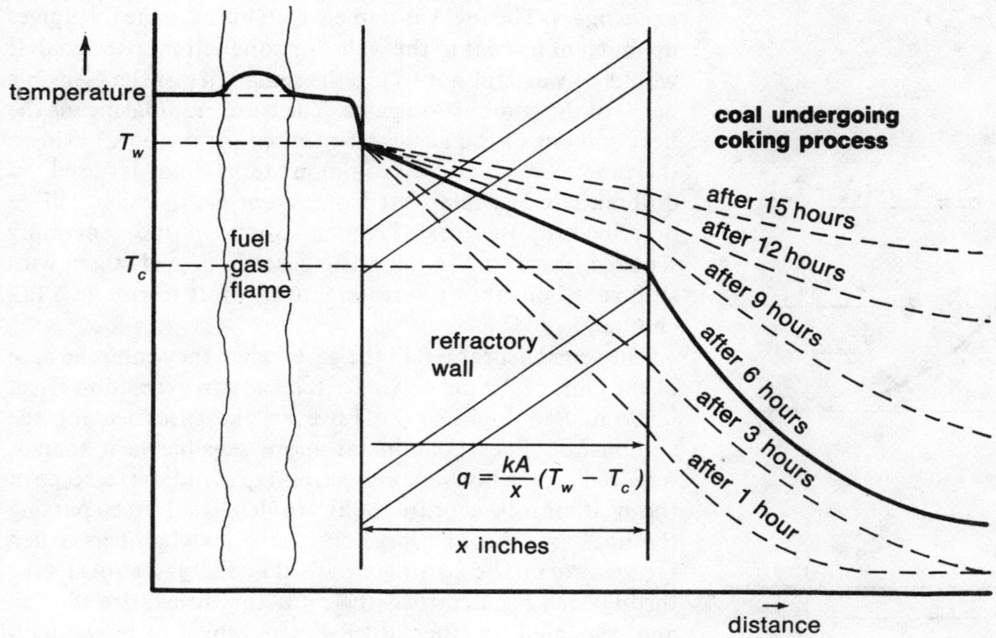

Figure 3-4. *Heat transfer through a coke oven wall.*

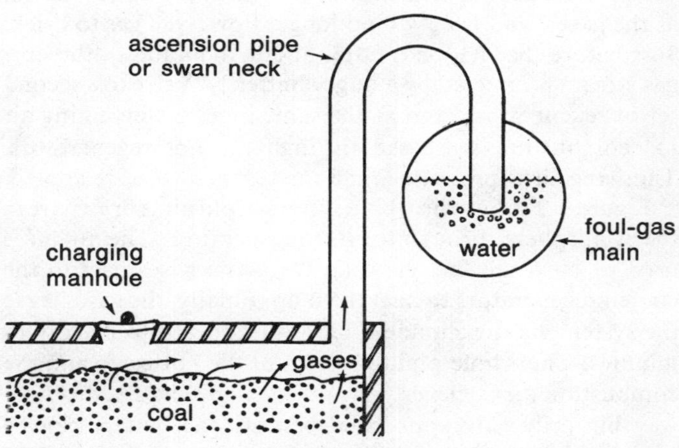

Figure 3-5. *The volatile matter leaving the coal.*

increases the heat conduction rate. At around 600°F the coal substance begins to break down. The volatile components of the coal begin to exude from the pores in the coal as a brown sulfurous vapor. Because the only way out of the oven is through a collection pipe which leaves the top of the oven (see Figure 3-5), the brown tarry vapors rise up through the bed, further helping to distribute heat throughout the bed. The vapors are sucked out of the oven and into a foul-gas main and then to a by-product recovery plant which separates this product stream into a wide range of substances.

From each ton of coal the by-products will include about 6,000 cubic feet of fuel gas with a heating value of around 500 Btu per cubic foot, 20 pounds of ammonium sulfate (a fertilizer), 8 gallons of tar, 24 gallons of ammonia and 3 gallons of an oily mixture which contains pyridine (a raw material for the pharmaceutical industry), gasoline, resins, benzene, toluene, and a variety of other chemicals in lesser amounts. These are separated into feedstocks for synthetic fibers, industrial solvents, inks, pharmaceuticals, dyes, perfumes, soaps, and literally hundreds of other consumer products.

Although this part of the steel plant is most fascinating and uses just about all chemical engineering skills in its design, I will leave out a description here since it is so similar to the plants described in Chapters 5, 6, and 7.

The fuel gas, however, is worth a mention. After it has been stripped of all its valuable by-products, it is used as a fuel gas in various parts of the steelworks. In particular it is used to heat the flues between each of the 100 ovens in the coke oven battery. A simple chemistry calculation shows that each cubic foot of typical fuel gas will need 5 cubic feet of air to burn and give up all its heat, assuming 5% extra air (called *excess air*) is supplied in order to be sure that every molecule of fuel gas meets up with a molecule of oxygen, thus preventing wastage of the valuable fuel gas. The chemical energy locked away in the carbon monoxide, hydrocarbons, and hydrogen in the fuel gas is released during combustion and transferred into the bulk of the combustion product gases. Some of the heat will be transferred to the walls of the flue, where it flows through to the coal.

The process of transferring the heat from the hot flame to the wall is a convection process. Figure 3-6 shows the temperature gradients from the bulk gas flow through to the coal. The equations in Figure 3-6 show the relationships

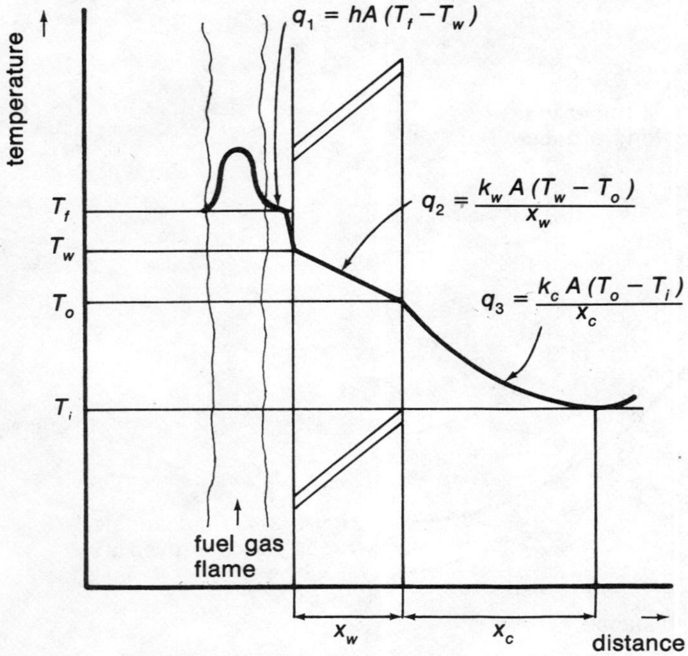

q_1 = heat transferred from flame to wall
q_2 = heat transferred through wall
q_3 = heat transferred into coal
k_w = thermal conductivity of wall
k_c = thermal conductivity of coal
A = area of path of heat transfer

Figure 3-6. *Heat transfer equations for a coke oven wall.*

between the temperature at various points, and the heat flowing. Notice that the thermal conductivity, k, of the coal and the wall material is required. The manufacturer of the wall material will provide a value of k at the temperature range of its service. The coal presents a bigger problem because of the air gaps and the complexity of the heat transfer through them.

Fortunately, many researchers have studied these systems and have published values of k for various coals under a wide range of conditions. The heat transfer coefficient, h, between the flame and the wall is also an important feature in the equations describing the heat transfer in this process. Again, researchers have determined values for heat transfer coefficients under such conditions.

The objective of the coke oven is to heat the coal as quickly as possible in order to gain the highest throughput for maximum efficiency. To achieve this, it is essential to have the fastest heat transfer through this wall and that means the highest flame temperature possible to steepen the temperature gradient down which the heat will fall. The flame temperature increases if the combustion air is preheated before being mixed with the fuel gas for combustion. This is done in *cyclic regenerative heat exchangers*. The fuel gas flame which burns in the flues gives up much of its heat to the walls for conduction to the coal. It would be wasteful not to use the remainder of the "sensible heat" of the combustion gases. The word *sensible* means the heat content of the gas can be sensed (or deduced) using a thermometer. It is a common term and is used to differentiate sensible heat from *latent heat*, which will be described in the next chapter. Some of the remaining sensible heat is left in the waste gases to provide them with sufficient bouyancy to create enough draft to rise up a tall chimney.

The sensible heat left in the gases when they enter the base of the chimney is put to work. It causes an extraction effect illustrated in Figure 3-7. However, the gases leaving the combustion flues contain far more sensible heat than is required for the bouyancy effect, and to avoid waste, some of this heat must be captured. The problem is solved by passing the hot gases through large checker-brick chambers called *regenerators*. The tortuous path that the gases must take through the regenerators (the refractory bricks are stacked and cemented together to leave gaps which form channels for gas flow) causes them to give up much of their heat to the bricks. After about 12 minutes, the bricks are almost as hot as the gases, and heat will no longer flow from gas to brick. Just before that happens (after about 10 minutes), the hot-gas stream is directed by a huge "butterfly" valve to a second set of regenerators, and at the same time, the incoming air for combustion is passed through the hot regenerators. Thus, the air is preheated, and the regenerators are cooled.

Figure 3-2 shows the whole system. Cold air is drawn from the atmosphere through the hot regenerators. The hot air is used to burn the fuel gas, and the waste gases pass to the cooled regenerators to heat them up. Finally, the gases leave the system via the chimney. The cycle repeats itself every 10 minutes. The whole piping system of the hot gases and the combustion air is cleverly arranged such that a set of two-way butterfly valves mounted beneath the battery can be operated by one, long, steel rod running the entire length of the battery.

After about 17 hours in the oven the coal is around 1,900° F. Its appearance and texture have changed considerably. It is now glowing red-hot, and it is a porous, brittle solid which when cooled down will have a steely-gray appearance. This is called coke, and of the 20 tons of coal put into the coke oven, only about 15 tons of coke remain. By the way, the coal's ash is contained in the coke.

At the end of the coking cycle both doors are taken off the ends of the oven, and a 40-foot ram (see Figure 3-8) is pushed through the oven, causing a spectacular cascade of red-hot coke to descend into a railroad car which whisks it away to be quenched with 2 tons of cooling water, of which 1.5 tons will be collected, filtered, and reused; the other half ton is lost as steam. This particular corner of the steelworks is most awe-inspiring, and even seasoned steel workers pause to watch a coke-push. The empty

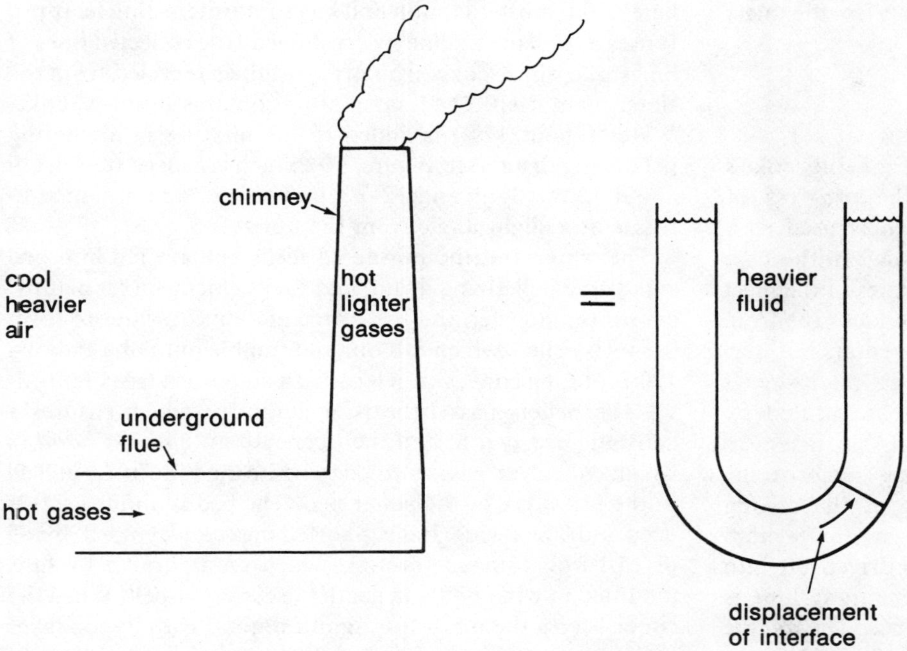

Figure 3-7. *The chimney effect.*

ovens are left open a few minutes to allow burnoff of any tars or pitch that have seeped into the seals around the door. Then the doors are replaced and the oven charged up from above as before.

When the oven is filled, it puts a huge heat load on that part of the battery. To avoid uneven distribution of heat, which could cause distortion of the expensive refractory and slow down the coking process in affected parts of the battery, the ovens are charged and pushed in a carefully planned order which spreads out the heat load. One way to schedule pushing is to number the ovens by omitting any number with a zero in it (thus, no ovens are numbered 10, 20, 30, etc.). Then the ovens are pushed in the following order. First to be pushed are all ovens with a number ending with one (1, 11, 21, etc.), then ending with three (3, 13, 23, etc.), and so on, until all the other odd numbers are pushed in sequence. Next, the same method of ordering is applied to the even-numbered ovens. In this way the temperature in any local part of the battery has a chance to recover after each oven in that part has been pushed.

The quenched coke is dumped from the coke car onto a *coke-wharf*, where it slides down to a shaking screen that separates out any chunks greater than 2 inches. These are crushed and recycled. The smaller coke (called *minus 2 inch*) is screened again to take out *fines* (coke dust which is minus 1 inch). The fines are taken to the *sinter plant*,

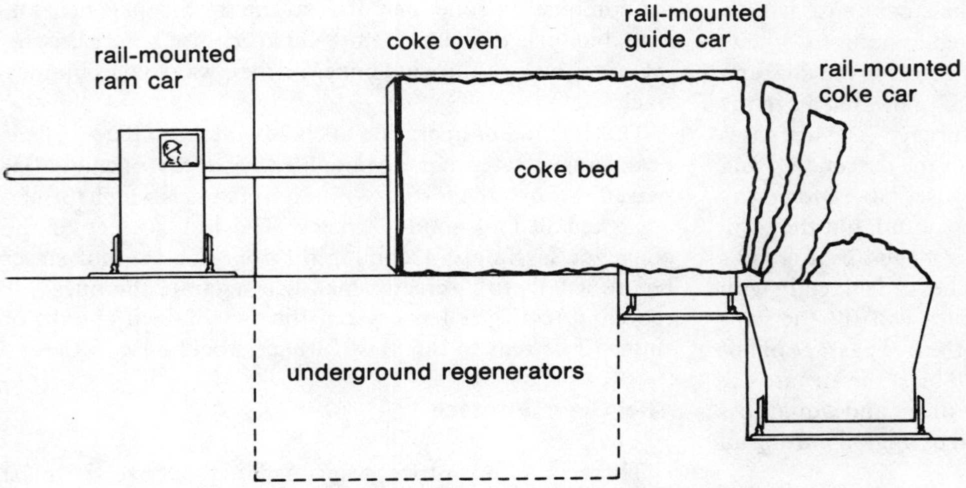

Figure 3-8. *Section through a coke oven during a coke push.*

and the minus 2 inch coke is sent directly to the blast furnace *stockhouse*.

The Sinter Plant

Let us now pause in the description of the coke's progress through the plant and trace the progress of another raw material—the limestone which is used as a *flux*. A flux is any substance that causes flow. In this case the limestone flux makes the molten impurities in the blast furnace run together and become separable from the iron.

The limestone (calcium carbonate) which enters the steel plant has been shipped from the mines in its raw state. It is crushed into three fractions which are separated by screening. The first fraction (plus 2 inch) is taken to vertical furnaces which heat it by burning some of the *coke oven gas* which was a by-product of the coking process. The calcium carbonate breaks down into calcium oxide (lime) and carbon dioxide which is driven off into the atmosphere as a harmless gas. The remaining lime is taken directly to the blast furnace stockhouse. The middle fraction of limestone (minus 2 inch plus 1 inch) is also sent directly to the blast furnace stockhouse. The third fraction (minus 1 inch) is the fines and is sent to the sinter plant.

The next raw material to consider is water. To make a ton of steel, 27,000 gallons of water are required. Some of it is recycled, but the rest is returned to rivers or the sea after first being cleaned to remove any material which may present a threat to the environment. Before the water can be used it has to be treated so it will not damage expensive equipment. (Water treatment will be discussed in the next chapter.)

Another raw material is air. Air is free in an ironworks. However, you will see later that we have to separate oxygen from the air in order to make steel by the most modern methods.

The final raw material is iron ore. Again, it is delivered in its raw state. After it is crushed, the larger chunks are sent to the blast furnace stockhouse, and the fines are sent to the sinter plant.

Eventually, all the ingredients in the stockhouse will be stacked into the blast furnace and heated in order to "smelt" out the iron. The source of the heat is the coke in the mixture, which burns when exposed to a blast of preheated air in the lower section, or hearth, of the blast furnace. The hot gases leaving the hearth zone rise up through the descending mix of limestone, iron ore, and coke. The gases have a long way to travel through this packed bed of material, but they will flow quite easily as long as the bed is made up of chunks between 1 and 2 inches. If, however, fines are mixed in with the blast furnace charge, then the fines will fill the gaps between the larger lumps and obstruct the free passage of hot rising gases. This means the fans which blast the air into the furnace would have to be bigger and stronger and would cost more to buy and to run. This is avoided by agglomerating all the fines in the sinter plant.

There are several types of sinter plants. The one described here is the most common and has some interesting features. It uses a feedstock made by crushing all the collected fines of limestone, lime, coke, iron ore, and fines recycled from the sinter plant itself. The final mixture contains about 4% coke. Water (about 12%) is added to the mixture in a rotating pelletizing drum (see Figure 3-9) which is a large steel drum about 12 feet long and 12 feet in diameter and mounted to rotate at a slight angle from the horizontal.

The water and the powdered feedstock are fed into one end of the pelletizing drum, and the "cement-mixer action" causes the powder and the scarce amount of water to form 0.5-inch pellets which fall out and tumble onto one end of a 120-foot long conveyor belt called a sinter bed (see Figure 3-9). The pellets pass beneath a guillotine which ensures a constant bed depth, and then beneath an *ignition hood* in which coke oven gas is burned to ignite the coke in the pellets in the top layer of the sinter bed. The bed is about 1.5 feet deep and 8 feet wide. It is supported on a conveyor belt made of cast-iron louvers through which air is drawn by fans mounted in wind boxes beneath the conveyor belt. When the sinter leaves the 6 x 8-foot ignition hood with its top layer ignited, combustion continues down through the sinter bed for the remainder of its 120-foot journey.

The sintering process is most interesting because it incorporates many basic fundamentals of chemical engineering. Figure 3-10 shows how, after 20 minutes of travel, the hot combustion zone has moved to approximately half way down through the bed.

Above the combustion zone is cooling sinter—a black shiny material formed into chunks which are quite strong, owing to the partial melting of the lime, iron, and the impurities in the mixture. Beneath the combustion zone are fresh pellets. Consider what happens to the air which is being drawn through the bed from the atmosphere (see Figure 3-10). First, it enters the hot sinter and becomes preheated itself as it cools the sinter. Because the air enters the combustion zone hot, it causes a higher flame temperature (2,600° F) which helps the melting or fusion process to occur, giving physical strength to the sinter. The hot gases leaving the combustion zone then preheat the fresh pellets beneath it, and in turn, the gases themselves are cooled before leaving the sinter bed via the wind boxes on their way to the chimney stack.

This method of operation takes advantage of the excellent heat transfer rates that are possible when a gas (or a liquid) is passed through the tortuous random channels which form in a packed bed of solid particles. The last 20 feet of the conveyor is simply a cooling-off zone for the hot sinter before it falls to a screen which will separate the fines (for recycling to the pelletizer) and the 1- to 2-inch chunks of sinter to be sent to the blast furnace stockhouse.

The Blast Furnace

There is a tiny office suspended high above the blast furnace stockhouse for the stockhouse controller, whose job

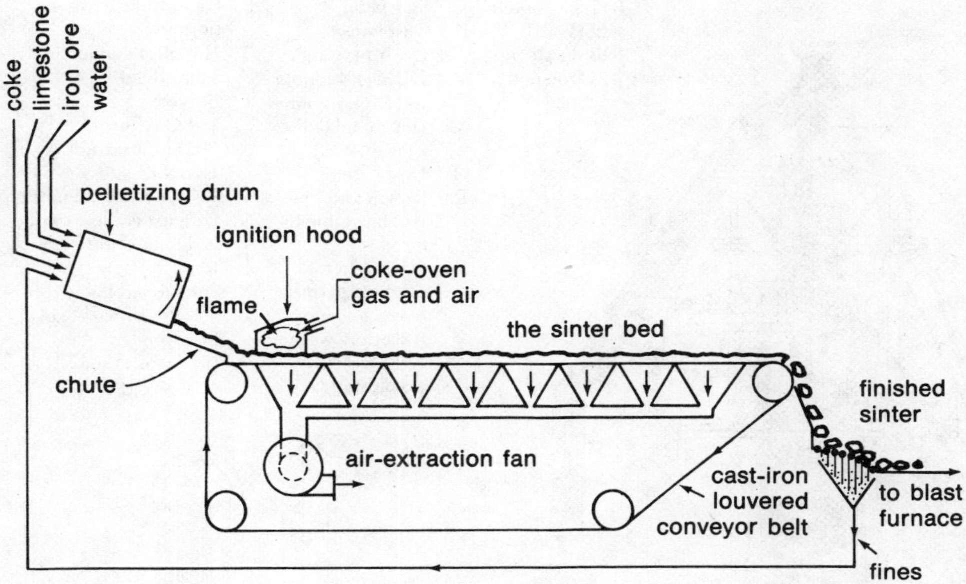

Figure 3-9. *A sinter plant.*

is to ensure a steady feed of the various iron-making ingredients to the blast furnace in the right proportion and small enough amounts that the bed of mixture, or *blast furnace burden*, contains no large pockets of any one component. Buttons at the controller's fingertips operate the discharge gates of the hoppers of materials which are situated above a chute leading to a skip conveyor. The skip conveyor hoists rotatable skips to the top of the blast furnace, where they turn over and dump their contents through a *double-cone hopper* to become blast furnace burden.

Figure 3-11 shows the schematic appearance of the blast furnace. The reason for the burden entering the blast furnace through the double-cone hopper is to prevent gases from leaving through this particular opening. Instead, the gases are forced to pass through another port at the top of the blast furnace.

The double-cone hopper works as follows: The smaller, upper cone is moved downwards, and the burden resting upon it drops into the space between the cones. Then the upper cone closes, and the larger, lower cone is lowered. In this way the fresh burden enters the top of the blast furnace but only a tiny amount of gas escapes.

Even an average-sized blast furnace will be 140 feet high. Its internal diameter at the widest part, around the hearth, is around 35 feet, and the walls at that point are about 3 feet thick, although the wall thickness at the upper, cooler part of the blast furnace will be 6 inches less. The furnace structure weighs about 20 million pounds (or 10,000 tons) and rises over 200 feet above the ground because of the gas ducts and skip conveyor equipment on top of the furnace. The whole structure rests upon a concrete foundation almost 40 feet in diameter and 11 feet thick—itself weighing nearly 2,000 tons!

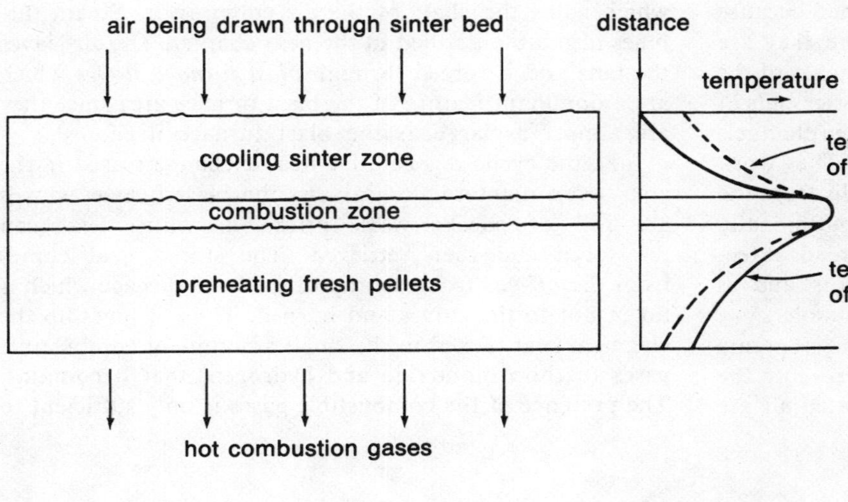

Figure 3-10. *Section through the sinter halfway along bed.*

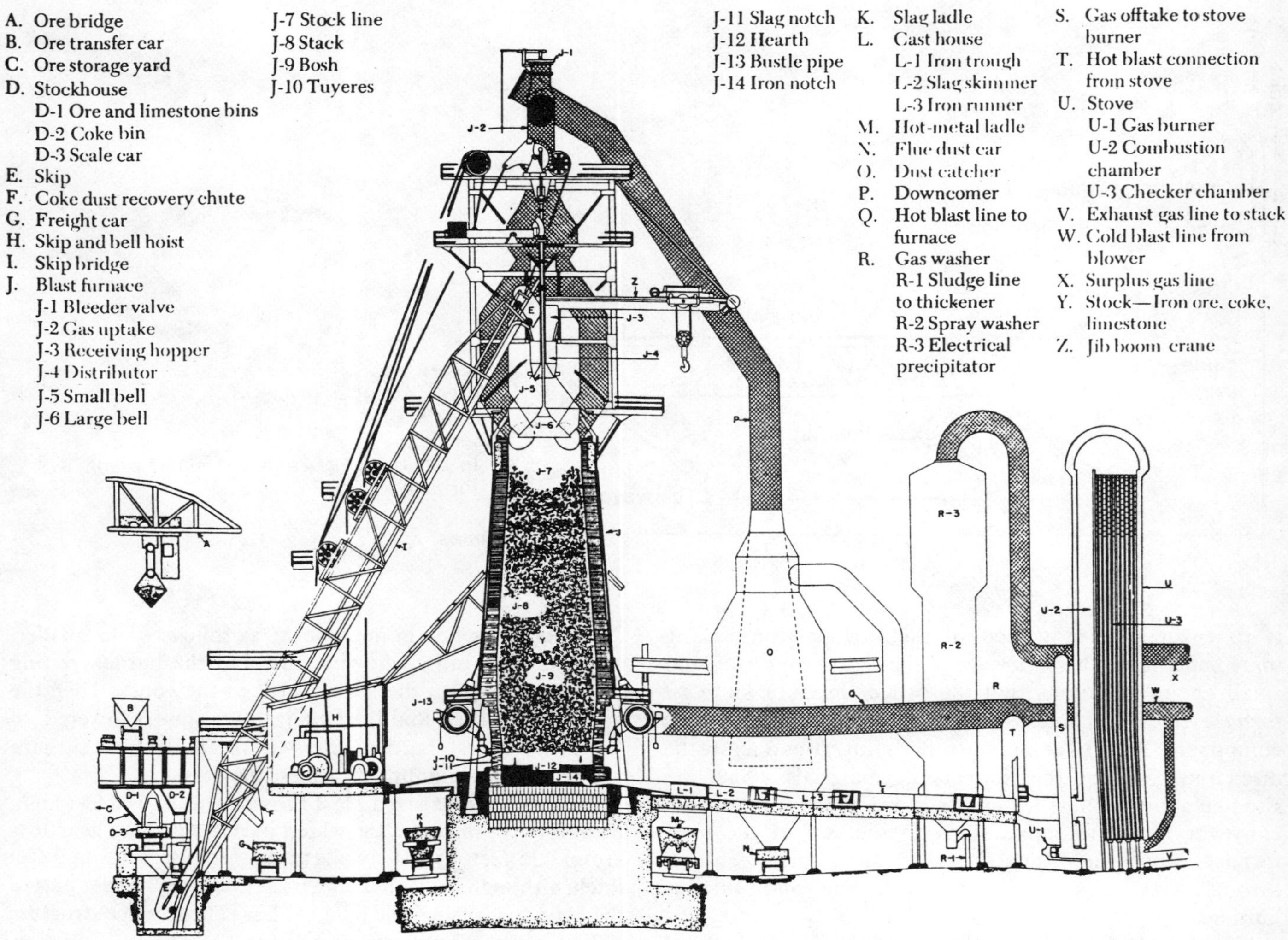

Figure 3-11. *A cross-section of a typical modern blast furnace plant. (© 1971 by U.S. Steel Corp. Used with permission.)*

The walls are made of refractory. Refractory has a longer life if kept at a lower temperature than the inside of the blast furnace (up to 3,000° F). To lower the wall temperature, *cigar coolers* are inserted into the wall at intervals of 3 to 4 feet in every direction. Cigar coolers are so named because they have the appearance of steel cigars before they are embedded into the refractory wall to within 6 inches of the inside wall surface. Water is pumped into the hotter ends of the cigar coolers, and it flows out to be collected in channels below the coolers on the outside of the furnace. They work by absorbing the heat which soaks into the walls from the inside. Just as a drop of ink spreads two-dimensionally across blotting paper, so does the heat spread three-dimensionally through the blast furnace walls and is captured for removal by the water in the cigar coolers.

Blast furnaces carry that name because they operate using a continuous blast of air through the *tuyeres* into the hearth zone, or *bosh*. About 4.5 tons of preheated air are needed to make 1.0 ton of iron. The air is compressed in giant turbo-blowers, driven by steam provided from a utility boiler (described in the next chapter). The high-pressure steam is forced through a turbine, the shaft of which spins the shaft of the air compressor. Steam turbines also are described in the next chapter. The air leaves the fans and is forced through *blast furnace stoves* which are a dominant feature of the blast furnace area since they are almost as large as the blast furnace itself.

Like the cyclic regenerative heat exchangers used in the coke oven plant to preheat air, the blast furnace stoves are full of checker-work bricks in which heat can be stored and then retrieved. The stored heat comes from the *off-gas* from the top of the blast furnace which is taken hot to the stoves and burned. Thus, it gives up the chemical heat locked in the small amounts of combustible gases (carbon monoxide and hydrogen) that it contains. The presence of the combustible gases is only sufficient to give a heating value of around 90 Btu's to the off-gas. Such a "lean" gas is not worth the expense of piping it to be used elsewhere in the plant. The roles of these two stoves are reversed every 10 to 15 minutes. Again, as in the coke oven plant, a valve device ensures a quick and efficient changeover. The preheated air is blasted through the tuyeres into the bosh where it burns the coke, causing it to be transformed into carbon dioxide and a remaining ash which melts in the high temperature in the bosh (about 3,400°F).

Notice that a *passenger* is introduced into the blast furnace. Although only oxygen is needed to cause the combustion reaction in the bosh, air is used which is about 21% oxygen and 79% nitrogen. The nitrogen serves no purpose whatsoever, and its presence carries heat away from the bosh, thus lowering its temperature. If we wish to achieve a higher temperature in this zone, then we can enrich the air blast with oxygen taken from the air separation unit of the steel plant.

As the hot gases leave the bosh and filter up through the packed bed of burden, they serve several purposes. First, they carry heat away from the bosh and use it to preheat the descending burden. Figure 3-12 shows a temperature distribution of the burden and the gas. A number of crucial chemical reactions occur. More carbon dioxide is produced as the remaining limestone becomes *calcined* (changed to calcium oxide), and the hot coke (which has not yet begun to burn) reacts with the carbon dioxide to give carbon monoxide. In addition the carbon monoxide reacts with the iron ore to leave iron and carbon dioxide. These equations are shown in Figure 3-13, but remember

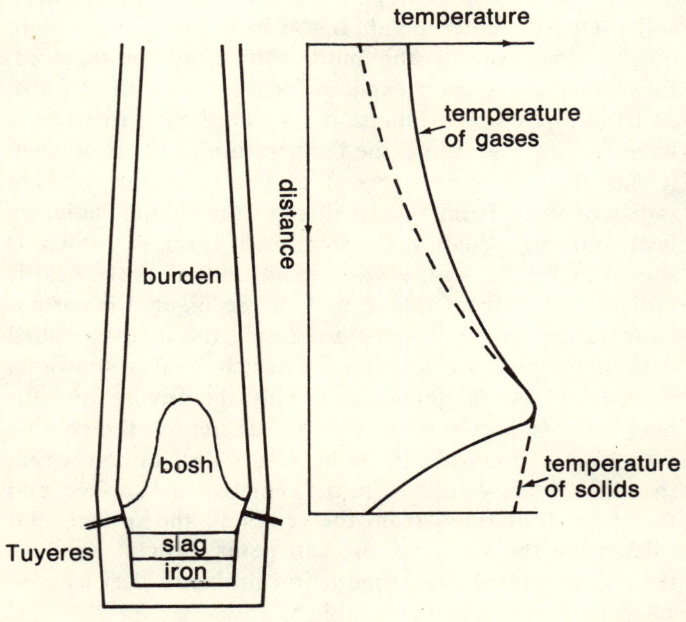

Figure 3-12. *Temperature distributions of burden and gases in a blast furnace.*

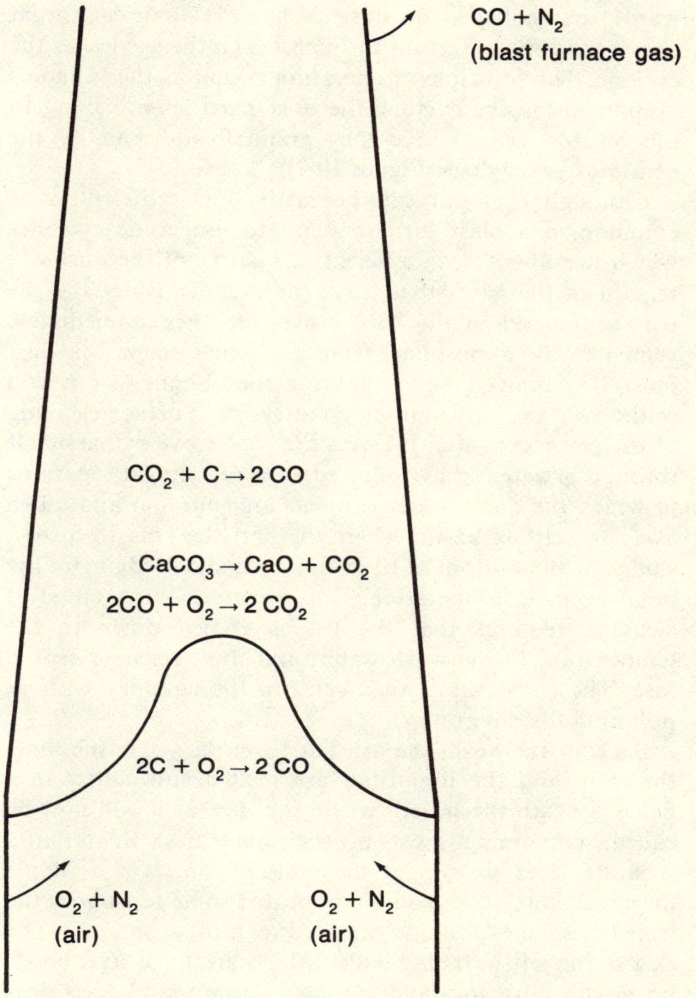

Figure 3-13. *Simplified chemical reactions in a blast furnace.*

that they are merely a simplification of far more complex chemistry which occurs in the burden. Although the reactions that actually occur are more complex, in the end their effect is the same as the reactions shown in Figure 3-13.

Again, notice the ideal heat transfer arrangement in the blast furnace as in the sinter bed. The descending burden is preheated by the ascending gas by receiving heat from the gas which no longer needs it. The gas, which contains a little combustible carbon monoxide, leaves the blast furnace through pipes projecting from the top of the furnace. This off-gas is carried down to whichever blast furnace stove is "on" and is burned for air-preheating purposes as you saw earlier. Since the gases rising in the blast furnace are moving at high speed, they pick up small particles of raw materials on their way to the exit pipe. A dust catcher removes this solid material by whirling the dust-laden gas around inside a cyclone. A cyclone is a vertical cylinder with a conical section pointing downwards that collects the dust; it has a cylindrical throat above the conical section through which the gas leaves the cyclone. The principle of operation is that as the gas spins around inside the cyclone the dust particles are flung to the outside walls where they gradually descend by the action of gravity (see Figure 10-7).

Although cyclones can be made quite efficient, it is common in a blast furnace plant to use crude cyclones which are about 75% efficient; a quarter of the dust will remain in the gases that leave the cyclone. Since the gas will be burned in the "on" stove and then immediately vented to the atmosphere from a chimney stack, this dust must be removed before leaving the chimney or it will settle over the surrounding countryside. Further cleaning of the gas occurs after it leaves the "on" stove by passing it through a water spray followed by a centrifugal separator in which the dusty water droplets are spun out and taken away to settling basins where the particles sink to form a sludge at the bottom of the pond. Periodic dredging of the pond keeps it in operation. Unfortunately, this method of cleaning requires that the gas is cooled down to the temperature of the wash water, and thus, sensible heat is lost. The spent gas is then sent up the chimney without polluting the environment.

Back in the bosh, the ash left from the coke, the lime, the iron, and the impurities are melted and collect in a basin beneath the hearth where the slag, as it will now be called, containing most of the impurities, floats in a separate layer on top of the molten iron. It is a simple matter then to have a slag tap located some feet above the iron tap so these liquids can be drawn off separately. The slag is run off into slag ladles which are huge steel bowls lined with refractory and mounted on railroad bogies. It is taken away, cooled, frozen, and crushed to be sold to cement manufacturers or to the construction industry for land fill or other purposes.

The iron is run off through an iron tap which discharges into a trench cut temporarily into sand. The trench leads to iron ladles which are refractory-lined steel bowls mounted on railroad bogies. The cascade of iron is stopped occasionally by damming the stream in order to move an empty ladle into position beneath it. The product at this stage is called *pig iron* because the iron from early blast furnaces was poured into temporary sand channels with side arms which were filled up on each *pour*. In between pours they were allowed to cool and freeze, and then, the bars of iron were broken up for remelting to steel.

In those early days of the industrial revolution, ironworkers came from a predominantly agricultural society, and the farmers with their newly acquired skills saw a resemblance between the main iron channels with its smaller side channels and a sow with a number of suckling piglets. Hence, they coined the name pig iron which has stuck ever since. Modern practice does not allow the pig iron to freeze (unless it is to be sold as merchant pig iron) since doing so would waste the sensible heat of the iron. Instead, the iron ladles are taken from the blast furnace site to the steelworks as quickly as possible.

So far, then, the plant has taken iron oxide and removed most of the oxygen and some other impurities, but the pig iron is still only a crude form of iron, and it contains carbon and other impurities. Some of the pig iron is taken molten to the steelworks. Some is sold as merchant pig iron—a name which signifies that it will be sold as an end product. For this latter use, it is frozen into small pigs weighing about 40 pounds each and then taken to foundries where it is remelted and cast into such products as pipes for waste or water transport, oil and gas burner parts, and automobile engine blocks. The list is endless.

Note the difference between the process which produces merchant pig iron for sale to end-users and the pig iron used for steel making. Merchant pig iron is made in small blast furnaces which are capable of a high degree of control over the composition of the final pig iron in order to meet the buyer's exact specifications. Merchant pig iron is emptied from the ladle into a channel which pours into small ingot molds mounted on a continuous conveyor belt. The molds pass under the cascade of iron at a rate which is controlled so that 40 pounds of iron will fall into the mold before it moves on and is replaced by the next one. A water spray cools the iron, and the molds move a long distance along the conveyor belt to ensure that the pigs freeze before they reach the far end where the molds overturn and drop the pigs into a railroad car.

The design of the conveyor belt is crucial. Imagine the waste, confusion, and danger to workers if molds were overturned into the railroad car before the iron had frozen solid. On the other hand, if the iron was cooled more quickly, then considerably more water would be required, thus causing a greater expense for the water, the pumps, and the pipes which handle it. We need then to strike a balance. Since we know the temperature of the iron when it fills the pig mold, we can calculate the rate of heat radiated away from the small ingot using the radiation heat transfer equation, a simplified form of which is shown in Figure 3-14. However, notice that as the heat is radiated away from the surface of the ingot, the surface temperature will fall, and thus, inside the ingot we shall have a temperature distribution which is also shown in Figure 3-14. As the temperature falls, the driving force for heat transfer becomes smaller, and therefore, the cooling rate drops. Of course, the temperature distribution within the ingot is "downhill" from the center to the surface, and therefore, heat flows from the center to the surface. But calculating the working rate can be as difficult as in all three-dimensional heat conduction problems. Let us consider how to solve the problem.

We know the surface temperature of the iron initially, but as soon as 1 Btu is radiated away from the surface,

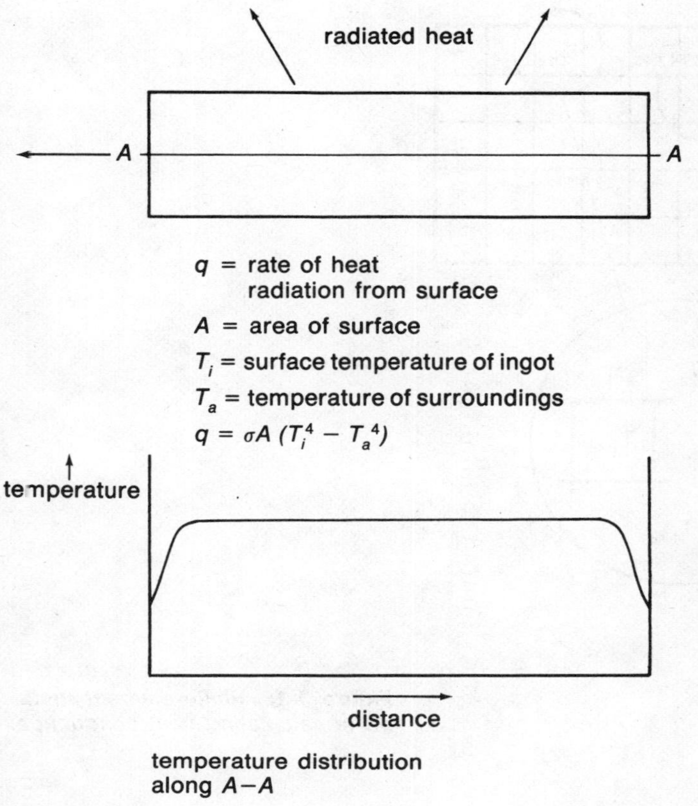

Figure 3-14. *Heat transfer from a pig-iron ingot.*

its temperature falls, and we can no longer calculate the rate because we don't know the new temperature. Normally, if a body is cooling and we know its specific heat, then we can calculate its new temperature after we know the amount of heat which has left it. However, in this case the heat is leaving the surface and thus tending to lower its temperature, and yet, fresh heat coming from the interior of the ingot tends to raise the surface temperature. It looks like a deadlock.

However, tools are available to the chemical engineer to calculate the cooling time of these ingots down to, and below, their freezing temperature. A simple way, which always works, is to imagine that the liquid iron is made up of thousands of tiny cubes of liquid iron stacked together to form the ingot. We can calculate the rate of radiation from the surface cubes. Assume that in the first tenth of a second of cooling, the temperature of the surface cubes remains constant, and therefore, the rate of radiative heat transfer remains constant over the tenth of a second period. We can then calculate the heat lost from the surface cubes. Knowing the heat lost from the cube, we can now calculate its new temperature. In the second tenth of a second we can do two things.

First, we can again calculate a heat loss assuming a steady temperature and radiative heat loss rate for the whole of a tenth of a second for each of the surface cubes. Second, we can calculate the heat flow which will occur from the cubes next to the surface cubes into those surface cubes, as a result of the tiny temperature difference between them. In this way we can calculate the new temperature of the surface cubes and the new temperature of the next layer in. This stepwise arithmetic operation can be repeated for all the cubes in the ingot from surface to center, and if we do this for hundreds of tenths of a second, we will find that the cooling mechanism works its way to the center of the ingot. If we keep this up long enough, we would reach the point where all the heat contained by the ingot has radiated from the surface of the cubes.

Figure 3-15 shows the basis for this method of calculation of heat conduction which is called the *finite-element* method because it uses the elemental cubes of a finite size. Study it and you will see that the equations are very simple, and yet, they do a difficult job. The other method of dealing with this difficult kind of problem is to use differential calculus. Using differential equations we can equate the *rate* of radiative heat loss to the *rate* of heat conducted from the inside of the ingot. Calculus is beyond the scope of this book and will not be discussed further.

The finite-element method of calculating heat conduction is simple but not trivial. We have to use such small cubes, or *finite elements*, in order to make the answers accurate that there are literally millions of simple arithmetic calculations to carry out. For instance, suppose we divide our ingot into 1,000 rectangular elements 10 x 10 x 10. Then suppose that it takes 5 minutes for the ingot to cool below the freezing point. Then the total number of calculations we have to do is 1,000 x 5 x 60 x 10 = 3,000,000 heat balances, each of which involves several subcalculations. This would take forever by hand. A computer can carry out over a million calculations like this in just a second or two. There are several proprietary computer programs which are general enough to solve any problem involving heating, cooling, conduction, radiation, and convection. One of the section B computer program problems in Appendix I is based upon heat conduction in a furnace wall and you will use a similar method for its solution.

Having calculated the time for the entire ingot to cool to below its freezing point using either of the above methods, and knowing the rate of pig iron production, it is easy to calculate the length of conveyor belt necessary to allow the ingot to cool for that calculated length of time. We must always take this amount of painstaking care when designing even such a simple plant parameter as the length of a conveyor belt. Too short, and there would be all sorts of terrible problems. Too long, and expensive capital equipment would be bought unnecessarily.

If the ironworks is integrated with a steelworks, the blast furnaces will be considerably larger than those used to make merchant pig iron, and the blast furnaces will be

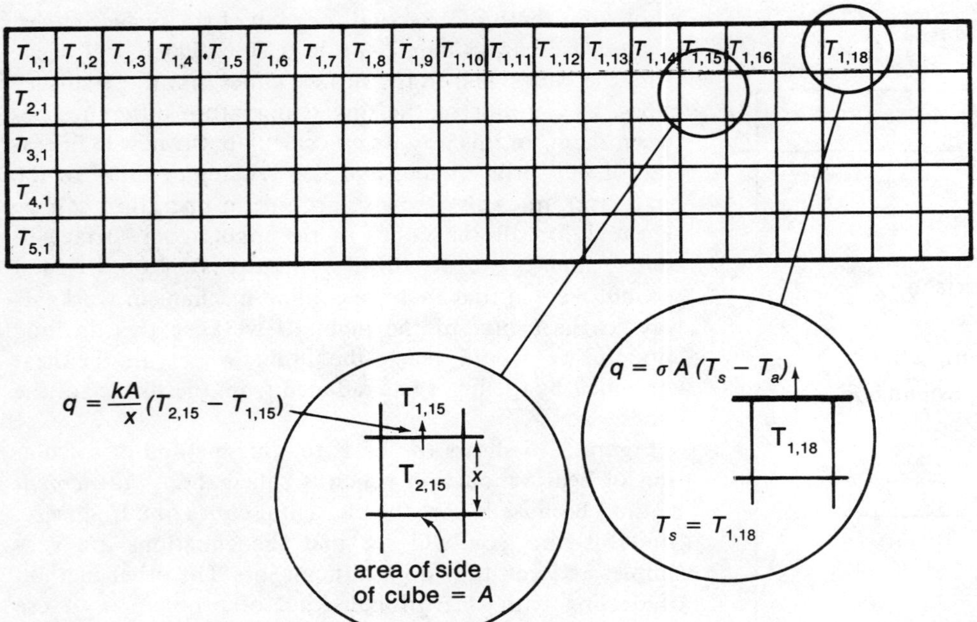

Step 1. First tenth of a second; only surface cubes lose heat.
Step 2. Second tenth of a second; surface cubes radiate more heat; next cubes in conduct heat to surface
Step 3. Third tenth of a second; surface cubes radiate more heat; inside cubes conduct heat outwards
etc.

Figure 3-15. *Finite-element method of calculating heat conduction.*

less capable of finely controlling the composition of the finished iron. The molten pig iron will be taken away by ladle either to a holding reservoir of hot molten iron called a *mixer*, or it will be taken directly to the steel-making furnace.

The Open Hearth Furnace

There has been a variety of ingenious means of heating the iron and removing all its impurities before adding the exact amounts of other ingredients which will turn the soft iron into the much stronger steel. The most well-known are the *open hearth furnace* and the *basic oxygen furnace*.

The open hearth furnace was invented in the early 1700s, but it was over 100 years later in 1865 that Sir Francis William Siemens patented the first truly practical design. Many, many changes have been made to the original design, and the ones in operation today are best represented by Figure 3-16. The fuel gas enters through a checker-brick regenerator and is preheated. The combustion air is preheated in another regenerator and mixed with the fuel gas in the burner which is mounted horizontally at the end of the furnace. The flame is established over the bath of hot metal and radiates heat to the furnace walls and ceiling as well as to the bath of metal itself. Most of the radiated heat which strikes the hot refractory walls and ceiling is reflected back towards the bath of steel. The remaining heat is either conducted through the refractory and lost to the surrounding environment by radiation and conduction from the outside of the furnace; or, it is carried out of the furnace in the hot combustion gases. The bath is 2 to 4 feet deep (depending on the size of the furnace) with molten pig iron straight from the ironworks or from the mixer (a refractory-lined reservoir of molten iron). Since the flame is sweeping over the surface of the iron, the excess oxygen in the gases helps burn off the remaining carbon in the iron. Of course, the residual oxygen in the iron also helps to burn the dissolved carbon.

Raw iron ore is added to the bath using a *charging machine* which looks like a 30-foot teaspoon. The oxygen in the ore reacts with dissolved carbon to release carbon dioxide bubbles in a vigorous boiling action called the *ore boil*.

Next, limestone is added to the bath using the huge charging machine. When the metal bath reaches a high enough temperature, the carbon dioxide is driven out of the calcium carbonate, causing a bubbling action called the *lime boil*. The ore and lime boils serve several purposes. First, the bubbles agitate or stir up the bath, causing more of the metal to be exposed at the surface to the oxygen in the furnace gases. Second, the carbon

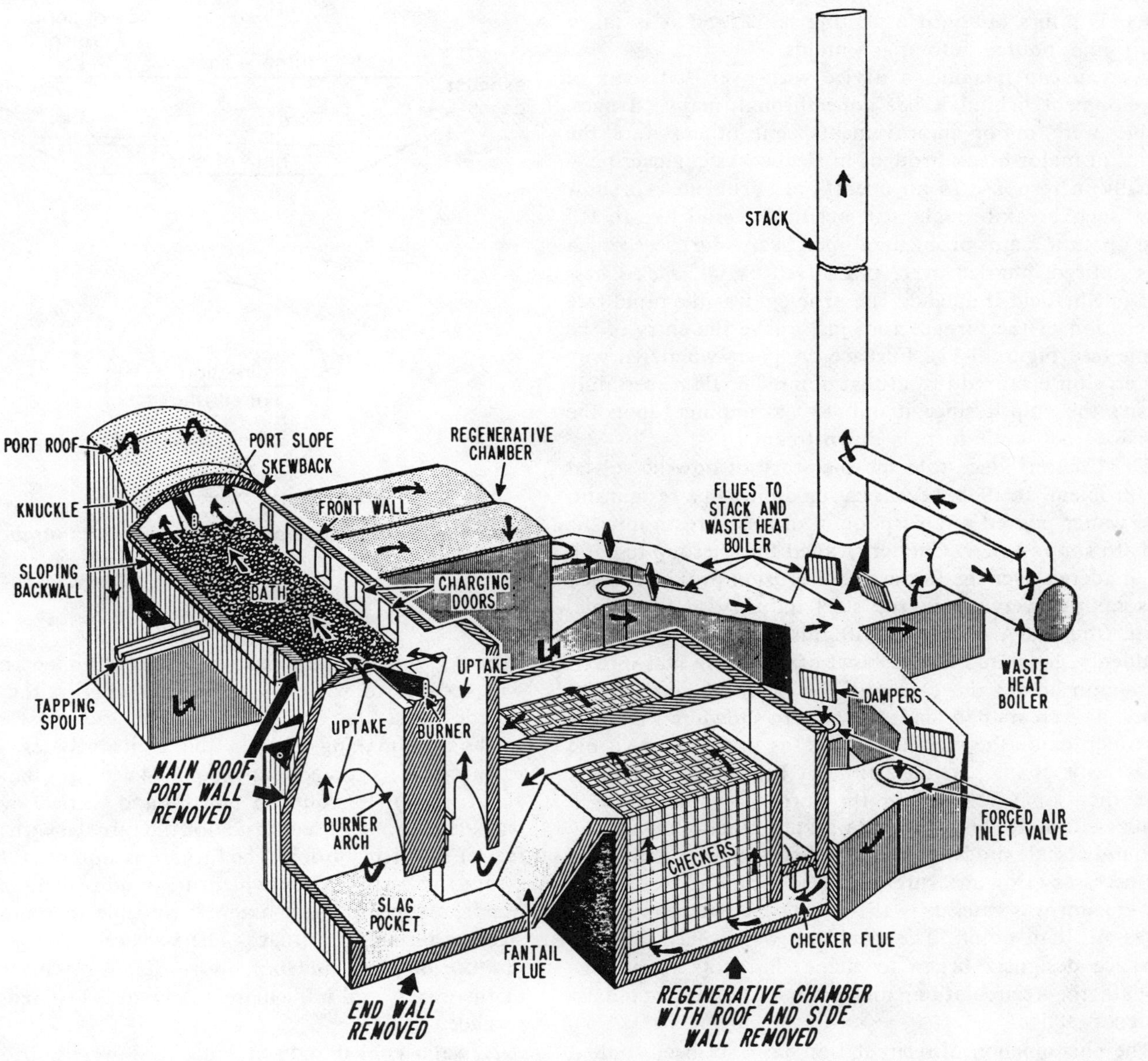

Figure 3-16. *The principal parts of an open hearth furnace. Heavy curved arrows indicate direction of flow of preheated air, flame, and waste gases when a liquid fuel is fired through a burner in the trench at the right end of the furnace. The five doors shown are in the front wall of the furnace, and the checkered chambers extend under the charging floor, which is not shown. When reversal of firing takes place, the function of the uptake is reversed, and it becomes the downtake. (© 1971 by U.S. Steel Corp. Used with permission.)*

dioxide reacts with carbon in the metal to form carbon monoxide which bubbles out of the metal, helping rid it of carbon. Third, the heat transfer from flame to metal is increased since the bubbles cause cooler metal to be moved to the surface. The cooler metal causes a greater temperature difference between the bath surface and the flame, and thus, the heat flux increases. Fourth, the bubbles burst at the surface, pushing away the thin slag layers and exposing more metal to the oxidizing furnace atmosphere.

When the steel is found by sampling and analysis to have the carbon content required by customer specifications, the charging machine is used to add other ingredients—such as manganese, chromium, tungsten, nickel, and scrap steel—to produce the required final composition.

After about 10 hours from the beginning of the *heat*, the flame is turned off, and the steel is poured (from a tap below the slag layer) by tilting the whole furnace about its horizontal axis. The 200 tons of molten steel at about

2,900°F pours out into a waiting ladle, and it is taken away and poured into ingot molds.

As you can imagine, a device with over 200 years of development behind it has gone through many changes. Some were minor improvements, but others were the result of major breakthroughs in steelworks engineering—usually in response to an operational problem. To show how such breakthroughs can occur, I would like to tell you about a major problem of open hearth furnaces which was noticed, puzzled over, and solved by Dr. Jack Chesters in Sheffield, England. The problem was the rapid rate of erosion of the furnace roof just above the entry of the flame (see Figure 3-17). Furnace designers wondered why the erosion occurred just at that point. The flame certainly wasn't the culprit since it didn't even impinge upon the furnace roof until further downstream.

Dr. Chesters once told me the story of how he solved the problem. In 1948, he was eating dinner in a restaurant. The waiter poured a tablespoonful of fresh cream into his tomato soup. As Dr. Chesters waited for the soup to cool, he pondered over the furnace roof erosion problem which was taxing everyone in the steel industry at that time. Absentmindedly, he toyed with his spoon in the soup. Suddenly, he stared at the bowl of soup. He had moved his spoon across the center of the bowl of soup several times. The cream had made the pattern shown in Figure 3-17 which made the recirculation paths in the soup visible as a result of his stirring action. It became clear to him that the flame velocity in the furnace would cause a similar effect. He could barely wait until the following day to build a scale model of the open hearth furnace and take the necessary flow measurements. He soon discovered that the erosion was caused by the violent recirculation of hot gases at that point. The problem was solved at last. Furnace designers began to shape the roofs to accommodate the recirculation, and soon it became standard design practice.

The phenomenon of recirculation has since been studied in detail. It is caused by another phenomenon called *entrainment*. The fast-moving molecules in the flame bump against the surrounding gases and set them in motion in the same general direction as the flame. As they move off, they leave a space which is filled up by more gas molecules from the "stagnant" area around the burner. At high velocities this process of entrainment causes rapid backward flow by the furnace roof sidewalls and the hot metal bath. It was this violent recirculation which particularly affected early furnace roofs.

The open hearth furnace has had its heyday. Gradually, it is being replaced by the basic oxygen furnace.

The Basic Oxygen Furnace (BOF)

The basic oxygen furnace is so simple that at first sight one wonders why it was not developed hundreds of years ago. As you read the following description perhaps you will be able to work out why it is only in the last few decades that the BOF has become the most common method of making steel in the United States.

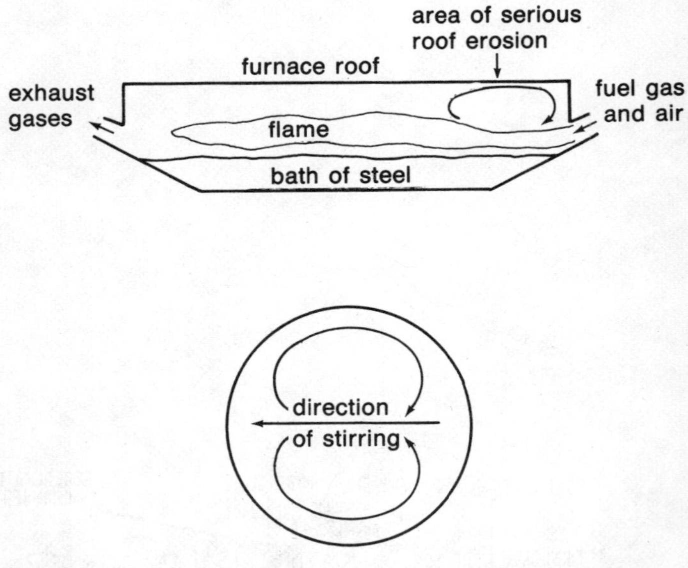

Figure 3-17. *Recirculation in an open-hearth furnace and in a bowl of tomato soup.*

The BOF is a steel cup about 24 feet in diameter (see Figure 3-18). Its rounded bottom and vertical cylindrical walls rise about 20 feet inside of the furnace with almost 3 feet of refractory brick. The furnace is supported by short, 1-foot-thick stub axles which are mounted diametrically opposite each other, making it possible to pour out the steel by tilting the furnace. The furnace is charged with a mixture of scrap and molten iron from the blast furnace or the mixer, and immediately, a hood is lowered over the furnace.

A water-cooled oxygen lance is lowered through the hood to about 5 feet above the surface of the iron, and high-pressure oxygen is blown straight at the metal surface. The pressure of the oxygen, which at 150 psi is over five times greater than the pressure in car tires, forms a deep depression in the surface of the molten metal. In fact, bubbles of oxygen actually may break away from the pit of the depression and bubble up through the iron. The violent agitation caused by the oxygen lance is beneficial because the stirring action keeps the bath composition uniform, i.e., no pockets of impurities will accumulate. The carbon, sulfur, and other combustibles in the iron react with the oxygen, which is just another way of saying they burn. When they do react, they produce heat just as if they were burning in, say, a piece of coal. The liberated heat is used in two ways.

First, it melts the scrap by providing what is called the *latent heat of fusion*. The latent heat of fusion of a

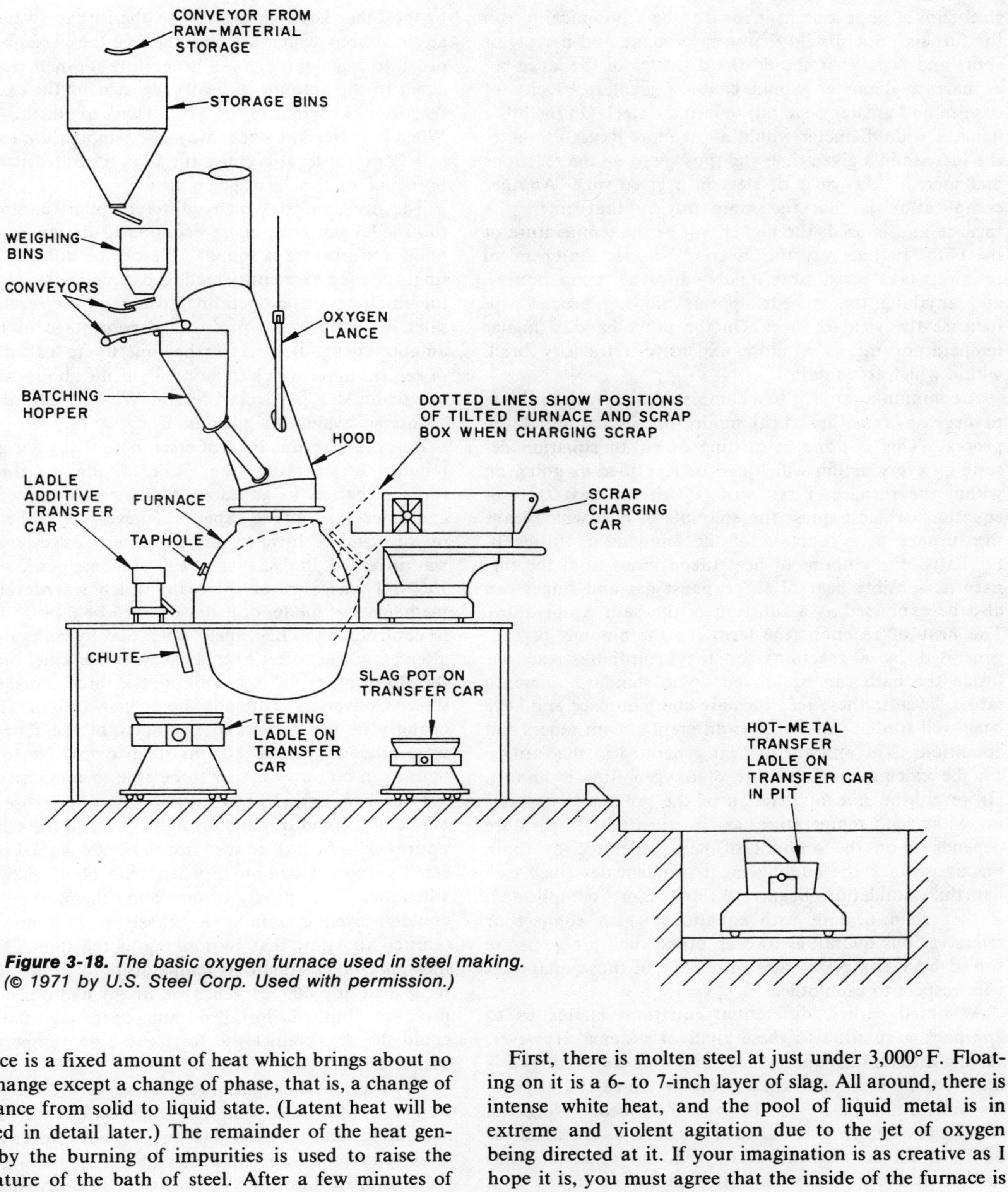

Figure 3-18. *The basic oxygen furnace used in steel making. (© 1971 by U.S. Steel Corp. Used with permission.)*

substance is a fixed amount of heat which brings about no other change except a change of phase, that is, a change of a substance from solid to liquid state. (Latent heat will be discussed in detail later.) The remainder of the heat generated by the burning of impurities is used to raise the temperature of the bath of steel. After a few minutes of being oxygen-blown, the entire contents of the bath are molten. Lime and other fluxes are added to collect any remaining impurities as a floating layer of slag. So, we now have a most interesting situation. Imagine what is going on inside that vessel.

First, there is molten steel at just under 3,000° F. Floating on it is a 6- to 7-inch layer of slag. All around, there is intense white heat, and the pool of liquid metal is in extreme and violent agitation due to the jet of oxygen being directed at it. If your imagination is as creative as I hope it is, you must agree that the inside of the furnace is an awesome sight.

Now turn your mind to the problems faced by the designers of furnaces like these. Designers must answer questions such as how big the vessel should be. From experience the designer knows that at least 300 tons of steel should be in one heat for it to be economical to run the furnace. But should the furnace be tall and narrow or short and fat? What should the diameter of the lance be? A narrow diameter would cause a greater velocity of oxygen and greater penetration into the steel. On the other hand, a wide diameter would allow more oxygen to enter the furnace in a given time and thus speed up the reactions and increase the yield of steel in a given time. Another complication is that the more oxygen that enters the furnace and is used, the higher will be the temperature of the bath. In one way this is good because the chemical reactions take place more quickly at higher temperatures, and anything that speeds up this blowing process will increase the yield of steel. On the other hand, a higher temperature means a more expensive refractory brick within which to contain it.

A common approach to a complex design such as this is to develop a mathematical model or simulation of the process. This is done by writing down an equation describing every action which can be identified as going on within the furnace. First, write down the heat transfer equations which express the amounts of heat which leave the furnace as a function of the temperature inside it. Similarly, the amount of heat taken away from the furnace as sensible heat of the exhaust gas and fumes can also be expressed as a function of the bath temperature. The heat of reaction (the term for the amount of heat generated by a reaction) for the identifiable reactions inside the bath can be looked up in standard reference tables, because these reactions are commonplace and have been well-studied, albeit under different circumstances and conditions. The amount of heat generated in the furnace can be calculated if the rate of oxygen flow is known. However, the rate of reaction of the pollutants depends upon the bath temperature, and in turn the temperature depends upon the amount of heat generated by those reactions. Like the cooling ingot problem described earlier this simulation begins to look more complicated. Rather than dealing with equations which connect or relate various quantities to each other, such problems are solved by relating the rates of change of those quantities with respect to each other.

As noted earlier, differential equations enable us to approach a solution to these kinds of systems. However, solving differential equations is often time-consuming. It is more common to program a computer to carry out process simulations such as the one described. Using the simulation, we can guess at such variables as the diameter of the lance, the feed rate of oxygen, the initial temperature of the iron, or the percentage of carbon in the iron. Then put these data into the model and allow it to reach a balance where the temperature of the bath settles to a steady state or equilibrium value. Then we shall know that the rate of heat loss in the exhaust gases and fumes added to the other heat losses through the furnace walls and to any reactions which are endothermic (i.e., require heat in order to react rather than generating it when reacting) is equal to the amount of heat generated by the exothermic reactions occurring in the bath. Think about this balance. When the "heat" is under way, the temperature settles out at a constant level because the heat generated is balanced or equal to the heat being lost.

The designer tries many different plant parameters in the model until the most economical design solution to build and operate is found. This can be done by calculating, for each case considered, the capital costs of building the equipment, the operating costs, and the revenue from sales. By calculating profit as a percentage of the total amount spent, the designer has one figure, called the *rate of return*, upon which to base judgments about whether or not to build a particular design. We will talk about this economic evaluation task in Chapter 11.

Returning to our bath of steel, note that after about 45 minutes of operation the flame leaving the top of the furnace changes its appearance. The change may be subtle and detectable only to experienced eyes. An earlier version of this kind of furnace, the Bessemer Converter furnace, was invented in 1851 and was still in operation in the 1960s. I was part of the team which was developing a mathematical model of a Bessemer to be used as the basis of controlling the Bessemer's operation by computer. Blow after blow, heat after heat, I sat with the "chief blower" in the "blowing cabin" overlooking the three enormous Bessemer Converters. Patiently, he explained to me the subtle changes in the color and character of the flame which were obvious to him, but invisible to me. He told me it had taken him over a year to be able to see these changes.

I never mastered the art of knowing when these changes heralded the end of the blowing period and the start of the pour. Our own task seemed hopeless. We wished to have a small computer determine when "end blow" had arrived, when we, the programmers and chemical engineers, couldn't even determine it ourselves. You may be interested to know that by looking at the flame, the chief blowers could estimate its temperature to within 10°F. No mean feat at 3,000°F! When we finally had our computer program finished, installed, and operating, the best it could do was predict the final end-blow temperature to within 20°F.

However, our program could usually predict the final weight of the steel and its composition a little more accurately than the chief blower. So, our efforts were not entirely in vain, and I had a computer terminal placed in the blower's cabin so he would have this information at all times.

Incidentally, it is most important that the temperature, weight, and composition of the bath are exactly to specification when poured. The temperature affects the cooling of the ingot, the composition affects its ability to meet the conditions for which it is intended, and the weight of the bath ideally should be an integral multiple of the weight of steel in one ingot. The next process that the ingot goes through, i.e., the cogging mill, cannot cope with partial ingots, and they have to be returned as scrap, thus wasting all the sensible heat they have retained and, of course, reducing the annual yield of steel.

When the end blow of the BOF is reached, the oxygen is switched off and the lance withdrawn so the furnace can be tilted over the waiting steel ladle. A tap hole on the side of the vessel is tapped and the steel escapes into the ladle. The slag floating on top of the steel remains in the furnace because the progressive tilt of the furnace causes it to be always above the tap hole. After pouring out the steel the slag is emptied into a slag ladle and taken away for disposal. The steel ingots are taken to the cogging mill. Immediately, unless minor refractory repair is required, the furnace is recharged with scrap and molten iron, and the process is repeated.

The Electric Arc Furnace

The predominant steel-making process of the future will be the electric arc furnace. Recycled scrap steel is sorted into its different types, and each "heat" of the furnace will be for a specific type of steel. For example, one heat could be for a certain alloy, and segregated scrap containing only that particular alloy would be charged in the furnace.

The electric arc furnace is a cylinder which can hold a bath of 300 tons of steel in its hearth (see Figure 3-19). Above the hearth are three retractable electrodes which have a 16-inch diameter and are 20 to 30 feet high. After charging the furnace the electrodes are lowered into position just above the steel and the current (a massive 80,000 kva) is switched on. An arc jumps between the electrodes and the charge, and back to another electrode, and so on. The temperature rises quickly and soon the bath of steel is molten. At this point, if the nature of the steel is to be changed by adding new components, they are introduced into the hearth and allowed to melt before the steel is poured from the tilting furnace into the ingot molds.

However they are made, the hot ingots roll out of the steel-making shop on low rail bogies. They have cooled somewhat by the time they arrive in the cogging mill, and the outside of the ingot may be frozen. However, the central core of the ingot will still be molten, and it is customary to strip the ingot from its mold, and then, using a huge pair of tongs, place the ingot in a heated chamber called a *soaking pit*, where the temperature of the ingot is allowed to become uniform at a temperature just below its freezing point. The red-hot ingot is now ready to be shaped into the desired product.

Huge machines with almost unbelievable strength will mold and ply the steel into such useful shapes as turbine shafts, steel strip, wire, reinforcing rod for concrete, and plate for cars and domestic appliances. Mechanical engineers control the design and operation of most of the remaining processes in a steelworks with, perhaps, some help and advice on heat transfer from chemical engineers.

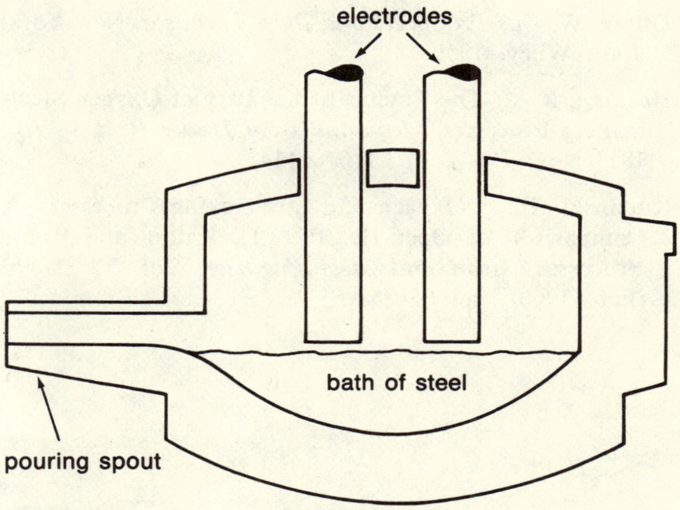

Figure 3-19. *The electric arc furnace.*

Summary

You will have become aware during the description of the ironworks and steelworks processes that these industries use and manipulate considerable amounts of heat. Perhaps more than in any other industry, the principles of heat transfer dominate both the design and operation of hardware. You have learned how heat is transferred by conduction, convection, and radiation, how heat transfer is exploited to preheat air in regenerators and packed beds of solids, and how it is used to heat up and smelt iron ore and make steel.

Further Reading

American Iron and Steel Institute. *The Basic Oxygen Process*. 1960.

Bray, J.L. *Ferrous Process Metallurgy*. New York: John Wiley, 1954.

Ess, T.J. "The Modern Arc Furnace." *Iron and Steel Engineer*. 7AF-58AF, (Feb. 1944)

Ess, T.J. "The Modern Open-Hearth." *Iron and Steel Engineer*. Vol. 25 0-19-0-66, (July 1948)

Gumz, W. *Gas Producers and Blast Furnaces*. New York: John Wiley, 1950.

Hacking, R.A. "The Practical Chemistry of Oxygen Steelmaking Processes." *Iron and Coal Trades Review*. Vol. 181 (Nov. 1960), pp. 1107–1114.

Kurzinski, E.F. "Oxygen and Steelmaking Processes—A Comparison of Open-Hearth, LD, Kaldo, and Rotor Processes." *Iron and Steel Engineer*. Vol. 37, No. 2 (Feb. 1960), pp. 65–75.

Oltrogge, A.R. "Large Electric Arc Furnaces." *Journal of Metals*. Vol. 18, No. 12 (Dec. 1966), pp. 1289–1292.

Sims, C.E. *Electric Furnace Steelmaking*, Vols. I and II. New York: Interscience, 1962.

United Nations. *Comparison of Steelmaking Processes*. New York: 1942.

United States Steel. *The Making, Shaping and Treating of Steel*. 1971.

Chapter 4

Steam and Electric Power Generation

Chemical plants need a constant and reliable supply of electricity and steam. In small plants it may be more economical to buy electricity from a local utility company and to produce steam on site by having a small steam boiler. However, in large plants it is more economical to build a large enough boiler plant to produce electricity, steam, and hot water. Electricity is used to drive motors, compressors, pumps, conveyor belts, skip conveyors, and electrolytic reactions, and to heat, cool, and light offices, laboratories, and lunchrooms. Steam powers steam engines which drive pumps and compressors, provides heat to other fluids, and creates partial vacuums if needed. Where the chemical reaction calls for water, steam may be supplied at an appropriate temperature and pressure. In this chapter you will learn how steam is made to generate electricity and how excess steam can be used to meet other process plant requirements.

First, as with the steel industry, consider the materials flow in and out of the boiler plant (see Figure 4-1). The boiler system requires a supply of water and fuel in order to produce electricity and steam. Consider first what happens to the fuel as it passes from storage into the furnace part of the boiler.

There are three types of fuel used to raise steam: natural gas, oil, and coal. Nuclear energy, which will play a much larger role in the future, is not considered in this book, even though the nuclear industry is a great employer of chemical engineers. An excellent description of nuclear energy can be found in *Steam* by the Babcock and Wilcox Company.

Natural Gas as Fuel

Natural gas firing is the cleanest and most convenient way to heat the boiler. Natural gas is found in association with crude oil, or petroleum, in huge underground reservoirs where it is trapped at high pressure. The light hydrocarbon gases such as methane, ethane, propane, and butane readily dissolve in liquid oils. A high pressure on the crude oil is usually maintained until it is out of the ground because the pressure keeps the light hydrocarbons in solution; they affect the petroleum by reducing its viscosity, making it easier and cheaper to pump. At the surface the pressure is released and the light hydrocarbons vaporize. Propane and butane are separated and sold pressurized in steel bottles. The methane with a little ethane in it is distributed by pipeline to customers.

There are also gas wells that contain small amounts of petroleum. These are tapped and the natural gas, which contains predominantly methane, is cleaned to remove dust, sand, water, and hydrogen sulfide, and then compressed into pipelines for distribution. It is good engineering practice always to run pipelines at their maximum load, or throughput. This means if the boiler plant using the gas has an hourly or daily fluctuation in demand for electricity, provision has to be made at the boiler plant site to store the delivered gas in slack periods in order to meet peak period demands.

There are many ways to store gas, the simplest being the familiar water-sealed gas holder (see Figure 4-2). The gas holder is an inverted cylindrical cup immersed in water which has a layer of tar floating on it to prevent evaporation. Gas is pumped into the gas holder which rises to accommodate it. Because these gas holders store the gas at a low pressure, the volume of the gas is huge.

From recognizing that much greater pressures would reduce both the volume of the gas and the need for large areas of land to be permanently committed to storage, the spherical pressure storage vessel was developed to store the gas at a very high pressure, thus occupying a fraction

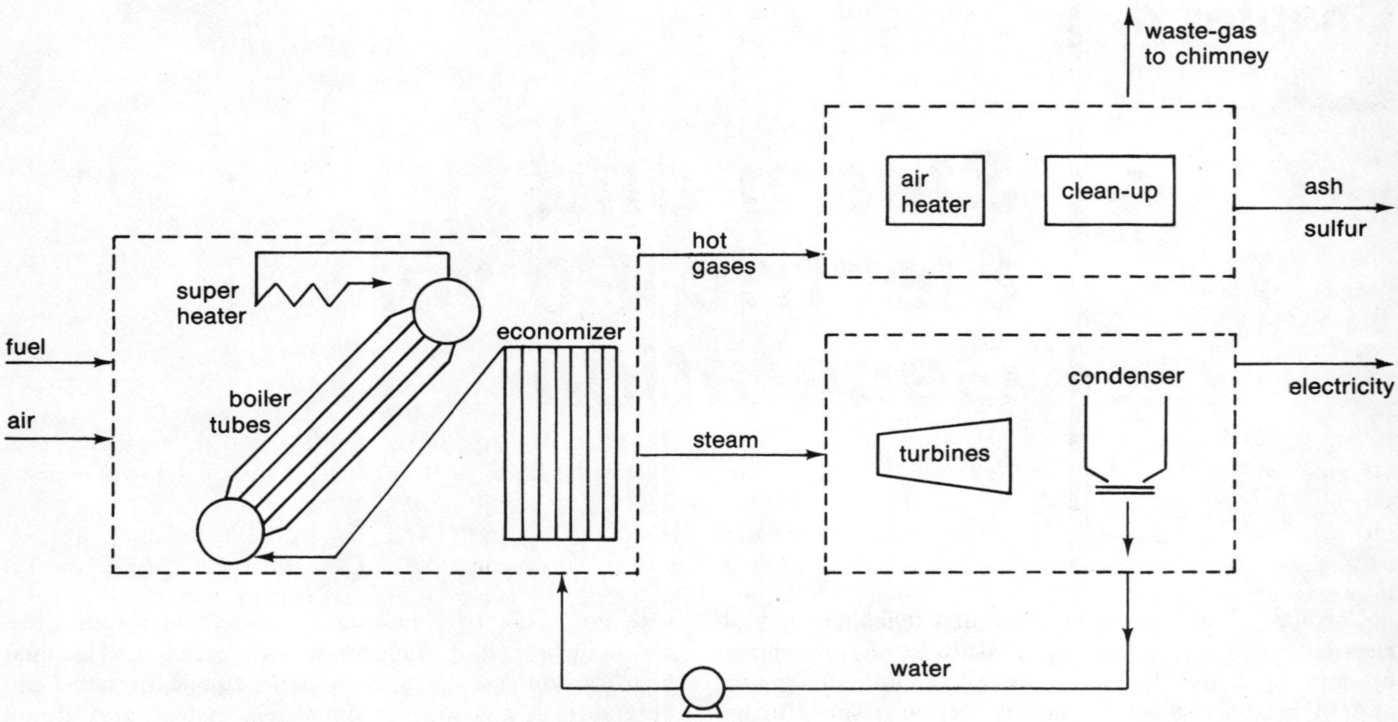

Figure 4-1. Materials flow in a utility boiler plant.

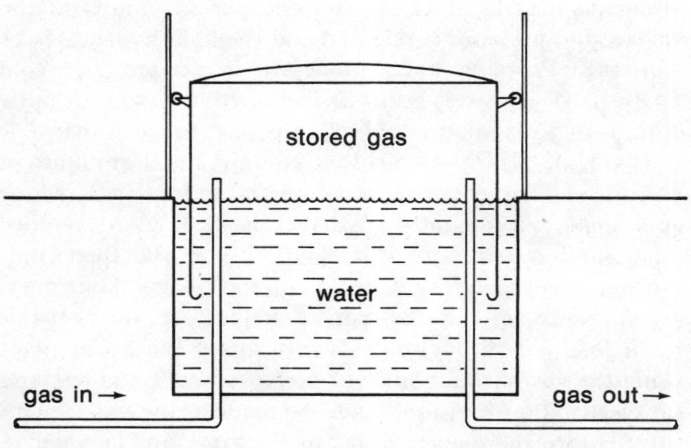

Figure 4-2. A simple gas-holder (shown half-full).

of the volume it would in a gas holder. An important advantage of high-pressure storage is that a compressor is not required for further distribution of the gas on the boiler plant site.

Recently, as natural gas wells have become depleted, gas engineers have found a way of using them as high-pressure underground storage reservoirs for natural gas. Obviously, this is a convenient means of storing gas only if it is close to the site where it will be consumed.

To burn natural gas all that is required is a means of mixing the gas with the correct amount of air which contains the oxygen necessary for combustion to occur. Table 4-1 shows a hand calculation to determine the amount of air required to burn sufficient gas to generate 4 million pounds of steam at 500 pounds per square inch (psi) pressure and 500° F. We can look up the amount of heat contained in steam at any given pressure and temperature in Steam Tables, a tabulation of steam and water properties found in the *Chemical Engineer's Handbook* and many other chemical engineering texts.

The calculation assumes that only 80% of the heat generated by combustion will be transferred to the steam. The rest of the heat will be lost as sensible heat of the exhaust gases, nitrogen, carbon dioxide, unused oxygen, and water; and heat will also be lost as latent heat of vaporization of the water in the exhaust gases. There will also be heat losses by conduction through the walls of the boiler and by convection and radiation away from the surface of the boiler.

The calculation shows we can use the chemical equation for the reaction between the methane and the oxygen (in the combustion air) as a basis to calculate the ratio of fuel gas to oxygen required for perfect combustion. The concept of a *mole* may be a new concept to some of you. A mole signifies an amount of a substance which is its molecular weight in pounds (or kilograms). In other words one mole of O_2, which is oxygen gas with a molecular weight of 32, contains 32 pounds of oxygen. Methane has a molecular weight of 16, so one mole of methane contains 16 pounds. A mole of *any* chemical species contains an equal number of molecules. Thus, the chemical equation in Table 4-1 shows that one molecule of methane reacts with two molecules of oxygen. The products of reaction are one molecule of carbon dioxide and two of water, whose molecular weights are 44 and 18, respectively. Then, since all moles contain equal numbers of molecules, we know that one mole of methane and two moles of oxygen give one mole of carbon dioxide and two of water. Therefore, 16 pounds of methane plus 64 pounds of oxygen will give 44 pounds of carbon dioxide and 36 of water. You can see how this is the basis for the calculation.

To start off with, it is assumed that the methane and air are mixed together so well that each methane molecule, in the short time period it is in the hot flame, can find two molecules of oxygen and carry out the reaction to form carbon dioxide and water. This would be much easier if methane and oxygen molecules were not forever bumping into the inert nitrogen molecules in the gas mixture. The following analogy demonstrates how this reaction takes place.

Imagine a crowd of 110,000 people completely filling a football field: 10,000 are called "methanes," 20,000 are "oxygens," and 80,000 are "nitrogens." A methane trying to find his way through the crowd looking for two oxygens is not only constantly buffeted by nitrogens who serve no purpose except to get in the way, but even when the methane has met one oxygen, another methane or nitrogen could bump into the pair of them and cause them to be separated. Also, imagine that when the methane has met up with two oxygens, a transformation occurs and from the three people, three more are made and their names are "carbon dioxide" and "water." The carbon dioxides and waters will now also get in the way of the dwindling numbers of methanes and oxygens looking for each other. When only 2% (200) of the methanes are left trying to collect two oxygens, they have to struggle through 110,000 people. This will be a lengthy process. Given enough time, all the methanes can find two oxygens and react, and eventually the stadium would only contain inert nitrogens, carbon dioxides, and waters.

However, during combustion, we do not have infinite time. In the flame, combustion must occur quickly to release the chemical heat while the gases are still in the furnace area, and to ensure that all the precious methane is used up and not wasted it is common practice to use a few percent of surplus air, or *excess air*. This is shown in the calculation in Table 4-1 as a 5% excess. The oxygen is accompanied in the air by a large amount of nitrogen. On a volumetric basis there is 21% oxygen and 79% nitrogen. You can see in Table 4-1 how this allows the calculation of the combustion air flow rate.

Since this calculation is important later on in a discussion about chimneys, let us look at the remainder of Table 4-1. For each mole of methane burned, the reaction produces one mole of carbon dioxide and two moles of water. Because we used excess oxygen and we also had accompanying nitrogen, there will also be oxygen and nitrogen mixed in with the exhaust gases. Adding the numbers of moles of products gives the total number of moles and, incidentally, allows us to calculate the percentage of each component in the exhaust gas stream.

It is conventional to calculate a percentage of just the three "dry" gases as though no water was present and to write them down as *percent dry* as in Table 4-1. The reason for this is that many measuring devices, which automatically sample the gas stream, dry it before analysis. When the equipment displays the percentage carbon dioxide on a dial in the control room, this is a dry percent figure. Since we can also calculate the *expected* dry percents of carbon dioxide and oxygen, we now have a means of diagnosis if trouble occurs in the plant by watching the *actual* values on control room instruments.

For instance, if in our example the percentage dry carbon dioxide falls below the expected 11.11%, we can find the trouble quickly by checking one or two of the other gauges. For instance we can check the fuel gauge. If the rate of feed of methane is still normal, then we know that for some reason not all the methane is burning, and we would suspect a fault has developed in the burner mechanism. Incidentally, this fault would also be reflected in the oxygen meter which would, of course, read higher than the expected 1.11% because less would have been used up in combustion. Another possibility indicated by the signs above would be an in-leakage of air somewhere in the boiler, thus diluting the carbon dioxide and increasing the oxygen in the output stream. However, this is most unlikely because the furnace is designed to operate at a pressure slightly above atmospheric pressure, and if any leakage does occur, it will be outwards and minimal.

The mole has another property we can exploit: one mole of any gas contains an equal number of molecules, and further, that number of molecules occupies just 359 cubic feet at what are called *Standard Conditions*, i.e., 32°F and one atmosphere (or 14.7 psi) pressure. Furthermore, using a combination of Boyle's law and Charles' law, which will tell how gaseous volumes change with temperature and pressure, we can calculate the altered volume at the conditions we have either in the feed or in the exhaust gas. For reasons beyond the scope of this book, the temperature scale used in the equation is not degrees Fahrenheit (°F) but degrees Rankine (°R), which is obtained by adding 460 to the temperature in °F. The result of this calculation shows we will have to provide a compressor (or a number of compressors) capable of handling 1.13×10^5 cubic feet of methane per minute. Similarly, the air fans will have to cope with 1.13×10^6 cubic feet of air per minute, and the chimney designed to

Table 4-1
Combustion Calculations for Natural Gas

Basis: 4 x 10^6 lbs/hr of steam at 500 psi and 500° F with heat content 1,231.2 Btu/lb (from *Steam Tables*)

Gas Requirement:

Heat contained in steam $= 4 \times 10^6 \times 1{,}231.2 = 4.925 \times 10^9$ Btu/hr

Heat released from gas (assuming 80% combustion efficiency) $= \dfrac{4.925 \times 10^9}{0.8} = 6.156 \times 10^9$ Btu/hr

Gas (1,067 Btu/ft^3) burned $= \dfrac{6.156 \times 10^9}{1{,}067} = 5.769 \times 10^6$ ft^3/hr

Moles of gas burned (1 mole = 359 ft^3) $= \dfrac{5.769 \times 10^6}{359} = 16{,}070$ moles CH_4/hr

Products Flow and Composition:

$CH_4 + 2O_2 \rightarrow CO_2 + 2H_2O$ (assume 5% excess air)

$1 + 2 \rightarrow 1 + 2$ molecules
$1 + 2 \rightarrow 1 + 2$ moles
$16 + 64 \rightarrow 44 + 36$ lbs

Reactant	Calculation	Moles	Product	Calculation	Moles	Percent Dry
CH_4	—	16,070	CO_2	—	16,070	11.11
O_2	$16{,}070 \times 2 \times 1.05$	33,750	H_2O	$2 \times 16{,}070$	32,140	—
N_2	$16{,}070 \times 2 \times 1.05 \times \dfrac{79}{21}$	126,950	O_2	$2 \times 16{,}070 \times 0.05$	1,607	1.11
			N_2	—	126,950	87.78
				Total dry	144,630	100.00
				Total wet	176,770	

Expected waste-gas composition (dry) = 11.11% CO_2, 1.11% O_2, 87.78% N_2

Fan and Chimney Ratings:

Methane flow rate $= \dfrac{16{,}070 \times 359}{60} \times \dfrac{(460 + 80)}{(460 + 32)} = 1.055 \times 10^5$ ft^3/min (at 80° F and 14.7 psi)

Air flow rate $= \dfrac{16{,}070 \times 2 \times 359}{60} \times \dfrac{100}{21} \times \dfrac{(460 + 80)}{(460 + 32)} = 1.005 \times 10^6$ ft^3/min (at 80° F and 14.7 psi)

Waste-gas flow rate $= \dfrac{176{,}770 \times 359}{60} \times \dfrac{(460 + 410)}{(460 + 32)} = 1.870 \times 10^6$ ft^3/min (at 410° F and 14.7 psi)

Waste-Gas Dew Point:

Partial pressure of water in waste-gas $= \dfrac{32{,}140}{176{,}770} \times 14.7 = 2.67$ psi

Therefore, from Steam Tables, dew point = 136° F

carry away 2.00 × 10^6 cubic feet per minute of exhaust gases.

Finally, the other useful information the calculation has produced is the amount of water in the exhaust gases. Not only can this figure be used to calculate the amount of heat loss due to the latent and sensible heats of the water, but also to calculate a property of the flue gas stream called its *dew point*, which is the temperature to which the exhaust gases would have to be cooled before the water would condense.

To visualize this, imagine a long glass tube carrying the exhaust gases away from the plant. As the gases flow down the tube they give up heat to the walls of the tube and become cooler. At some point, a temperature is reached when the first droplets of water condense out of the gas stream onto the walls of the glass tube. Looking at the glass tube from the outside, we would see the sudden appearance of a film of dew—hence the name dew point. Now, if only steam is present in the combustion gases, then it would obviously condense at its boiling point, 212°F. But this is not so. Also present are nitrogen, oxygen, and carbon dioxide, and although the total pressure exerted by all these gases (including the water) on the surroundings may be, say, 14.7 psi (which is atmospheric pressure), they each only contribute what is called their *partial pressure* to the total pressure.

Imagine in one mole of mixed gases (in our case, 0.091 moles of carbon dioxide, 0.0091 moles of oxygen, 0.718 moles of nitrogen, and 0.182 moles of water) the contribution of any one species to the total pressure is proportional to its fractional presence, or its *mole fraction*. So, in our case the partial pressure of the water is 2.67 psi (see Table 4-1).

There is a complicated relationship between the pressure of water and its boiling point. As the pressure is increased, the boiling point goes up. Whereas at atmospheric pressure (14.7 psi) the boiling point of water is 212°F, at 2.67 psi its boiling point is 136°F. With a partial pressure of 2.67 psi in the exhaust gas, the water condenses to dew when its temperature falls below 136°F. Figure 4-3 shows that if this terrible state of affairs occurs somewhere between the air preheater and the top of the chimney, liquid water will form.

Furthermore, if we use a fuel gas which contains sulfur traces, then there will be small traces of sulfur oxides in the exhaust gas which dissolve in the water and form mild acids. Over the years, the sulfurous and sulfuric acids would eat their way through the expensive hardware. Therefore, designers must go to great lengths to ensure that the exhaust gas remains well above its dew point until it clears the chimney.

Oil as Fuel

Let us retrace our steps and imagine that the fuel we chose was oil. Crude oil is taken from the ground and separated into many different products (see Chapter 5). One of these is the "heavy fuel oil" used to generate steam for electricity. Heavy fuel oil is difficult to burn. To produce the same output of heat as the natural gas did in the last example, the oil feed must be 3.4 × 10^5 lbs/hr as calculated in Table 4-2. Imagine what would happen if that much oil is pumped through a hole in the furnace side. It would certainly begin to burn, but the stream of oil would form a puddle on the furnace floor, where it would continue to burn. However, it would burn so slowly that shortly the puddle would completely fill the boiler—with obviously terrible consequences.

Burning oil is more difficult than you might think. Imagine we wish to produce 4 million pounds of steam an hour. That much steam contains 4,925 million Btu's. Thus, if the boiler plant is 80% efficient, i.e., it takes 100 Btu's in the chemical energy of the oil to give 80 Btu's to the steam, then the oil must provide 6,156 million Btu's each hour. Since the heating value of the oil may be around 18,100 Btu's per pound, we need to burn 340,000 pounds an hour, or around 680 gallons a minute.

The next problem would be trying to burn 680 gallons a minute. You know how big a 50-gallon drum is. Imagine taking the tops off 14 of them and igniting them. Even assuming that with heavy oil you could get the flame going, it would take considerably longer than one minute to burn it all. You would do better to spill the 680 gallons of oil onto the ground and then try to ignite it. It would burn more quickly but still last considerably longer than a minute. To get it to ignite you have to hold the match at the surface long enough for the surface layers of oil to become hot enough to evaporate so they will catch fire. As the surface oil burns more oil evaporates as the flame heat radiates towards cooler parts of the oil surface. Once ignited, the burning is self-propagating.

By spreading the oil on the ground, the area exposed to radiative heating from the flame is increased, more evaporation occurs, and burning progresses more quickly. But to burn the 680 gallons in one minute, we clearly have to expose considerably more surface by some means. So, we need to come up with a way of making sure the oil burns far more quickly. Current engineering practice is to atomize the oil feed into a spray of tiny oil droplets, each with a large surface-area-to-weight ratio. The droplets are so fine they are dispersed into the furnace looking something like a heavy fog, and, because of their small size, they heat up very quickly and evaporate.

The oil droplets are spherical. They leave the burner jet and are immediately exposed to the radiant heat coming back from the burning oil downstream. Because they have an immensely large surface-area-to-weight ratio, the droplets quickly pick up radiant heat and vaporization starts. If the 680 gallons of oil came in one large spherical droplet, its diameter would be 5.58 feet. The surface area of this droplet would be about 98 square feet. If the 680 gallons could be made into one-inch diameter spheres,

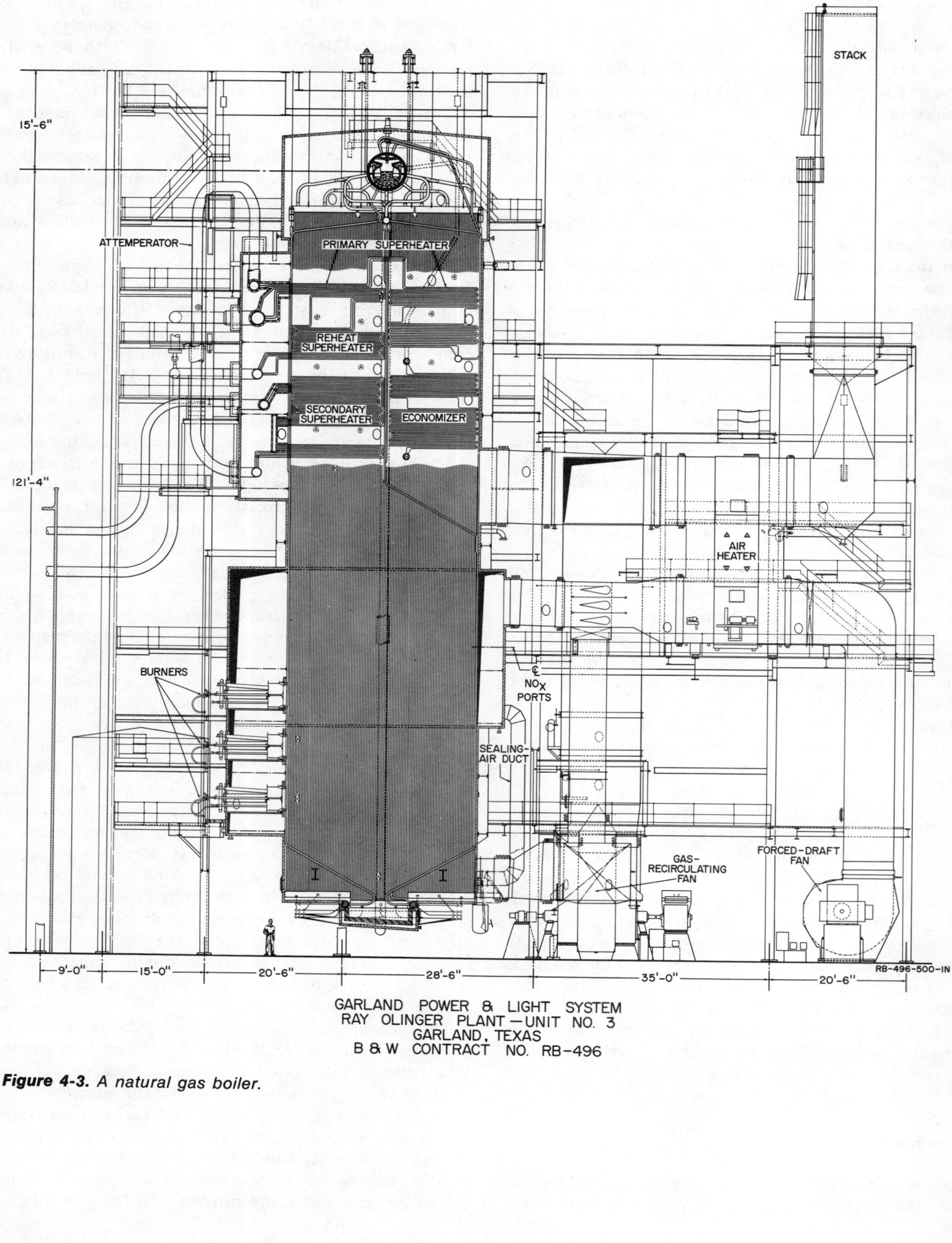

Figure 4-3. *A natural gas boiler.*

Table 4-2
Combustion Calculation for Oil

Basis: 4×10^6 lbs/hr of steam at 500 psi and 500°F with heat content 1,231.2 Btu/lb

Oil Requirement:

Heat contained in steam $= 4 \times 10^6 \times 1{,}231.2 = 4.925 \times 10^9$ Btu/hr

Heat released from oil $= \dfrac{4.925 \times 10^9}{0.8} = 6.156 \times 10^9$ Btu/hr

Oil (18,100 Btu/lb) burned $= \dfrac{6.156 \times 10^9}{18{,}100} = 3.40 \times 10^5$ lbs/hr

Moles of oil burned $= \dfrac{3.4 \times 10^5}{204} = 1{,}667$ moles/hr
(assuming $C_{15}H_{24}$)

Products Flow and Composition:

$C_{15}H_{24}$	+	$21O_2$	→	$15CO_2$	+	$12H_2O$		(assume 10% excess air)
1	+	21	→	15	+	12		moles
204	+	672	→	660	+	216		lbs

Reactant	Calculation	Moles	Product	Calculation	Moles	Percent Dry
$C_{15}H_{24}$	—	1,667	CO_2	$15 \times 1{,}667$	25,005	14.42
O_2	$21 \times 1667 \times 1.10$	38,507	H_2O	$12 \times 1{,}667$	20,004	—
N_2	$38507 \times \frac{79}{21}$	144,860	O_2	$21 \times 1{,}667 \times 0.10$	3,501	2.02
			N_2	—	144,860	83.56
				Total dry	178,370	100.00
				Total wet	198,370	

Expected waste-gas composition (dry) = 14.42% CO_2, 2.02% O_2, 85.56% N_2

Fan and Chimney Ratings:

Air flow rate $= \dfrac{21 \times 1667 \times 359}{60} \times \dfrac{100}{21} \times \dfrac{(460 + 80)}{(460 + 32)} = 1.095 \times 10^6$ ft^3/min (at 80°F and 14.7 psi)

Waste-gas flow rate $= \dfrac{198{,}370 \times 359}{60} \times \dfrac{(460 + 410)}{(460 + 32)} = 2.10 \times 10^6$ ft^3/min (at 410°F and 14.7 psi)

Waste-Gas Dew Point:

Partial pressure of water in waste-gas $= \dfrac{20{,}004}{198{,}370} \times 14.7 = 1.48$ psi

Therefore, from Steam Tables, dew point = 109°F

it would form 300,000 droplets, and their total surface area would be 6,545 square feet. With a fine atomizer reducing the droplet size to 1/100 inch, the 680 gallons would make 300 billion (3.0×10^{11}) droplets, and the surface area available to pick up radiant heat transfer would be 654,500 square feet. Clearly, this is the way to go with the oil-burning problem.

The vaporized oil mixes quickly with the air so that the situation becomes similar to the gas flame described above. Atomization is quite simple and there are several ways to achieve it. The obvious one is to blast the liquid oil through a tiny orifice. Exactly like an aerosol spray, the oil breaks up into tiny droplets. To adopt the pressure jet method the oil must first be warmed so it flows more freely into the jet, and then have other small jets (see Figure 4-4) of air or steam that impinge on the jet of oil and break it up even further. The steam or air atomizer is the most widely used method of burning oil. Furthermore, steam is nearly always the atomizing agent because on a boiler plant site it is readily available at high pressure, whereas air would have to be raised to a sufficiently high pressure, which means the purchase of a compressor.

There are other kinds of atomizers which are generally termed "mechanical" atomizers because they do not require an extra agent such as steam. In these devices the oil is pumped to such a high pressure that when it is ejected into the furnace over mechanical obstructions, the tiny droplets needed are formed. Because of the extra equipment needed to handle the oil at the high pressure, mechanical atomizers are seldom used except in installations (e.g., on ships) where the steam is too valuable to use as an atomizing agent.

Table 4-2 shows the calculation for the volume of air required and the volume of gases produced when burning this much oil to produce the same amount of steam as the example with natural gas.

Coal as Fuel

Let us for the last time retrace our steps through the combustion process, but this time with coal as the fuel. Coal was formed by the constant pressure of overlying rocks upon the vegetable matter from primeval forests which for many thousands of years produced beds of decaying vegetable matter thousands of feet thick. As the earth grew older, continents moved and created new mountains, rivers, and oceans, and in the general reshuffle some vast beds of compost were then relocated at the bottom of deep seas with a constant stream of sediment settling upon them as the products of land erosion washed down from the continents in streams and rivers. The weight of the oceans combined with the ever-increasing weight of the thickening bed of sediment imposed a huge pressure on the compost.

In addition, because the compost was closer to the molten core of the earth, its temperature began to rise. These two effects caused the lighter elements in the compost to be squeezed out and a new material to be formed. Because of the compression, strong molecular bonds were formed and the new substance developed the consistency of a soft rock. Its name is *lignite* or *brown coal.* There are huge deposits of this material in North and South Dakota, Texas, Germany, and the Soviet Union.

Most lignite, however, was subjected to more development as the earth continued to change. Great seams of lignite and compressed compost were sunk deeper into the

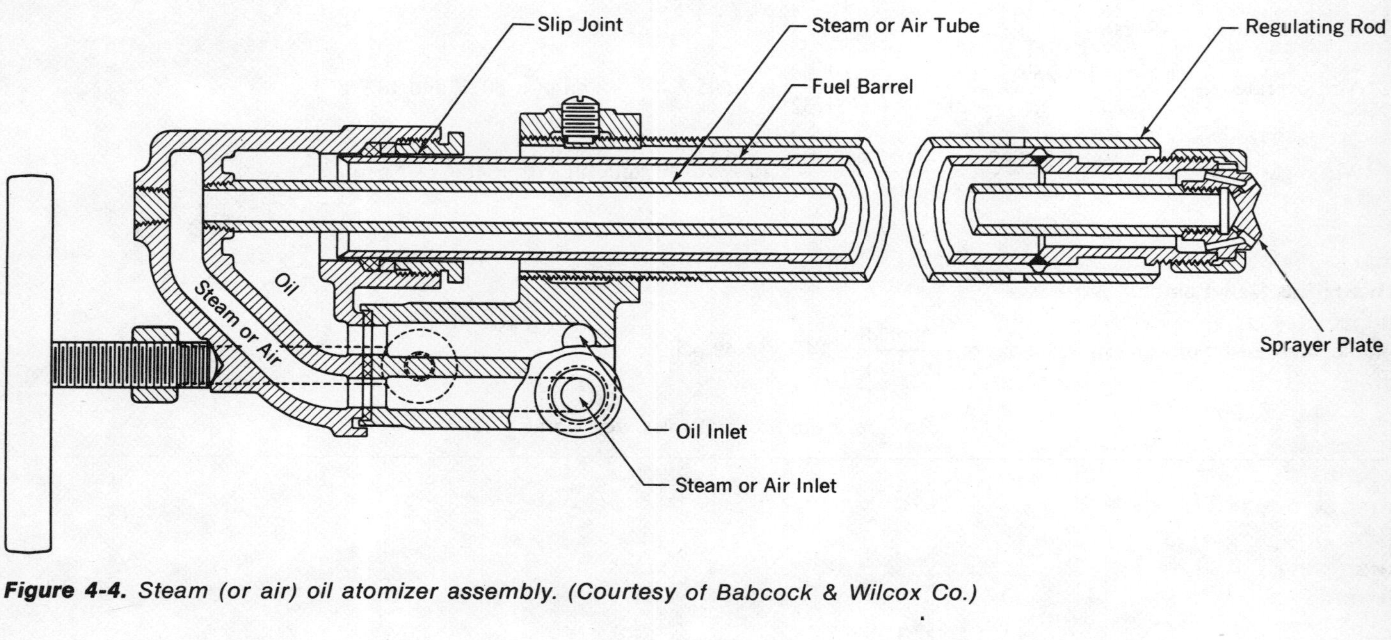

Figure 4-4. *Steam (or air) oil atomizer assembly. (Courtesy of Babcock & Wilcox Co.)*

earth and subjected to even greater pressures and temperatures. The processes that had already begun continued, and as the seams were compressed, more of the hydrogen was squeezed out, thus increasing the percentage of carbon. At any point in this process there may have been changes in the earth's structure that moved the seam to cooler regions closer to the surface. In these cases further change was prevented.

Gradually, as the new mountains above were eroded by wind and rain and rivers, the solid fuels became more and more accessible to mining. The seams of coal which have been subjected to these processes of change the longest are now called *anthracite*—a hard, steely-gray coal containing a high percentage of carbon and little else. Its high "heating value" (for example, at 16,000 Btu's per pound) is up to three times higher than the heating value of lignite. To give you some idea of the pressure and temperature conditions necessary to form anthracite, imagine taking an average garden compost heap of 2′6″, and pressing it until it is 1″ thick. The first 5 inches would be easy, but what about the remaining 2 feet?

There are coals called *bituminous coals* somewhere in age between lignite and anthracite. They vary in carbon content and heating value according to their age. Bituminous coals are the most common form of solid fossil fuel. Some bituminous coals form a strong, steely-gray material called *coke* when subjected to high temperatures. We saw in Chapter 3 how this process, called *carbonization*, is used to make blast furnace coke. All kinds of solid fossil fuel from the anthracites to lignites are burned to raise steam for power. Obviously, there are differences in the design of the hardware to meet the special requirements of each of the fuels.

What burning rate of coal would give 4 million pounds of steam an hour? Table 4-3 shows how it works out to be 520,000 pounds an hour or 8,667 pounds a minute. This would be a 4.6-foot cube of coal. We have 1 minute to burn it. If we manage to get it going and keep it going, it would probably burn for many days. Clearly, we will have to resort to something to bring the burning time down to a minute.

Let's study the way the coal catches fire and burns. Imagine we could see the molecules of oxygen and nitrogen in the air and the molecules of combustible material in the particle of coal as shown in Figure 4-5. An oxygen molecule that is jostled by the surrounding air molecules until it reaches a sufficient velocity that it strikes a carbon atom may react with it to form carbon dioxide. Since this is a gas molecule, it immediately leaves the surface and further adds to the congestion of molecules around the coal particle. More oxygen molecules may bounce off the carbon dioxide molecules and never reach the coal surface.

However, we can improve the probability of oxygen striking the surface by heating the air to give individual

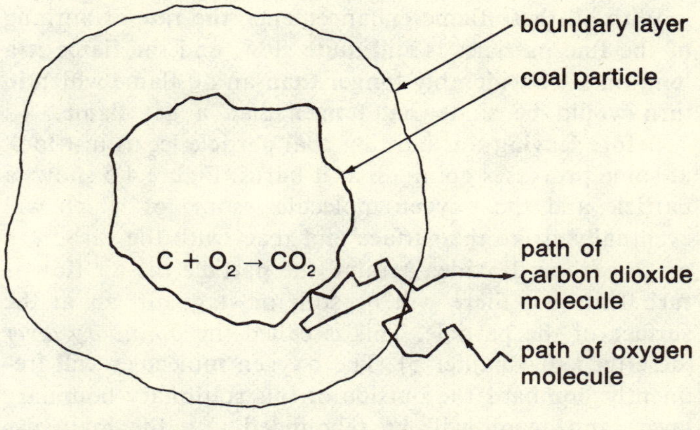

Figure 4-5. *Combustion of a single particle of coal.*

molecules greater velocities. We can also increase the probability of an oxygen molecule striking the particle by increasing its surface area. As we saw with oil droplets the way to increase surface area is to break down the coal into smaller and smaller pieces. A third way is to give a higher degree of turbulence to the air stream so that the oxygen molecules are randomly mixed throughout the gas and once again more likely to strike against the coal surface. A fourth way is to move the coal particles around relative to the air in order that the coal surfaces meet up with a greater volume of air, hence having a greater probability of meeting up with an oxygen molecule.

This study, in fact, gives us a clue as to how coal is burned at such large amounts per minute. First, it is finely crushed in grinding machines called *pulverizers* or *mills*. In just a moment you will see schematically some of the more common types of mills used to prepare pulverized fuel. The 4.6-foot cube containing 8,667 pounds of coal can be reduced to 168 trillion (1.68×10^{14}) cubes with a side of 1/1000 of an inch. Assuming the particles are spheres rather than cubes, then the surface area now available would be 3.7 million square feet.

The second way of speeding up combustion is by increasing the temperature of the combustion air. The hot combustion gases leaving the boiler still have sufficient heat to warm up the combustion air. This is usually done by an *air heater* (described later in this chapter). As the air enters the furnace, it is heated some more by the radiation from the flame and the walls. Also, the recirculating hot furnace gases surrounding the jet of fuel and air are drawn into the flow and cause the air to be heated still more. Finally, the coal particles themselves receive radiant heat from the flame and some of this is convected from the hot particles to the air. This also has the effect of heating the air.

The third improvement that can be made is the promotion of turbulence. This is produced by swirling the fuel—air mixture as it leaves the burner at high velocity.

With all these flame enhancements, the rate of burning of the fine particles is still quite slow, and the flame is a long one—considerably longer than an oil flame which in turn would be somewhat longer than a gas flame.

Before leaving this burning coal particle let us just look at some processes going on as it burns. Figure 4-5 shows a particle and the oxygen molecules, some of which will eventually strike the surface and react with the carbon to make carbon dioxide. Around the particle the air flow is turbulent, but there will be still, or stagnant, air at the surface of the particle. This is called the *boundary layer* (described in Chapter 5). The oxygen molecules will frequently bombard the outside of this stationary boundary layer, and many will be rebounded into the main gas stream. Others will be shouldered around inside the boundary layer and perhaps eventually find their way to the coal particle surface.

After the oxygen molecules react at the coal surface, carbon dioxide, carbon monoxide, water molecules, and any other gaseous products which are generated will leave the surface and add to the general congestion as they move outwards across the boundary layer. An analogy would be as follows.

Imagine a sold-out football game that ends at 6:00 p.m., and some shortsighted reservations clerk has planned a concert to begin in the stadium at 7:00 p.m. The stadium has only one large entrance for spectators, and this entrance is on a busy street. Picture the chaos at 6:30 p.m. Concert fans would be heading inwards through the gate; football fans would be leaving. Passers-by would be moving across both of these flows. The concert fans are the oxygen; the football fans are the combustion products; the passers-by are the nitrogen molecules in the air. However, the analogy breaks down. In the analogy both sets of fans have an objective, and if a concert fan is pushed back by the football crowd, as soon as he becomes untangled from the mob, he will again head toward the gate. This is not so with oxygen molecules migrating through the gas film. If a particularly energetic carbon dioxide molecule comes bounding away from the surface and strikes the oxygen molecule, it will be bounced away and will *not* head back for the coal surface unless jostled to do so by other molecules in a random way. This process is called *diffusion*, and it obeys the laws of mass transfer (described in Chapter 6).

Next, consider the hardware used to burn coal and make its chemical energy available for raising steam. The historical development of coal combustion is fascinating.

The simplest way man could find was to build a coal fire on a cast-iron grate enclosed by brick such that the hot gases from combustion rose upwards past the boiler surfaces and into a chimney which was high enough to give sufficient draft to draw air under the bed and through it, thus perpetuating the process. Occasionally, a furnace door in the brickwork was opened and fresh coal was thrown in by the stoker who also had the occasional job of poking the coal ashes away from the flame into an ash pit where they could be collected for disposal.

As industry geared up to an ever-increasing production capacity, the demand for power and steam grew. To feed in coal at the rate necessary for a large modern utility boiler (say 2,000 MW) would require the stoker to work at the rate of over 100 shovelfuls of coal every second. So, to increase the size of the furnace a mechanical stoker had to be found.

The composition of naturally occurring solid fuels and most artificial ones is nonuniform, being a mixture of compounds of an uncertain chemical and physical nature. This phenomenon causes the design engineer a number of problems. First, it is difficult to predict the best values for combustion parameters such as air-to-fuel ratio, combustion-zone temperature, and the overall reaction rate. Second, the combustion process does not occur in an intimately mixed (the technical term is *homogeneous*) phase as with gaseous or vaporized liquid fuels, where the air supply needs only to be increased a few percent over the stoichiometric proportions necessary to cause all the fuel to come in contact and react with the oxygen. Instead, combustion reactions occur on the surface of the fuel (or *heterogeneously*). Thus, no matter how abundant the oxygen supply, the reaction can only proceed where there is an exposed, or vacant, site on the surface of the solid.

These two constraints on solid fuel combustion have three important implications for the designer of hardware to burn solid fuel. First, the surface area available for reaction must be made as large as possible by (1) grinding the fuel as small as possible and (2) arranging for turbulence of the furnace gases in order to sweep the solid surface clear of reaction products.

Second, the reaction rate will obviously be governed by the rate at which new surface is exposed by reaction. Thus, it is important to allow sufficient time for all the fuel to be exposed for reaction. Third, the reaction itself must be made more rapid in order that the turnover of reaction sites is higher. This can be achieved if the temperature in the combustion zone is increased by reducing the heat load in that part of the furnace: the higher the temperature, the faster the coal burns.

These three design considerations are often referred to by combustion engineers as the three T's: Time, Temperature, and Turbulence. In all three cases, the designer of combustion equipment must seek the best value without jeopardizing other aspects of the design. If we examine each of the three T's, we will see the implications for the designer.

First, consider temperature. As stated before, it is favorable in combustion kinetics to operate at as high a temperature as possible. However, an upper limit is set by other factors. The cost, reliability, and availability of construction materials which can operate safely at high temperature are major constraints. Also, higher temperatures may help to form undesirable pollutants such as the nitrogen oxides. Higher temperatures may cause the coal ash to melt and stick to the steel heat transfer surfaces, causing resistance to heat transmission and exposing the steel to corrosive attack by the chlorides, sulfates, and vanadium compounds in the coal ash. Of course, in slagging combustors such as the cyclone furnace (which will be discussed shortly), it is *intended* that the ash will melt, and provision is made to minimize the impact of the corrosive compounds on the expensive water tubes and other surfaces.

Second, consider the promotion of turbulence, the objective of which is to create relative motion between the combustion air and the fuel particles. This turbulence will scrub away reaction products and make fresh combustion sites available.

One obvious method of promoting turbulence is to cause the reactants to follow a tortuous path through the furnace. However, this is not practical due to the high cost of installing and maintaining baffles which can withstand high temperature and the erosive and corrosive effects of the gas-borne residues. Baffles cause a resistance to the flow of gases, and extra large and more expensive air fans would be necessary. A common solution is to channel the air flow through a bed of fuel in which the small interparticle channels cause high gas velocities.

Then, to increase the reaction rate or burning rate, it is a simple matter to turn up the air feed rate. But this can only be done up to a certain limiting air velocity. At that point, the air blows up through the bed so fast that it lifts the coal off the grate and carries it through the boiler and up the chimney. Such pollution is unthinkable and the waste of fuel is bad engineering. There is no way around this limit on the burning rate of coal on grates, and so, for large, faster burning systems designers go to pulverized fuel hardware which pulverizes the coal into extremely small particles (average size 1/1000 inch). With such fuel systems, the fuel–air mixture is swirled in the combustion chamber, and the resulting turbulence has the desired effect.

With temperature and turbulence optimized, the last of the three T's must be considered. Sufficient time must be provided to ensure that a reasonable utilization of the fuel occurs. Ideally, all the combustible material reacts and gives up its heat, but in most cases 100% combustion would require infinite time and capital. A trade-off is reached between the combustor size and the percent loss of unburned fuel.

There have been many ingenious designs for systems to burn solid fuels. Some of those which have withstood the test of time are reviewed below.

Originally, solid fuel beds were fed or "stoked" manually, and it is easy to see that some mechanical stoking devices are attempts at imitating the manual stokers in a controlled way. The spreader stoker (Figure 4-6) is an example. A mechanical shovel is charged with raw fuel in its "home" position. It is then automatically operated to

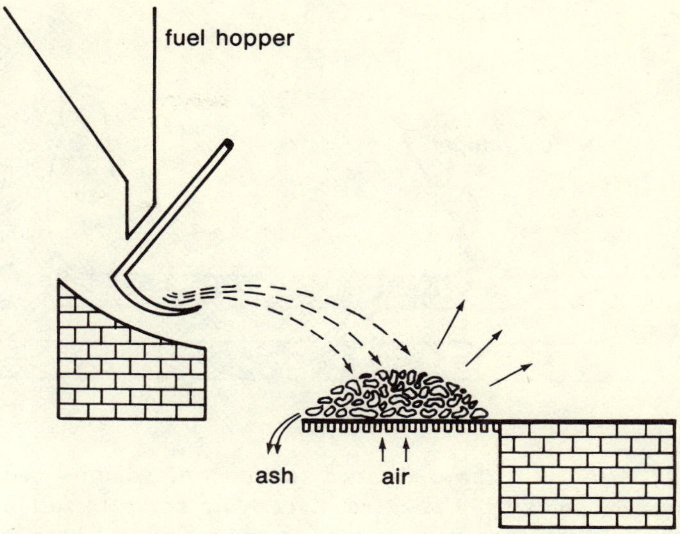

Figure 4-6. *The spreader stoker.*

throw the coal onto the burning fuel in the combustion zone, before being returned to its home position empty. Many enhancements have been developed for this type of stoker. Some systems allow variation of the length and direction of the shovel's swing according to a repeated preprogrammed schedule. In one system the shovel swings in a direction which is purely random in the hope that the bed will be served equally. This device is called a *drunken stoker*.

Another common enhancement is to move the grate horizontally to allow the ash to be carried away from the combustion zone to an ash pit. This can be achieved using a continuously-moving grate made of hinged cast-iron louvers, similar to, but considerably smaller than, the sinter equipment in Chapter 3. Alternatively, an arrangement of reciprocating or vibrating grate bars may be used. The combustion air is passed upwards through the fuel bed, and this may cause a problem of solid "carry-over" if fines are present. Furthermore, carry-over increases with ash content, and grit emission into the flue may be a problem, especially when running at high load. The quenching effect of falling raw fuel, especially when it is wet, which is usually the case, may cause unburned volatiles to escape to the stack, causing smoke emission.

The sprinkler stoker works on essentially the same principle as the spreader—the main difference being that the fuel is gravity-fed through a drop tube immediately above the fuel bed. Ash builds up and is removed manually as required, or, in more modern designs, automatically at a frequency set by the combustion rate.

The underfeed stoker (Figure 4-7) operates without a grate to support the fuel bed. A screw feeder (see Figure 4-7 and work out how this operates) delivers the raw fuel to the base of the fuel bed and forces it up towards the combustion zone where tuyeres provide air. Secondary air

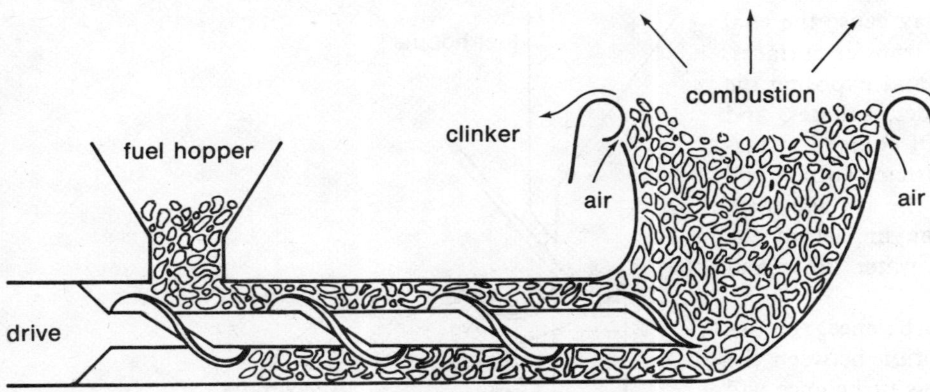

Figure 4-7. The underfeed stoker.

is introduced above the bed to burn off volatiles and reduce smoking. Underfeed stokers are cheap to install and can be inexpensive to operate and maintain. They are mostly used in small boilers up to 1 megawatt, where a major benefit is the lack of a requirement for skilled supervision. Unfortunately, underfeed stokers tend to make smoke (which is illegal in most parts of the world) and can suffer serious screw-feeder erosion problems with hard abrasive fuels.

A far more serious problem is the tendency to form *coke trees* in the bed. If the fresh coal becomes plastic on its way up to the combustion front, it begins to carbonize, and a central solid mass of coke forms. Since it is largely impermeable to air, and remote from the air tuyeres, a coke tree will not burn and has to be broken down manually before it rises to a great height over the bed and topples to cause considerable damage. Obviously, in an installation with little supervision this can never be allowed to happen. It is prevented by burning coals which are known not to form a sticky plastic mass when exposed to heat. The preferred fuel size is between 1 and 2 inches. Fines are kept to a percentage of the fuel weight because they could contribute to an impermeable central core and promote coke trees. Since ash removal at best is only semiautomatic, low-ash fuel is preferred, and the combustion temperature should be arranged so the ash just melts and fuses into large chunks rather than powdery ash that may be carried over to cause grit and dust emission problems.

The coking stoker (Figure 4-8) is arranged so fresh fuel can drop from a hopper into a position in front of a ram that pushes the fuel across a hot grate area where volatiles are released to burn over the fuel bed. The remaining coke is further pushed across the grate by a fresh charge of fuel pushed by the reciprocating ram. The grate may be static or arranged to perform some motion which will assist the movement of the ash towards the ash pit. These stokers lend themselves well to automatic control of the ramming frequency, length of stroke, grate movement, and air rate.

Only fuels which do not form sticky clumps should be used, thus avoiding formation of huge, impermeable coke masses. The fuel size should be 1 to 3 inches, with a minimum of fines. As with all stokers with a grate, fines will be lost to the ash unless the fuel is damped before combustion to cause the fines to stick to larger fuel pieces long enough, hopefully, for them to burn. Also, in common with other grate-supported beds, the ash content of the fuel should be high enough to form a layer on the grate. This prevents overheating and damage to the grate.

The above stokers all place the coal onto a fixed bed on a fixed grate. The range of *moving*-bed stokers is as wide as human ingenuity. Development was aimed at increasing combustion rates, and modern moving-bed stokers burn coal at least twice as fast as fixed-bed stokers.

In the traveling-grate stoker (Figure 4-9) fuel falls from a hopper onto the grate made of cast-iron louvers which carries it beneath an adjustable guillotine, thus forming an even, moving bed. The speed of the grate and the height of the guillotine are geared to and fixed by the air flow rate. The fresh bed passes beneath an "ignition arch" which is a short tunnel of refractory bricks that glow bright red with heat. The ignition arch is positioned to receive radiant heat from the combustion zone and reradiate to the fresh fuel feed heating it up to its ignition temperature. After ignition the bed burns away over a wind box fitted with

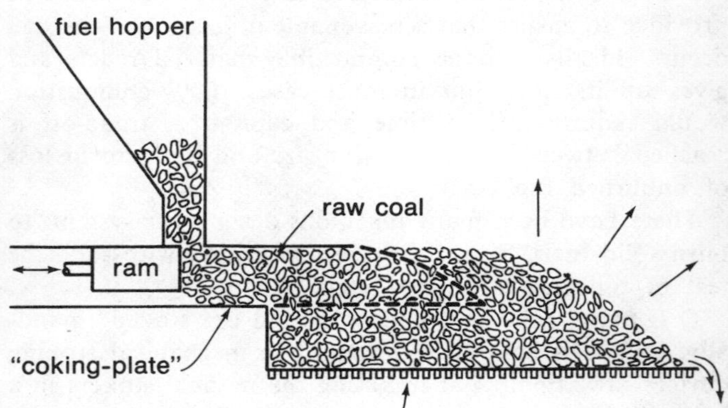

Figure 4-8. The coking stoker.

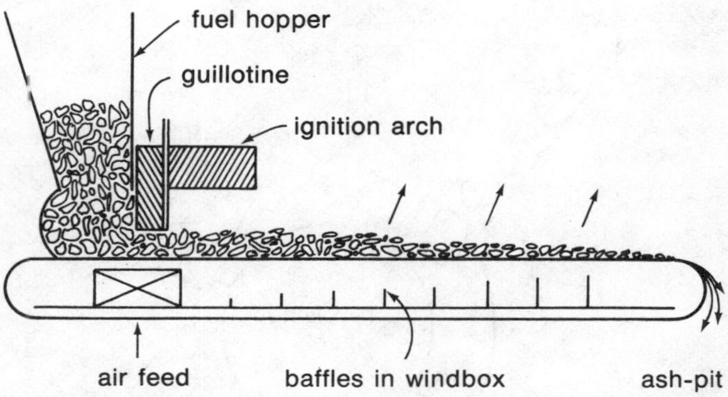

Figure 4-9. *The traveling-grate stoker.*

baffles to distribute the air, providing more flow through the middle combustion areas and less to the ignition and ash zones. Ash discharge is automatic from the end of the moving bed. It is normal practice to damp the feed to stop the fines being blown clear of the bed and out of the boiler.

The stoker can handle just about any fuel except anthracite, which burns with an unstable bed due to difficulties in lighting anthracite initially. To increase burning rates the grate can be speeded up with a corresponding air feed rate. However, a limit is imposed by high air rates blowing the bed off the grate, causing serious grit and smoke emission and a loss of combustible material.

Traveling grates have been used in utility boilers operating at up to 70 megawatts. A rapid response to changes in load makes this stoker a favorite in plants where the daily steam demand undergoes large changes. Other moving-bed stokers have been developed that rely on vibrating, louvred grates or traveling grates for moving the fuel from input to ash pit over a flow of combustion air. Most can achieve a high level of automatic operation, safely remain "banked" overnight, and burn a variety of fuels with a wide range of ash properties. It is unlikely, however, that moving-grate stokers will be used in anything but shell-type boilers in the future due to competition from the newer methods of fuel firing described below.

It was discovered that a mixture of finely powdered coal and air would burn almost like a gas when blown through a burner port. The principle was developed to lead to the huge, modern pulverized-fuel-fired systems used so widely in the power-generation industry. Almost any fuel can be burned in a pulverized fuel furnace, although coals which become plastic on heating can cause problems when the particles occasionally swell up to form tiny bubbles called *cenospheres.* They are up to 100 times the original particle diameter and float out through the flame and leave the furnace unburned.

The pulverized fuel flame is highly radiant. The design of the boiler must be carefully matched to take up the radiant heat in order to cool the hot gas and ash mixture before it impinges on convective heat transfer surfaces. Doing so, the ash will solidify in the gas stream and not adhere to the steel and cause corrosion.

The fuel must be prepared as follows. Drying can be carried out prior to crushing and pulverization, or the air feed to the mill can be preheated to around 550°F to act as a drying agent. It is dangerous to use higher temperatures because of the high risk of explosion with finely ground powders, particularly coal. For the same reason the fuel is ground immediately before it is needed to avoid storage and the risks of a serious fire and explosion.

The coal is first crushed using one of the types of crushers shown in Figure 4-10 to reduce its size to minus 1

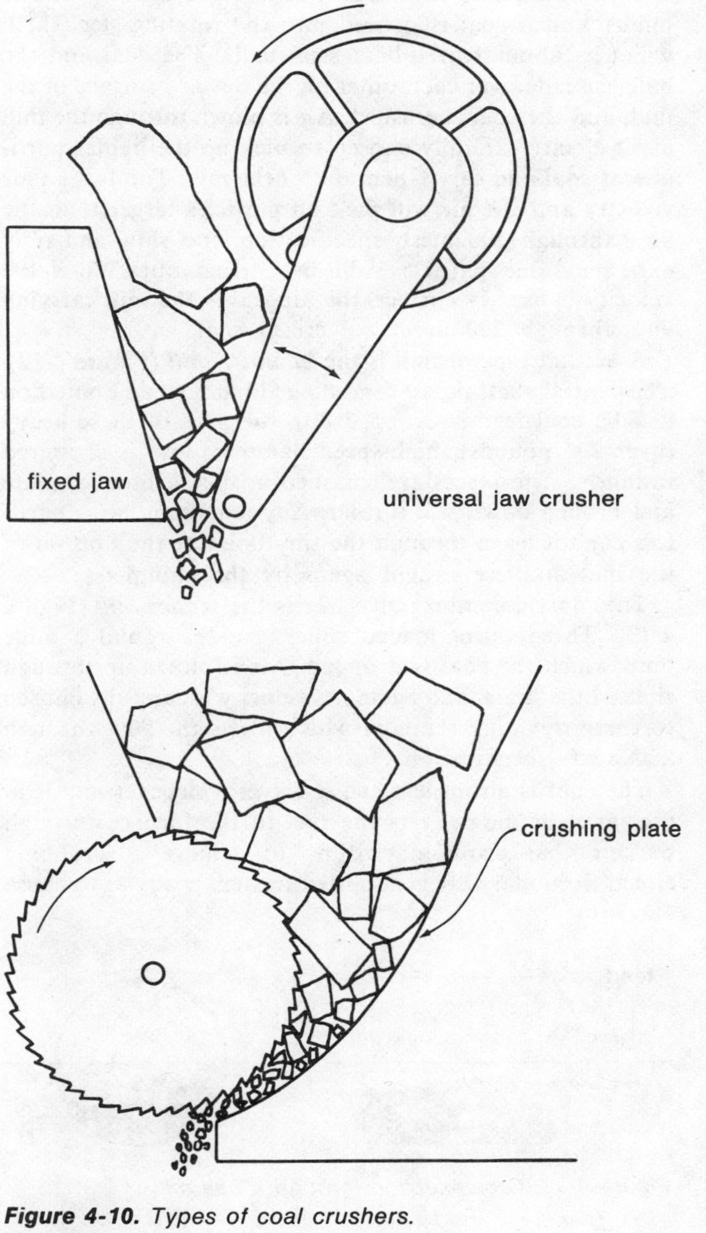

Figure 4-10. *Types of coal crushers.*

inch: this step may not be necessary if the delivered coal is already small in size. The next stage of *comminution* (the technical word for size reduction) is pulverization down to about "90% through 200 mesh." The meaning of this phrase is simple. The American Society of Mechanical Engineers (ASME) has standardized sieve sizes and given each standard a number. The test sieve with the number 200 has openings which are 0.00354 inches across. Thus, the phrase used above means that 90% of the powder would pass through a 200 test sieve mesh. The size could also be stated, less conveniently, as 90% minus 0.00354 inches and 10% plus 0.00354 inches.

There are a number of pulverizer types which will reduce coal to this small size (it looks exactly like black flour when it is so small). Figure 4-11 shows a cross section through a "ball mill," which works as follows. The minus 1 inch coal is poured into the rotating steel shell, which is about half full of steel balls. The coal and the balls cascade over each other on the upward surface of the mill, and the coal is crushed. Air is blown through the mill at a velocity carefully chosen to pick up the lighter particles of coal and carry them out of the mill. Too fast an air velocity and the air will pick up particles larger than the 90% through 200 mesh specification: too slow and only extremely fine particles will be carried out. When the velocity is exactly correct, the air leaves the mill carrying 90% through 200 mesh pulverized coal.

A second type of mill is the *hammer mill* (Figure 4-12). The central shaft has several hinged hammers mounted on it. The coal feed is dropped into the path of these heavy (over 200 pounds), high-speed hammers and is shattered instantly. The pieces are smashed against a breaker plate and broken down still further. Any remaining large particles cannot leave through the tiny holes at the bottom of the mill and are caught again by the hammers.

The most common pulverizer is the *roller mill* (Figure 4-13). Three spring-loaded rollers rotate around a table onto which the coal is dropped. Air is blown up through slots in the table, and again, its velocity is carefully chosen to carry out only the coal which meets the 90% through 200 mesh specification.

The coal is airborne when it leaves whichever mill is in use, and the air delivers the fuel to the furnace through burners that promote swirling to achieve a scrubbing effect. Secondary air is admitted in such a way as to cause

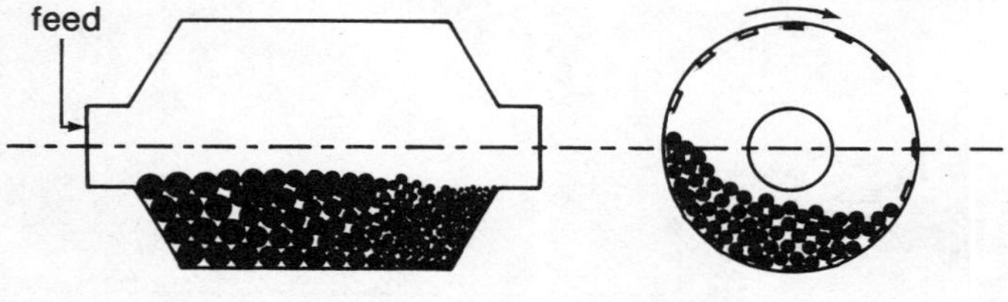

Figure 4-11. *Cross-section through a ball mill.*

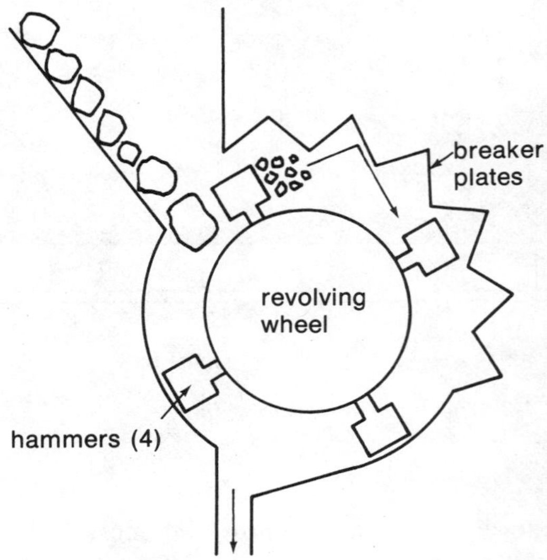

Figure 4-12. *Cross-section through a hammer mill.*

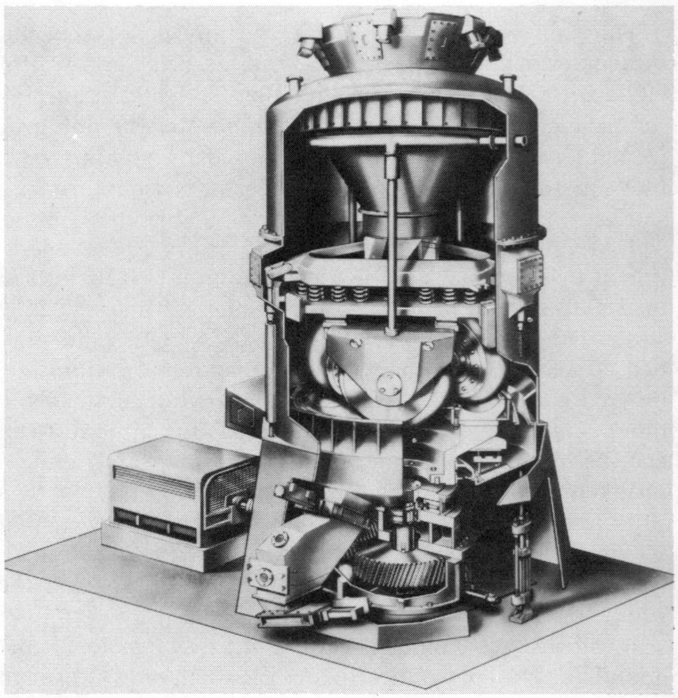

Figure 4-13. *The roller mill. (Courtesy Babcock & Wilcox Co.)*

more turbulence and various configurations of burners, and secondary-air inputs have been developed and proven in practice. (See Figure 4-14).

Notice in Figure 4-14b that the flame is directed downwards towards a high-temperature zone. The ash melts and runs out of the furnace as a liquid. Under these

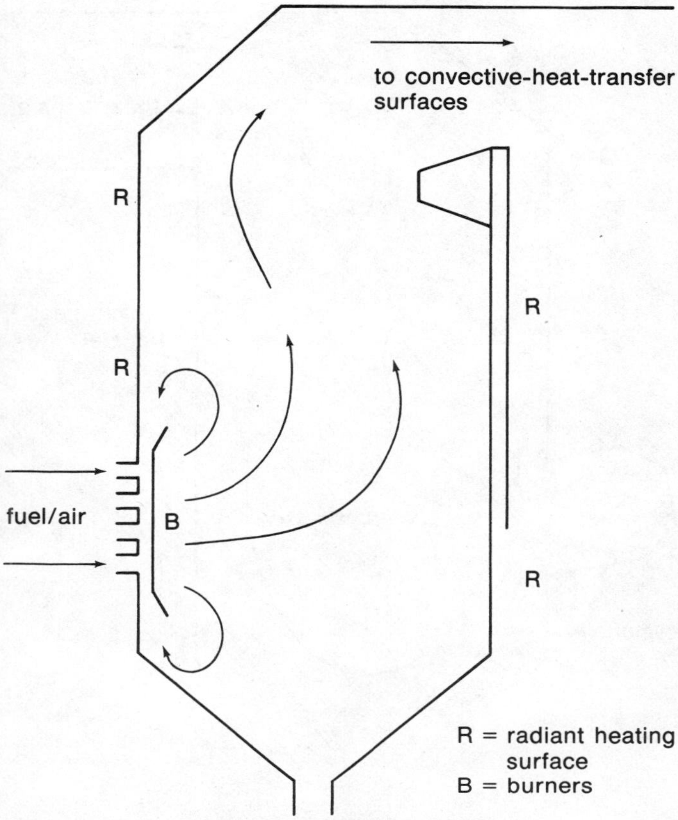

Figure 4-14a. *Front wall firing.*

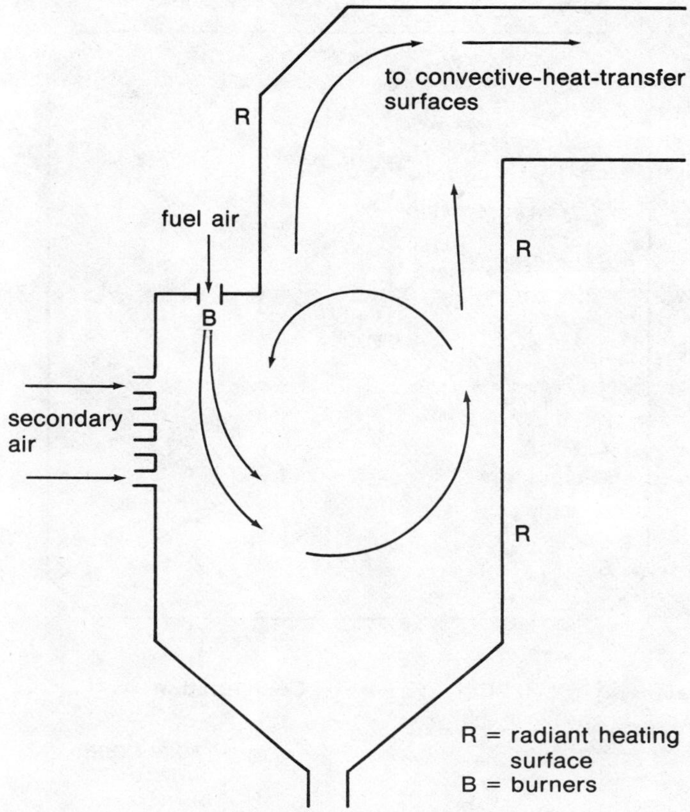

Figure 4-14b. *Down firing.*

circumstances a coal whose ash has a low melting point is obviously preferred. These kinds of boilers are called *wet-bottom boilers* because the ash liquifies. All other "dry" boilers must be fired with a coal whose ash melting point is high enough that no melting occurs or else the fine ash particles (called *fly ash*) would stick to the water tubes and cause poor heat transfer and tube corrosion. This phenomenon means the designer must restrict the flame temperature and, unfortunately, a lower temperature means a slower burning rate.

In the search for high-intensity, i.e., rapid, combustion at high temperatures and conditions of extreme turbulence, developers are limited by the availability of materials capable of containing the high-temperature combustion zone, and the availability of fuels with high-melting ash. The cyclone furnace (Figure 4-15) reconciles these problems by deliberately allowing the ash to melt and form a protective liquid layer over the water tube walls of the cyclone. The layer of molten ash (often called *slag*) experiences a temperature distribution across its thickness such that next to the steel water tubes it is frozen, but where it is exposed to the flame it remains liquid. The centrifugal effects in the cyclone cause larger particles to be flung to the wall where they are held by the sticky slag and scrubbed by the turbulent tangential gas flow until they burn away. The molten slag layer continuously runs down the walls to a slag tap that runs into a water pond where it freezes and is collected as a granular, glass-like material useful as roadfill or aggregate for concrete.

The outlet throat between the cyclone and the boiler is designed to reverse the axial flow of the gases and force them back into the cyclone before they exit, causing additional turbulence effects. Figure 4-16 shows where the cyclone furnace is placed in the boiler. The cyclone furnace is usually fired with coal crushed to less than 0.25 inches, which is considerably larger than needed for pulverized fuel firing and, accordingly, involves lower fuel preparation costs. Cyclones were used in installations up to 1,000 megawatts but have fallen from favor in recent years due to economic reasons and the higher emission of nitrogen oxides caused by the high operating temperature.

Before leaving a description of coal as a fuel, and in order to compare it to oil and gas, consider Table 4-3, which shows a calculation of the amount of a particular coal which is required to generate 4 million pounds of

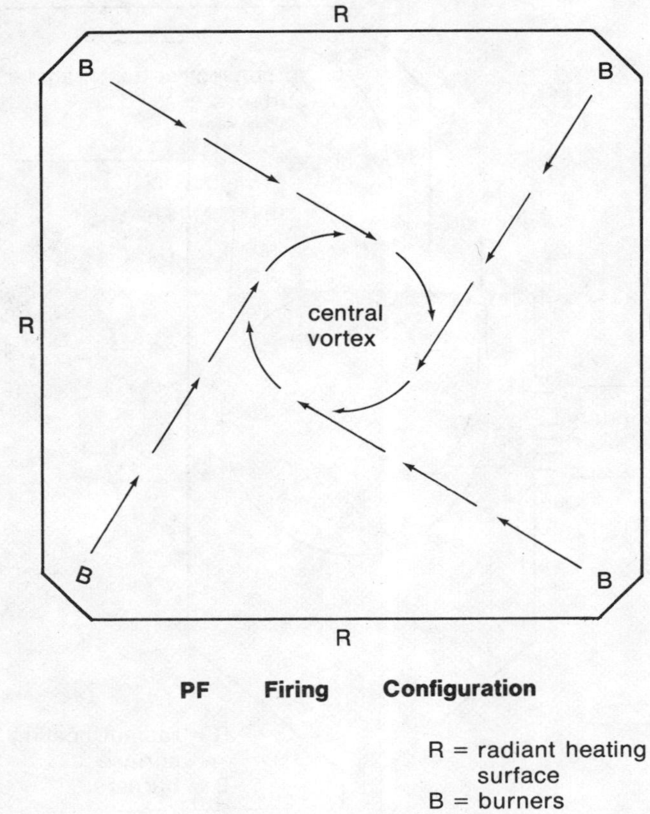

Figure 4-14c. Corner firing.

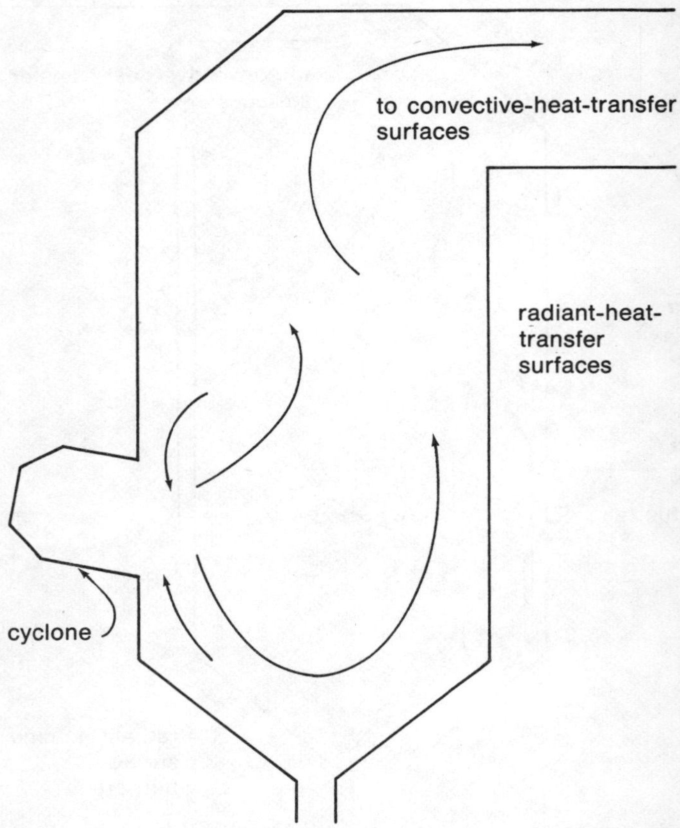

Figure 4-16. The cyclone furnace principle (siting the combustor).

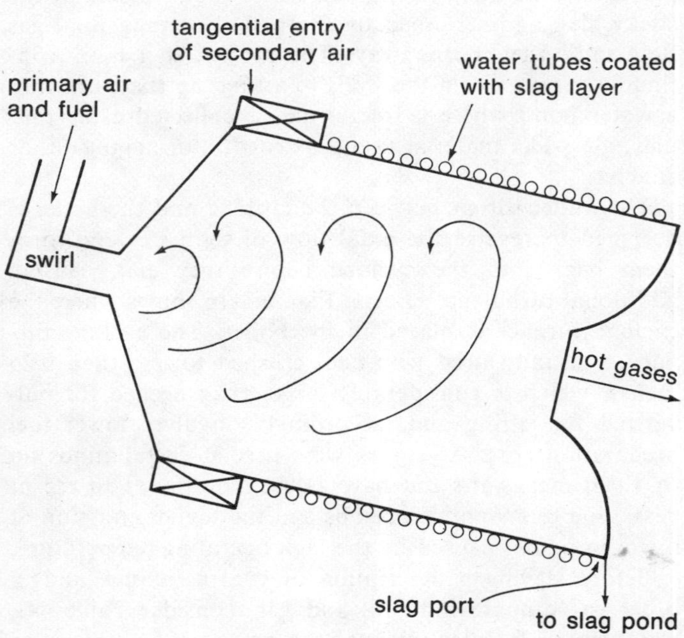

Figure 4-15. The cyclone furnace principle (the combustor).

steam an hour at 500 psi and 500°F. Follow the calculation carefully. You can see this time how it is necessary to include other fuel constituents (oxygen, nitrogen, water, and sulfur) in the calculations. Finally, to help you with the comparison of the data generated by the calculations in Tables 4-1, 4-2, and 4-3, a Table 4-4 is presented to compare the most important combustion data from the three cases. Make sure you understand all three calculations before reading further.

The Flame Temperature

Now, for the third time we have reached a point in the furnace where the fuel has given up its heat to the combustion gases. Predicting the temperature of the gases is a mathematical puzzle similar to the one described when discussing the basic oxygen furnace. Consider the following method of calculating the flame temperature.

Let's work back from the end product—the temperature. To calculate it we must first determine the total heat released by the fuel. That's easy. We know the heating value of each pound of fuel from laboratory tests, and we know the proposed feed rate of the fuel. Now, since this

Table 4-3
Combustion Calculation for Coal

Basis: 4×10^6 lbs/hr steam at 500° F and 500 psi with heat content 1,231.2 Btu/lb

Coal Requirement:

Heat contained in steam = $4 \times 10^6 \times 1{,}231.2 = 4.925 \times 10^9$ Btu/hr

Heat released from coal = $\dfrac{4.925 \times 10^9}{0.8} = 6.156 \times 10^9$ Btu/hr

Coal (13,150 Btu/lb) burned (ash-free basis) = $\dfrac{6.156 \times 10^9}{13{,}150} = 4.68 \times 10^5$ lbs/hr

Coal (10% ash) burned = $\dfrac{4.68 \times 10^5}{0.9} = 5.2 \times 10^5$ lbs/hr

Products Flow and Composition:

Assume coal composition (ash free) is: 92.4% C, 4% H, 1.4% N, 0.9% S, 1.3% O

Coal Constituent	Percent in Coal	Moles Used	Moles Oxygen Used	Products Formed
C	92.4	$\dfrac{0.924 \times 4.68 \times 10^5}{12} = 3.60 \times 10^4$	3.60×10^4	3.60×10^4 moles CO_2
H	4.0	$\dfrac{0.04 \times 4.68 \times 10^5}{1} = 1.87 \times 10^4$	0.47×10^4	0.94×10^4 moles H_2O
N	1.4	$\dfrac{0.014 \times 4.68 \times 10^5}{14} = 468$	—	234 moles N_2
S	0.9	$\dfrac{0.009 \times 4.68 \times 10^5}{32} = 132$	132	132 moles SO_2
O	1.3	$\dfrac{0.013 \times 4.68 \times 10^5}{16} = 380$	−190	—

Total oxygen used = $(3.60 \times 10^4) + (0.47 \times 10^4) + 132 - 190 = 40{,}642$ moles
(Assume 30% excess air)
Oxygen supplied = 52,835 moles O_2
Accompanying nitrogen = 198,759 moles N_2

Product	Calculation	Moles	Percent Dry
CO_2	—	3.6×10^4	14.56
H_2O	—	0.94×10^4	—
SO_2	—	132	0.0533
O_2	52,835-40,642	12,193	4.93
N_2	198,759 + 234	198,993	80.46
	Total Dry	247,318	100.00
	Total Wet	256,718	

(Table 4-3 continued on next page)

Table 4-3 (continued)
Combustion Calculation for Coal

Fan and Chimney Ratings:

$$\text{Air flow rate} = \frac{52{,}835 \times 359}{60} \times \frac{100}{21} \times \frac{(460 + 80)}{(460 + 32)} = 1.65 \times 10^6 \text{ ft}^3/\text{min (at } 80°\text{F and } 14.7 \text{ psi)}$$

$$\text{Waste flow rate} = \frac{256{,}718 \times 359}{60} \times \frac{(460 + 410)}{(460 + 32)} = 2.72 \times 10^6 \text{ ft}^3/\text{min (at } 410°\text{F and } 14.7 \text{ psi)}$$

Waste-Gas Dew Point:

$$\text{Partial pressure of water in waste-gas} = \frac{0.94 \times 10^4}{256{,}718} \times 14.7 = 0.54 \text{ psi}$$

Therefore, from Steam Tables, dew point = 82°F

heat will be absorbed into the combustion gases, we must calculate the weight of combustion gases that will be present. This also is easy. You have just seen such a calculation three times for three different fuels. All that remains, then, is to find the specific heat or heat capacity of each component in the gas. The average heat capacity of the combustion gases is made up by taking a molar average of the individual heat capacities, that is, by multiplying the heat capacity of each gas by its mole fraction and adding them as shown in Table 4-5.

Incidentally, the mean heat capacity is a measure of the temperature rise obtained each time 1 Btu is added to one mole of the gases. So, look in *The Chemical Engineer's Handbook* for the heat capacities of our individual gases. They are tabulated according to temperature. In other words the specific heat changes with temperature according to the equation shown in Table 4-5. Since we don't know the temperature, deadlock!

To get out of our dilemma, guess at the value of the flame temperature. Then, look up the heat capacities, and work out the average heat capacity; this will give the temperature rise associated with the heat release from the fuel. The resulting flame temperature will almost certainly be different from the one assumed earlier.

However, because the new temperature will be somewhat more accurate than the original guess, it can be used to go back to the heat capacity tables and start again . . . and again . . . and again . . . until the improvement in the value of the temperature is only in the first decimal place. This process is called an *iterative calculation*. The iterations can be speeded up if you are working by hand by watching how successive calculations change the temperature and then trying to leapfrog forward towards the answer.

The temperature we finish up with is called the *adiabatic flame temperature*, which means we have assumed there was no heat loss as the flame developed and burned. This, of course, is not true. There is a considerable amount of heat radiation away from the flame. The boiler designer's job is to place water tubes in the path of this radiation that capture the heat into the water and steam. By the time combustion is complete and the flame becomes a stream of hot gases, the actual temperature will be much less than the adiabatic flame temperature. How-

Table 4-4
A comparison of gas, oil, and coal combustion

Basis: The production 4 x 10^6 lbs/hr of steam at 500°F and 500 psi

					Waste-Gas Composition			
Fuel	**Feed Rate**	**Excess Air**	**Air Required (ft³/min)**	**Waste-Gas Flow Rate (ft³/min)**	**%CO_2**	**%O_2**	**%N_2**	**Dew Point**
Natural gas	5.769×10^6 ft³/hr	5%	1.005×10^6	1.870×10^6	11.11	1.11	87.78	136°F
Heavy oil	3.40×10^5 lbs/hr	10%	1.095×10^6	2.100×10^6	14.42	2.02	83.56	109°F
Coal	5.2×10^5 lbs/hr	30%	1.650×10^6	2.720×10^6	14.56	4.93	80.46	82°F

Table 4-5
Calculation of Adiabatic Flame Temperature

Basis: Combustion data from Table 4-1.

Heat released by 16,070 moles of CH_4/hr = 6.156×10^9 Btu.

$$\text{Heat released per mole of } CH_4 = Q = \frac{6.156 \times 10^9}{16{,}070} = 383{,}070 \text{ Btu/mole.}$$

From Table 4-1, products of combustion are: 1 mole CO_2, 0.1 mole O_2, 2 moles H_2O, and 7.9 moles N_2.

$$\text{Adiabatic flame temperature} = \frac{\text{heat released}}{\text{heat capacity of combustion gases}} = \frac{Q}{Cp_m}$$

However, Cp_m is temperature-dependent and temperature is unknown. Guess $T = 4000°$F. Can now prepare iteration table.

Component	Number of Moles	Molar Heat Capacity at 4,000° F	Total Heat Capacity at 4,000° F	Molar Heat Capacity at 3,983° F	Total Heat Capacity at 3,893° F	Molar Heat Capacity at 3,943° F	Total Heat Capacity at 3,943° F
CO_2	1.0	13.11	13.11	13.09	13.09	13.10	13.10
O_2	0.1	8.48	0.85	8.46	0.85	8.47	0.85
H_2O	2.0	10.58	21.16	10.54	20.08	10.57	21.14
N_2	7.9	8.01	63.28	7.99	63.12	8.00	63.20
Total	11.0	Cp_m =	98.40	Cp_m =	97.14	Cp_m =	98.29
		$\therefore T = \frac{Q}{Cp_m}$ = 3,893° F		$\therefore T$ = 3,943° F		$\therefore T$ = 3,897° F	

Adiabatic flame temperature = 3,900° F

ever, there is still a considerable amount of sensible heat which can be extracted from the gases before they are cleaned and sent off to the chimney. First, some of their heat is recovered by suspending the superheater tubes in the hottest part of the gas flame.

On the inside of the superheater tubes, steam produced in another part of the boiler picks up sensible heat which raises its temperature well above its boiling point. This process is called *superheating the steam*. Further along in the process the hot gases are made to pass over *convection heating surfaces* that are steel tubes with an outside diameter of about 3 inches. (You will learn more about what happens here in just a moment when the steam water cycle is described.) The hot gases are now considerably cooler than when they were in the flame. They continue their journey out of the boiler and into the economizer, which is simply a water preheater. (This too will be described in a moment.)

The gases enter the economizer at just over 800° F, and after giving up some heat to the boiler feedwater, leave at around 630° F. They now pass through an air heater that further reduces the temperature to 410° F, while preheating the air flowing inside the steel tubes to around 340° F. Since this air is to become the air for combustion, the air heater is a means of recycling some of the waste heat of the hot gases. Remember that we cannot take all the heat from these gases. We have to leave the temperature sufficiently high that after the inevitable further cooling in the duct work to the chimney and the chimney itself, the final temperature at the top of the chimney is still well above the dew point of the gases.

The combustion waste gas is not passed in all cases straight to the chimney in this way. If the original fuel had contained more than just a trace of sulfur, by federal law the percentage of sulfur oxides in the combustion gases must be reduced to below a safe, acceptable level. Any remaining fly ash must also be removed (only present when coal or some fuel oils are being burned) and in the future, nitrogen oxides will also have to be removed before the gases are introduced to the chimney and the atmosphere. In Chapter 10 the removal of sulfur oxides and dust is discussed.

The Steam Cycle

Having completed the description of the journey of the fuel and air through the boiler system to the waiting ash disposal cars and the atmosphere, let us look at the steam side of the system and see what happens to the water after it enters the plant. We shall see how the process of evaporating the water into steam, superheating it, expanding it through a turbine system, condensing it, and pumping it back into the boiler is a cyclic process. However, there will be some losses such as the boiler blowdown which will be described, and there will also be leaks from the steam pipe system.

In addition, if the boiler plant is part of an integrated chemical plant, then there will be a ready use for low-pressure steam to carry out hundreds of jobs around the plant site. This is called *process steam*, which is extremely useful to a process plant. To make up for all the water (or steam) losses, there must be a constant supply of fresh boiler water, which is called *makeup feedwater*.

At the design stage the company makes a decision on the water source for the plant. City water is expensive but relatively clean, which means it contains a small percentage of dissolved solids. Natural river or lake water will contain a higher amount of dissolved solids, and their removal means higher treatment costs. There is no way of knowing which water source to use until calculations are made for the cost per 1,000 gallons to fill the boiler system with water initially and then provide makeup water for its lifetime.

The makeup water and the returned condensate are pumped into the boiler, and as the water evaporates, the dissolved solids are left behind in the boiler, increasing the concentration and causing the possibility of deposits on the insides of the water tubes. This condition is called *scaling* and can lead to clogged pipes, poor heat transfer efficiencies, and, ultimately, explosions and serious damage. Obviously, treatment of the boiler feedwater is essential. The amount of cleanup required depends upon whether a blowdown (defined later) is possible (you will see later why in some designs a blowdown cannot be operated).

As the water evaporates within the boiler, leaving behind the dissolved solids, their concentration rises. Even water which looks sparklingly clean and pure can contain up to 100 parts per million (ppm) of various dissolved substances. This may not sound like much, but is equivalent to about one level teaspoonful in 12 gallons of distilled water. However, since big boiler systems evaporate anywhere between 200,000 and over 4 million pounds of steam an hour, the solids left behind in the boiler build up quickly unless they are removed before the water enters the system or somehow removed from the boiler itself. In practice both processes are used.

The first precaution is to pass all water through what is called the *water treatment plant*. The actual treatment administered to the water depends, of course, on what the available local water contains. Water treatment is only necessary for makeup water and the initial fill of water in the system. It is not necessary to treat the condensate returned from the turbine house for reuse. If some of the steam is used as process steam or to drive machinery, however, and its condensate is returned, then it may be contaminated and must go through a water treatment plant.

In addition to dissolved solids there may be traces of oil, dissolved gases, and suspended solids such as mud, clay, silt, and metallic oxides. There may even be organic matter such as bacteria, algae, and other microscopic cellular organisms. The damage they could cause inside the boiler is unthinkable.

Dissolved solids become more and more concentrated until they precipitate out onto hot surfaces. This, as you will see, gives an extra resistance to heat flow and reduces the efficiency of the system. In addition, dissolved solids, oil, and organic matter tend to thicken the water and cause it to foam as the steam is driven off. At high steaming rates this can be disastrous because the foaming steam bubbles burst violently and spray the concentrated boiler water into the steam which is rising so quickly away from the water surface that it entrains, or carries off, the dirty water. Then, as the contaminated steam passes through the superheater tubes, some solids will deposit onto the walls, causing eventual burnout and a potentially dangerous explosion. The undeposited solids remain suspended in the steam and may be deposited in the steam pipes, turbine, or condenser. In any of these places they are a nuisance, but the turbine would suffer most because it is a fast-moving, highly engineered piece of equipment that would deteriorate rapidly if its finely made blades were constantly sprayed with solid particles.

The second cleaning-up method is called *blowdown*. It works by arranging to have a small but constant flow of boiler water through an outlet situated at the lowest point in the water drum of the boiler. The flow rate of the blowdown can be calculated very simply using a material balance (see Figure 4-17). In low-pressure boilers the concentration of the total dissolved solids in the boiler water itself is never allowed to go over 3,000 ppm. The appearance of a glass of blowdown water with 3,000-ppm solids would perhaps surprise you if you compared it with a glass of treated feedwater. The former would look like a glass of milk, the latter would be crystal clear. Incidentally, this means of removing an unwanted substance from a cyclic process is called a *purge*, and only in this particular instance of a cyclic boiler system is it called a blowdown.

The raw water from a lake, river, well, or reservoir enters the plant and is stored in holding ponds to ensure a steady water supply. Then it is filtered through finely packed beds of sand (see Figure 10-4). This removes all the suspended solids and organic material. Only dissolved contaminants will pass through the filter. Coagulating chemicals such as alum and iron sulfate are added to the raw water prior to filtration. These cause the small particles in the water to agglommerate into larger ones which settle

Basis: 4×10^6 lbs/hr of steam

Required data: Maximum allowable solid content of water.
Assume it is specified to be 3,000 ppm.
Solids content of boiler feedwater.
Assume it is measured to be 0.04 ppm.
Solids carried over in steam.
Assume it is specified to be 20 ppb.

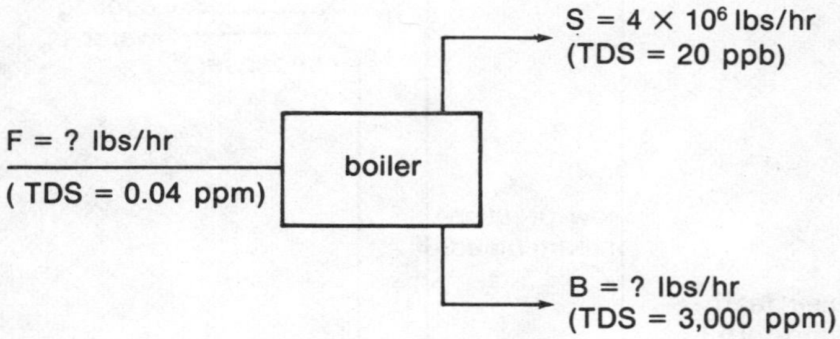

Mass balance on Total Dissolved Solids (TDS)

$$F \times \frac{0.04}{10^6} = (S \times \frac{20}{10^9}) + (B \times \frac{3000}{10^6})$$

$$4F = 2S + 300{,}000B \qquad \text{Equation (1)}$$

Mass balance on total material

$F = S + B$
$B = 267$ lbs/hr = blowdown flow rate
(= 32 gal/hr of blowdown)

Figure 4-17. *Calculation of blowdown flow rate.*

to the bottom of the holding ponds or are filtered out by the sand filter.

Sometimes, in areas where the water source may include surface runoff water, there may be unusually large amounts of organic material that ordinary filters will not remove. In these cases it is common practice to include activated-charcoal filters to remove the final traces of organic material. Incidentally, charcoal filters also remove any chlorine that may have been used initially to kill microscopic life. After filtration the water still contains dissolved material.

The next stage is called *sodium zeolite softening*, and it utilizes resin materials that exchange the hardness constituents—such as calcium and magnesium salts—for sodium. The resin is coated onto a stack of plates over which the water cascades. The zeolite would rather be attached to calcium or magnesium than chemically bonded to sodium and so the ions are exchanged. The process is a special type of ion-exchange system. When all the available sodium ions have been exchanged for calcium or magnesium ions, the resin is said to be exhausted, or spent. Just before this happens the flow of water is diverted to a standby ion exchanger whose resin has been freshly regenerated, and the spent ion exchanger is then regenerated by passing brine, a solution of sodium chloride, through it.

After leaving the sodium zeolite softening process the water could be used as boiler water, and often is used in conjunction with blowdown to remove the highly soluble sodium salts from the boiler drum. However, with high-pressure systems and high steaming rates, sodium salts can still cause foaming and carry-over of solids as described earlier. If sodium salts are still present, extra treatment is required for the water. In this case a hydrogen zeolite softener is used to exchange the sodium ions in the water for hydrogen ions, and the spent resin is regenerated by washing with acid.

Similarly, the nonmetallic ions (bicarbonate, carbonate, sulfate, and chloride) are replaced by hydroxide ions in an ion exchanger containing a specially prepared resin saturated with hydroxide ions. After being spent, this resin is regenerated by using a solution of caustic soda. In addition to all the water treatment undergone so far, extra processes may be needed. Dissolved oxygen concentrations must be less than 0.007 ppm or 70 parts per billion (ppb). Above this level oxygen can cause corrosion, or pitting, of the boiler tubes. To avoid hazards, pitted tubes must be replaced more often than if undamaged, which causes expensive downtime for the boiler. Small amounts of an oxygen-scavenging material are added to the feedwater and are present in the system at all times. A usual agent for this purpose is sodium sulfite.

Do not think the above description of a water treatment plant is the only one available. In fact, there are many different ways of treating water, and although the one described is the most common, there are often local water conditions that require different kinds of treatment. The design team responsible for a boiler plant will spend a great deal of time, money, and effort planning their water treatment needs with a water consultant and experts on local water conditions.

The treated water is added to the condensate return and passed through a succession of low-pressure bleed heaters that pass the water through tubes surrounded by steam bled away from the low-pressure turbine at various points (see Figure 4-18). Then, after passing through the boiler feedwater pump (also driven by steam, but this time bled from the high-pressure turbine) the water passes through high-pressure bleed heaters that are heated by steam from the high-pressure and reheat turbines. The water leaving the last high-pressure bleed heater enters the economizer at around 220° F, where sensible heat is taken from the hot combustion gases, thus reducing their temperature from 800° F to around 630° F.

The water leaves the economizers at around 290° F, but it is still liquid because at this high boiler pressure (say, 500 psi) its boiling point is 467° F.

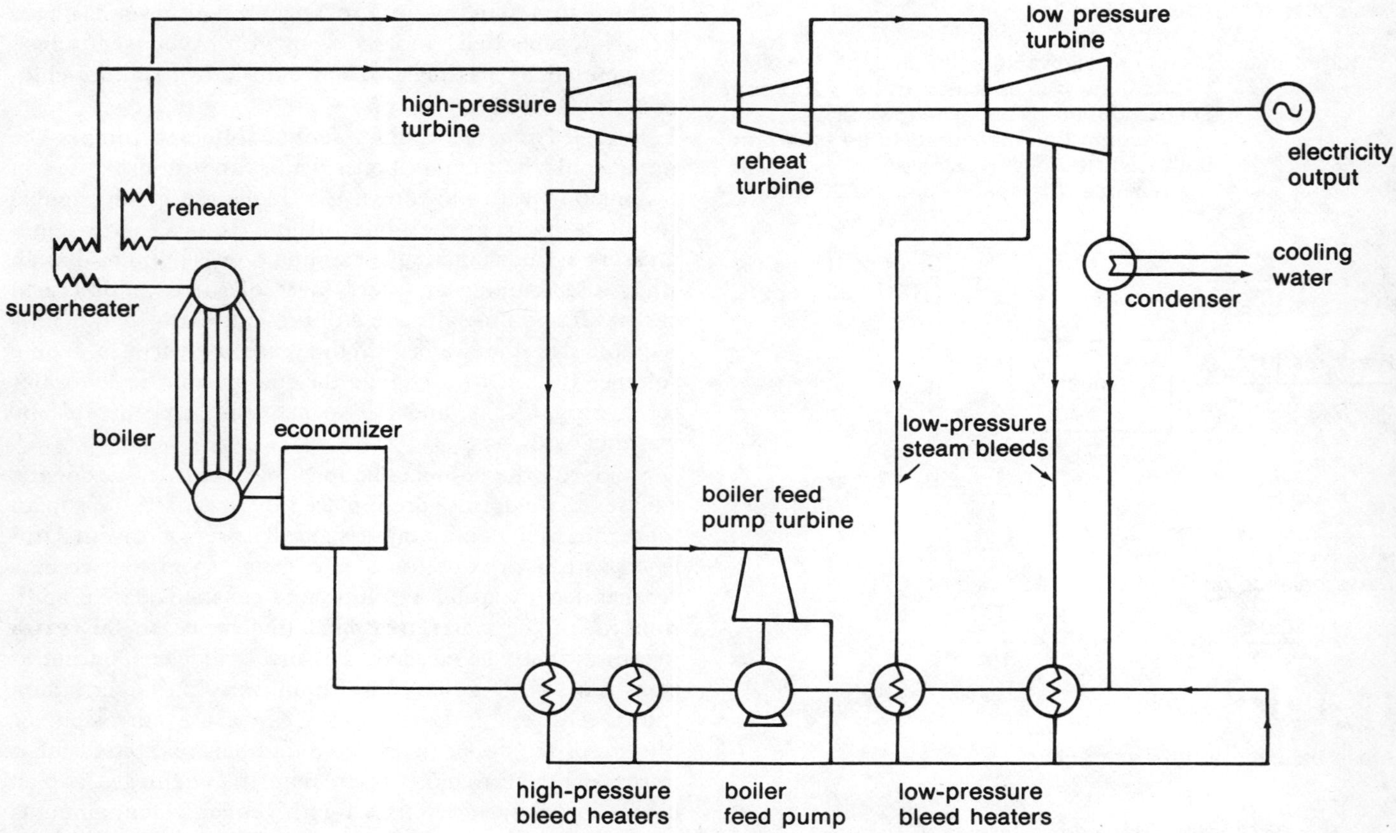

Figure 4-18. *A typical steam cycle.*

Figure 4-19 shows the cross section of an economizer tube. The outside diameter is around 2 inches, and the wall thickness is just over 0.125 inches, but even thicker and stronger for higher-pressure systems. Also, a thin layer of scale will build up between cleanings of the economizer tubes.

The hot gases pass over the outside of the tubes in the direction shown in Figure 4-19. Although the velocity of the gas stream is quite high, there will be some drag effect at the surface of the tubes, causing the maintenance of a thin boundary layer. For the sensible heat of the flue gases to enter the steel, it has to fall down a temperature gradient. Figure 4-19 shows the temperature gradient which establishes itself from the bulk flow of hot gases, across the slow-moving and stationary film of gases on the outside of the tube, through the tube itself, then the scale, and finally through a film of slow-moving water on the inside before it gets to the bulk of water flow.

Let us follow the progress of the heat down the temperature gradient. In the bulk of the gas stream the temperature gradient is almost flat. This is because internal turbulence in the gas stream causes rapid mixing, which means the temperature throughout the bulk is kept uniform. Notice what happens to the temperature close to the steel wall of the tube. The slower and slower moving layers of the gas are less well mixed, and the heat that escapes to the steel tube is not replaced so quickly by internal turbulence. Therefore, the temperature drops sharply near the wall surface. The gas molecules actually at the surface are stationary because they cling to the surface. This means that heat transferred through those last few molecules must be by conduction rather than the convective method, which involves molecular movement on a grand scale. The rate at which heat can be transferred from the bulk gas flow to the steel tube against this gas film resistance is called the *outside wall heat transfer coefficient* (denoted by h_o in Figure 4-18).

To know at the design stage the value of h_o that can be expected, a method is required for its calculation. We could try to relate it in some way to the thermal conductivity of the gases in the mixture, but it's not all conduction. Some heat is transferred through the film by convection and radiation. Even if we could come up with a value for thermal conductivity that includes the effects of convection and radiation, we still would have the impossible problem of finding the thickness of the boun-

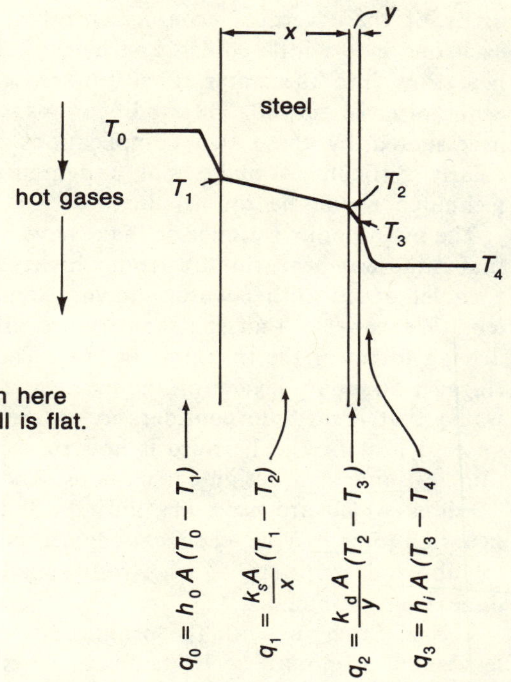

Warning:
Equations given here assume the wall is flat.

Typical Data:

h_0 = outside heat transfer coefficient = 5.0 BTU/ft^{3}hr °F

h_i = inside heat transfer coefficient = 600 BTU/ft^{2}hr °F

k_s = thermal conductivity of steel = 612 BTU/ft^{2}hr °F/ft

k_d = thermal conductivity of scale = 0.86 BTU/ft^{2}hr °F/ft

T_0 = hot gas temperature = 700°F

T_4 = water temperature = 260°F

x = tube thickness = 0.25 inches

y = scale thickness = 0.01 inches

Solve four equations to give:

T_1 = 266 °F

T_2 = 265.9 °F

T_3 = 263.8 °F

Figure 4-19. *Heat transfer through a steel wall.*

dary layer, which varies with the temperature and velocity of the gases. It turns out that this approach yields meaningless results.

Instead, it is normal to look up in heat transfer books such as McAdams' *Heat Transmission*, the values of h_o, under the expected conditions. The authors of such reference books have brought together the results of hundreds of experiments carried out with many gas and water flows and different arrangements of tubes. In addition to being able to look up the value of h_o, there are some calculation methods available. These equations are *empirical* because they are based on correlations of measured data. A *theoretical* equation, on the other hand, has been predicted by a theoretical consideration of the system under study and then confirmed by a series of carefully chosen experiments. Empirical relationships are used when the theoretical studies have revealed far too complicated a system to develop a suitable equation.

An example of an empirical equation in everyday life can be drawn from the experience of airline operators. They find that at flight check-in a percentage of people cancel their reservation or simply do not show up. To predict this percentage on a theoretical basis is impossible because of the vast number of variables and the even larger number of reasons why the "no-shows" fail to appear. However, over the years, the airlines have discovered the percentage of no-shows is around 10%. Thus, an empirical equation is:

$$\text{number of no-shows} = \text{number of bookings} \div 10$$

The empirical equation for h_o involves the values of the bulk velocity, density, viscosity, the thermal conductivity of the gas, and the diameter of the tube. The equation for heat transfer from the combustion gases to the economizer tubes is given in Figure 4-19.

Even though the heat transfer coefficient, h_o, can be looked up or calculated from empirical equations, and the bulk gas temperature is known, we cannot yet work out the actual amount of heat that flows across the boundary layer because the temperature at the wall surface (T_l in Figure 4-19) is unknown. However, let us consider the heat flow through the steel in the hope of finding some way to determine T_l. For the steel wall itself the heat transfer equation is easy. Heat conduction through a flat wall is given in Figure 4-19.

In this equation, A refers to the area of the wall at right angles to the direction of heat flow. A is constant from outside to inside for a flat wall. In the economizer tube this is not so: the area on the outside is obviously greater than the area on the inside for any length of tube. To accommodate for this changing area, some modification is necessary to the equations in Figure 4-19, which are for a flat wall. Although the modifications are quite simple, I would run the risk of masking the points I want to make here, so I have chosen to consider only the hypothetical case of a flat wall.

Across the wall, then, with outside temperature T_l and inside temperature T_2, the heat transferred is given by the equation in Figure 4-19. You can see that the temperature gradient will be quite shallow because of the excellent thermal conductivity of steel. However, since we do not know either the inside temperature T_2 or the outside temperature T_l, we still cannot calculate the heat flow. Let us go a little further and see whether any new information will enable solution of this problem.

Between the steel wall and the boundary layer on the

water side of the tube, there will be a thin scale buildup. This generally is calcium carbonate or a similar substance and has a low thermal conductivity. The heat flow across the scale is treated in exactly the same way as through the steel wall. We look up the value of k_d for boiler scale (very similar to the value for calcium carbonate) and use the equation in Figure 4-19. Between the inside scale surface and the bulk of the water flow in the economizer tube, there will be a boundary layer or film similar to the gas film on the outside. It, too, will present extra resistance to heat flow, and hence, there will be another steep temperature drop as shown in Figure 4-19. The inside water film resistance is treated in a similar way to the outside one because the rate of heat transfer is governed by a heat transfer coefficient, h_i. Again, we use reference tables or calculate it using empirical equations. Thus, the heat transferred through the film to the bulk (well-mixed and flat uniform temperature gradient) water flow is given in Figure 4-19.

In trying to solve this problem we have created four equations between the heat flux through each part of the composite wall and the temperature differences across the wall. Now, let's look at the heat fluxes, q_1, q_2, q_3, and q_4. They must be the same. Otherwise, if, for instance, q_1 is greater than q_2, more heat would be arriving at the outside wall surface than leaving it. If the temperature at that surface remains constant, as it will in practice, then clearly, heat doesn't build up, and the heat flow q_1 must be equal to the heat flow q_2. For similar reasons, $q_1 = q_2 = q_3 = q_4$, or let's say they are all equal to Q. Now we have four equations with four unknowns (Q, T_1, T_2 and T_3).

Using algebra, simultaneous equations such as these can be solved if the number of unknowns is no more than the number of equations. In Figure 4-19 typical values have been inserted for the thermal conductivities of the steel and scale and the various properties of the water and combustion gases. Notice how the temperature drop across the steel is only 0.1° F, whereas across the scale, the temperature drop is 2.1° F, even though the scale is only 1/100 of an inch thick. Also notice that the temperature drop due to the gas-side resistance is an enormous 434° F, whereas the water-side resistance is much lower and results in a temperature drop of 3.8° F.

Remember that in the vertical steel tubes in the economizer, there are several differences to this ideal composite wall case. First, there is the difference we have already discussed, i.e., the area term is not constant throughout the thickness of the wall. A trivial modification takes care of this. Next, as the water passes up through a single tube its temperature increases as it picks up the heat that has been transferred inwards. Thus, the driving force of the temperature difference from bulk gas stream to bulk water stream is reduced, and hence, the heat flow is also reduced.

Furthermore, the gas stream passes over literally hundreds of these vertical economizer tubes. After leaving each one it is a little cooler, and hence, the potential for heat flow into the water tubes towards the top of the economizer is reduced. The modifications to the equations necessitated by these two complications are not particularly difficult to make, but a demonstration of the technique would be too lenghty.

The main points I wanted to make have been made. The first is the way heat will flow from a higher temperature to a cooler one, in other words, down a temperature gradient. We can say that the temperature difference is the driving force for the transport of heat. The second is the sluggish boundary layers (on the inside and outside of the walls) that contribute considerably to the overall resistance to heat flow. The third is how to account for these film resistances in design calculations. The further modifications which are necessary must be left until the appropriate time in your chemical engineering education. If the above description has whet your appetite, then I have done what I intended.

Let us catch up with the progress of the water as it leaves the economizers. Its temperature is a little under 300° F, which is still below its boiling point at these high pressures. The hot water enters the boiler proper. It is convenient here to differentiate the two main kinds of boilers, which are very different in design. The first and older design is the Two-drum Sterling boiler. The second is the Universal-Pressure boiler.

I shall describe only the Two-drum Sterling boiler and the phenomena that occur within it (see Figure 4-20). The water from the economizer is circulated in water tubes which entirely cover the surface of the walls around the radiant, or visible, part of the flame. The main process occurring here is radiative heat transfer from the flame to the outer walls of the steel water tubes. In addition to radiant heat falling on the tubes there is a small amount of convective heat transfer due to the presence of hot gases circulating inside the furnace section of the boiler. The water is heated to its boiling point in these radiant tubes and enters a *water drum* mounted at the back of the furnace. Projecting from the top of the water drum are hundreds of water tubes which pass upwards into the *steam drum*, a similar drum above the furnace. As the boiling water is forced up the water tubes towards the steam drum, the heat transfer processes are similar to the ones occurring in the economizer. Of course, the outside bulk gas temperature is much higher than it will be when it gets to the economizer, and this produces some interesting effects on the inside of the tube.

Let us consider the heat transfer through this tube with the same amount of detail as before. Figure 4-21 shows a composite wall (again, flat for simplicity) and the temperature distribution across it. Also in Figure 4-21 is a curve showing the relationship between the heat flowing across the wall and the temperature difference driving it.

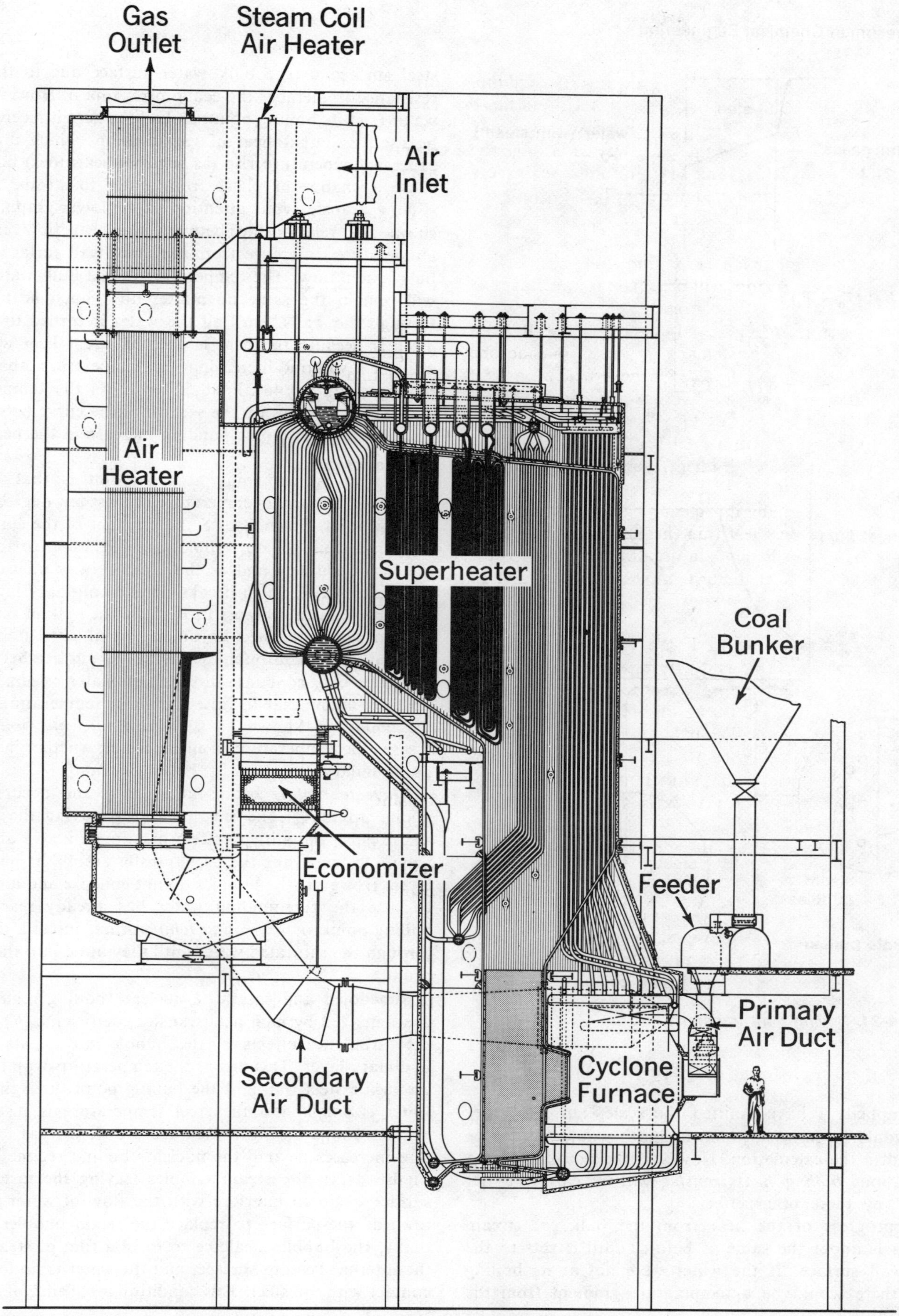

Figure 4-20. *The Two-drum Sterling boiler. (Courtesy of Babcock & Wilcox Co.)*

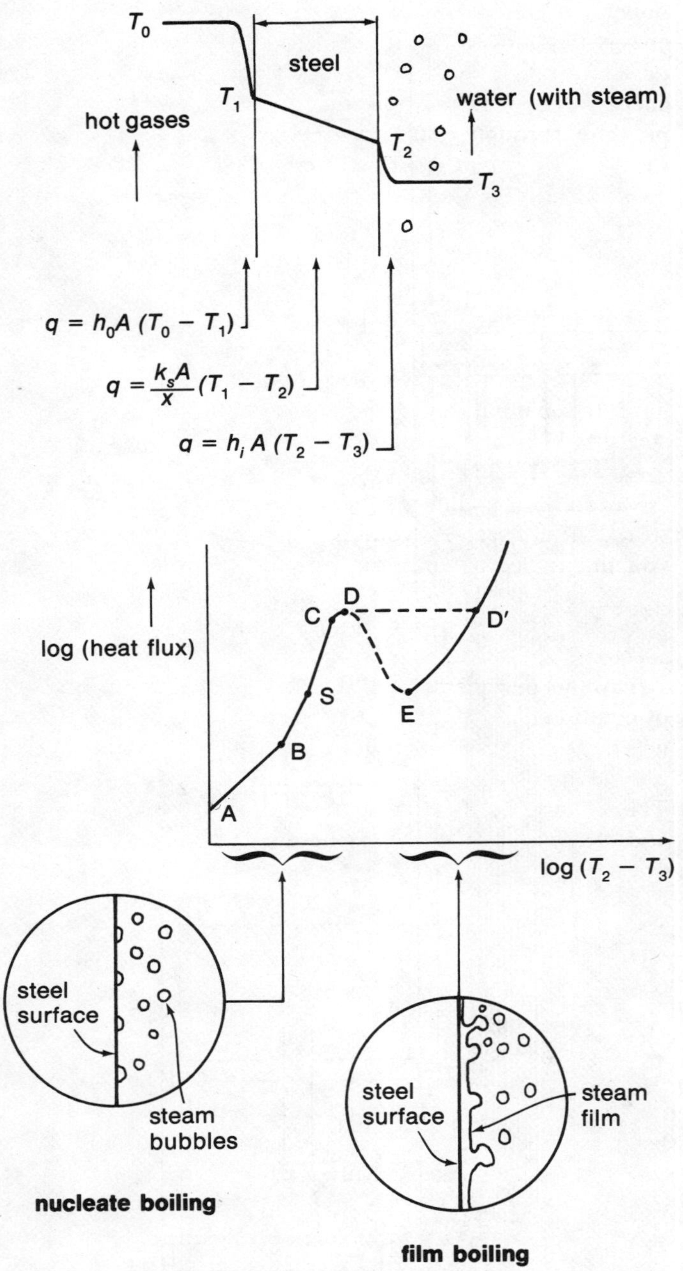

Figure 4-21. Boiling heat transfer.

For simplicity I have omitted the scale, but remember there would be some in practice, and it would have to be included in the calculation. However, the point I want to make about *boiling heat transfer* will be clearer if not masked by these other effects.

The progress of the heat from the bulk gas stream inwards is about the same as before, until it gets to the inner wall surface. If the water were not at its boiling point, there would be a temperature gradient from the steel surface to the bulk water surface due to the film resistance (remember the economizer tubes). However, the water *is* at its boiling point, and all the heat it receives will be used as latent heat of vaporization, which does not cause a temperature rise (as sensible heat does) but does cause a change of phase from water to steam.

Now, transfer your attention to the lower graph, which shows the relationship between the heat flow and the temperature difference from the inner wall surface to the bulk water flow. The temperature in the bulk water flow will remain the same no matter how much heat comes through the walls, until all the water is turned to steam. That is because the heat is used as latent heat which is absorbed without increasing the temperature above the boiling point of 467°F at 500 psi. As the temperature difference increases on the section of the curve between A and B, there is a corresponding increase in the heat flow rate.

All that is happening at this point is that we are increasing the water temperature the last few degrees to its boiling point. From B to S the gradient of the curve gets steeper because steam bubbles are beginning to form on the inside surface, and as they break away they cause a little turbulence which breaks up the boundary layer and helps heat transfer across it. Because the bulk of the water has not quite reached its boiling point until point S is reached, the steam bubbles leaving the surface between B and S quickly condense in the main water stream, giving up their latent heat to raise the water temperature. Thus, the condition known as *subcooled nucleate boiling* increases the temperature of all the water, whether bulk flow or boundary layer, to its boiling point.

Nucleate boiling begins at point S and occurs at all points up to C. The steam bubbles leaving the surface again cause the stirring effect mentioned above and contribute to increasing the heat transfer across the boundary layer. However, the bubbles do not collapse and disappear because the surrounding water has already reached its boiling point, or *saturation temperature*. Instead, they rise through the saturated water until they burst into the steam drum.

Subcooled and saturated nucleate boiling regimes are characterized by high heat transfer coefficients (h_i), due to the turbulence effects of the bubble movements on the boundary layer. This means the temperature drop between the inside steel wall and the boiling point throughout the water is small, and the steel temperature is just a few degrees above the water temperature. If the gas temperature increases beyond the nucleate boiling regime (B-C in Figure 4-21), the steam bubbles leaving the inside tube surface begin to interfere with the flow of water moving towards the surface to replace the steam bubbles. Eventually, the bubbles coalesce to form a film of steam over the internal heating surface, and the water is no longer in contact with the steel. This condition is called *film boiling*.

From D to E the film is unstable and sometimes breaks down into nucleate boiling, but as soon as it does, the heat flow increases again, with a corresponding return to film boiling. From E to F the film boiling regime is completely stable, and the film of steam acts as a constant resistance to the heat that must flow through the film to reach the water. In a boiler where a local heat flow (such as through a thin spot in the steel wall caused by erosion or slag corrosion) exceeds that which corresponds to point D in Figure 4-21, the temperature difference between steel and steam will rise quickly along the horizontal dotted line in Figure 4-21 to reach point D′. Look at the great increase in temperature difference in the distorted logarithmic scale.

Since the water and steam temperature will remain at boiling point, the steel temperature must rise quickly, just as it would if it suffered a total loss of cooling water. The steel temperature may rise above its melting point and may begin softening. Since there is a high internal pressure, any softening will be immediately followed by tube rupture and an explosion as the jet of high-pressure steam and water is pumped into the hot furnace. This condition, if it occurs, is called *burnout*, and D is called the *burnout point*. Point C, which is usually close to D as in Figure 4-21, is called the *point of departure from nucleate boiling* or the *critical heat flow*.

Because there is a temperature drop across the film in the film boiling regime, the temperature of the steam at the wall surface is above its boiling point or saturation temperature. Thus, the steam is superheated. At the interface between the film of steam and boiling water, of course, both the water and the steam are at the same temperature, which is the saturation temperature or boiling point at that pressure.

Here then is another hazard for the boiler tube designer. For economy and efficiency, the heat flux from bulk gas flow to bulk water and steam flow must be as high as possible to minimize the necessary length of expensive steel water tubes. On the other hand, the designer *must* avoid any chance of burnout occurring.

Now, consider the steam bubbles leaving either the steel surface or the film interface. They will rise through the saturated water until they reach the water surface which is in the lower half of the steam drum. As they burst through the surface they splash water, which may be entrained in the flow of steam leaving the steam drum, and, as we saw earlier, may cause trouble in the superheaters or the turbines. This is minimized in several ways: one approach is to minimize the amount of substances which would cause foaming by treatment of the boiler feedwater. Another approach is to add an antifoaming agent to the boiler water, and the geometry of the steam offtake from the top of the steam drum is arranged to cause water droplets to be thrown back into the bulk of the water.

The steam leaving the steam drum is, of course, saturated, i.e., still at the boiling point corresponding to the boiler pressure. The steam now contains two kinds of energy: latent heat and the energy due to its compression, or *pressure energy*. We can recover all of the pressure energy in the steam by expanding it down to a lower pressure through a turbine. However, if we attempt to recover the latent heat as well, then the steam would condense into water in the turbine. Turbines are extremely expensive because of the fine-tolerance engineering which goes into their manufacture. They are designed to withstand high temperatures and pressures and to operate with high velocities of steam passing over their surfaces.

However, at the speeds involved, water droplets would cause erosive damage to the turbine blades. If you find it hard to imagine water droplets affecting the high-grade steel used in turbine blades, remember the high velocities involved. Imagine sticking your hand out the window of an airplane traveling at 600 mph through a heavy rainstorm. The effect on your hand would be the same as if you put it out of your car window when traveling at 55 mph and being hit by a piece of flying gravel. Just think what that gravel would do to your windscreen at that speed!

Turbine designers usually specify a maximum amount of condensation of 15%, i.e., the steam leaving the turbine will have given up about 15% of its latent heat and 15% of the steam will have condensed into tiny water droplets. These considerations mean the amount of heat recoverable and turned into work in the turbine is limited to the pressure energy and just 15% of the latent heat. To increase the heat recovery, more heat can be put into the steam by superheating it.

A turbine transforms the superheat into work without forming moisture, and thus, all the superheat is recoverable in the turbine. This is done by taking the steam from the top of the steam drum into another set of tubes hanging down into the hottest part of the hot gas flow, i.e., just after the end of the visible part of the flame. The heat transfer processes which occur inside and outside the superheater tubes are similar to those in the other tubes of the boiler system except that the inside material is all steam rather than water or a mixture of steam and water, and so it has a gaseous film on the inside (the steam) as well as the outside (hot gases). The superheated steam leaves the top of the boiler from the superheater section and is taken straight to the turbine hall, where things get really complicated. Years of experience of operating steam cycles have led to many innovations that would, if described here, cloud the more important issues I want you to understand. I have chosen a simple turbine arrangement that is similar to those in common usage, but it does not represent the latest in turbine-systems design.

Superheated steam enters the first turbine, which is called the *high-pressure turbine*, at high pressure and a temperature well above its boiling point. There are many wheels of blades within the turbine. They are arranged in pairs of two different types. The first wheel of blades is free to spin and is called a *rotor*. As the steam passes through the angled curved blades it imparts a circulating motion to the ring; the steam itself is deflected tangentially. The next ring is called a *stator ring*, which is fixed and not able to spin. Its blades are angled the opposite way to the rotor blades, and its purpose is to turn the direction of the fast-flowing steam back to being parallel with the axis of the turbine. The steam then goes through another rotor ring and again through a stator, and so on, through the turbine until it exits at a lower pressure than when it went in, because it has lost some pressure energy as well as superheat.

The high-pressure turbine is arranged to convert sufficient pressure energy and sensible heat of superheat into work such that the steam leaving it is at a lower pressure and just a few degrees above its boiling point. Since taking any further heat from the steam would soon bring it down to saturation temperature and form water droplets, the steam is taken back through a steam pipe from the turbine hall into the boiler through tubes known as *reheater tubes*. These are located between the superheater tubes and the ordinary boiler tubes.

Since the steam goes in at a temperature above its boiling point, the reheater tubes add superheat only. The reheated low-pressure steam is now taken back to the turbine hall and expanded through a *low-pressure turbine*, which is similar in operation to the high-pressure turbine except that, at the exit end of the turbine, water droplets are beginning to form as the 15% of its latent heat is taken away and converted to work. Let us pause at the exit of the low-pressure turbine to see some other features of the high- and low-pressure turbines.

First, consider what happens to the pressure energy and sensible heat that has been converted into work. Figure 4-22 shows how to calculate the work outputs of the turbines. Notice that we look up the total heat content of the steam (called its *enthalpy*) in Steam Tables. The values for the example are shown alongside the steam flows in Figure 4-22. Also, there are some conversion factors shown. For instance, 778 foot-pounds of work are equivalent to 1 BTU. This is called the *mechanical equivalent of heat*. One *horsepower* is a rate of working that uses up 33,000 foot-pounds of energy per minute.

Knowing these conversion factors, which very soon in your career will be inscribed indelibly in your memory forever, we can calculate the horsepower available at the turbine shafts. In fact, it is impossible to recoup all the energy given up by the steam. Some heat is lost through the casing of the turbine and is radiated and convected away. Nothing can be done about this loss except by insulating the turbine to minimize it. In addition, friction will generate some heat in the bearings of the turbine, but this heat will be returned to the steam as it flows over the steel bearings. The horsepower at the shaft is converted

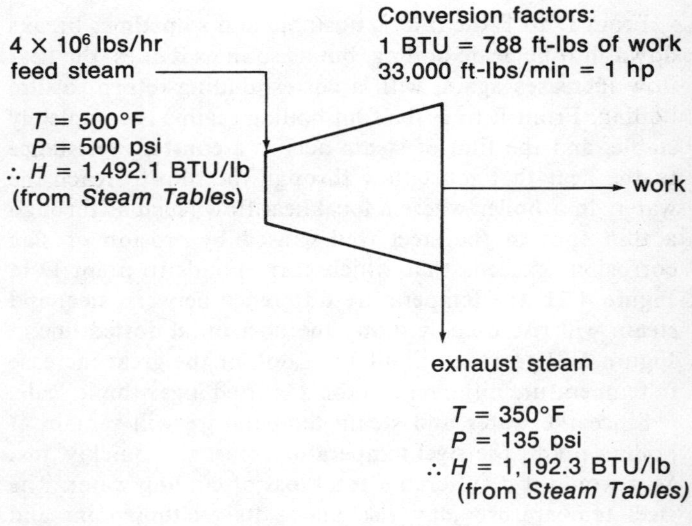

heat converted to work in turbine

$$= 4 \times 10^6 (1{,}492.1 - 1{,}192.3)\ \text{BTU/hr}$$
$$= 1.199 \times 10^9\ \text{BTU/hr}$$
$$= \frac{1.199 \times 10^9 \times 778\ \text{ft-lbs/min}}{60}$$
$$= 1.555 \times 10^{10}\ \text{ft-lbs/min}$$
$$= \frac{1.555 \times 10^{10}\ \text{hp}}{33{,}000}$$
$$= 471{,}200\ \text{hp}$$

Figure 4-22. *Calculating the horsepower of a steam turbine.*

into electrical energy by spinning a generator, similar in principle but considerably greater in size than the one in your car. Here then is the first useful product of a boiler plant: the power or electricity output.

There are some other interesting features in the turbines. In the high-pressure turbine it is common practice to arrange at two points in the turbine (see Figure 4-18) to bleed off a small amount of steam and pass it through a heat exchanger to heat up the boiler feedwater. There is a detailed description of heat exchangers in Chapter 5. The heat exchangers here are given a special name and are shown in Figure 4-18 as high-pressure bleed heaters.

Similarly, at two or even three points in the low-pressure turbine a similar bleed is taken off and used to heat the boiler feedwater in the low-pressure bleed heaters. In the high-pressure turbine the first bleed is at a higher pressure than the second and is also used to drive the feed pump that forces the boiler feedwater along its path into the boiler. This is done using a smaller version of a turbine in which the shaft is connected directly to the boiler feedwater pump. Pumps and compressors will be discussed in detail in Chapter 5. Some steam from the downstream or second bleed point in the high-pressure turbine may be added to the steam returned to the reheater (see Figure 4-18).

The spent steam from the low-pressure turbine, the high- and low-pressure bleed heaters, and the feed pump turbine is collected and taken to the condenser. Here, the steam condenses around the outside of túbes which contain cooling water. As it condenses, it creates a partial vacuum, which has the beneficial effect of helping to draw the steam through the turbines. The condenser is subject to the same rules of heat transfer as any other tubular system. However, there are some interesting phenomena that occur on the condensation side of the tubes. Let us look in detail at the condensation process.

The bulk flow of steam is mixed by turbulence and has a flat temperature profile (see Figure 4-23). Again, a boundary layer of steam is formed next to the surface, with the usual temperature drop across it. The steam actually next to the surface gives up its latent heat to the steel wall (to be conducted through into the cooling water), and the steam condenses. Consider the way this condensation occurs. It forms a film on the surface of the tubes. This is similar in principle to the films of condensation that occur when we breathe heavily on a mirror.

The film itself forms yet another barrier to heat transfer and thus slows down the condensation. However, some condensation does continue, and the film gets thicker until gravity causes it to flow downwards. As it gets thicker, heat transfer is reduced, the temperature gradient across it increases, and the temperature of the outside of the film is higher than that on the inside. The viscosity of water decreases as temperature increases, and the decreased viscosity causes the water film to flow downwards more quickly. But the quicker the water film flows away from the cooling surfaces, the thinner the film will be, and hence, the *better* the heat transfer will be. A balance will be reached between these two effects.

There is another effect to consider. In terms of heat transfer efficiency it is desirable to keep the film of water to a minimum thickness. One way of doing this is to make all the condensation surfaces vertical and to arrange for the steam flow also to be vertical in a downward direction. Thus, this has a "scrubbing" effect on the water and pushes it off the cooling surfaces. The condenser designer knows all these facts, and the design calculations will arrange for the conditions that create the highest heat transfer rate possible to minimize the initial cost of the water tube condensers.

The operation just described is called *filmwise condensation* for obvious reasons, but another kind of condensation can occur. Imagine that the cooling surfaces are made of plastic or perhaps steel coated with wax. As the steam condenses, it forms droplets. The majority of the cooling surface is left uncovered by film or droplets, and the heat transfer is much higher than it would have been if a film had formed to cause extra resistance. This operation is called *dropwise condensation* and is obviously preferable to filmwise condensation because of the higher heat transfer rates possible.

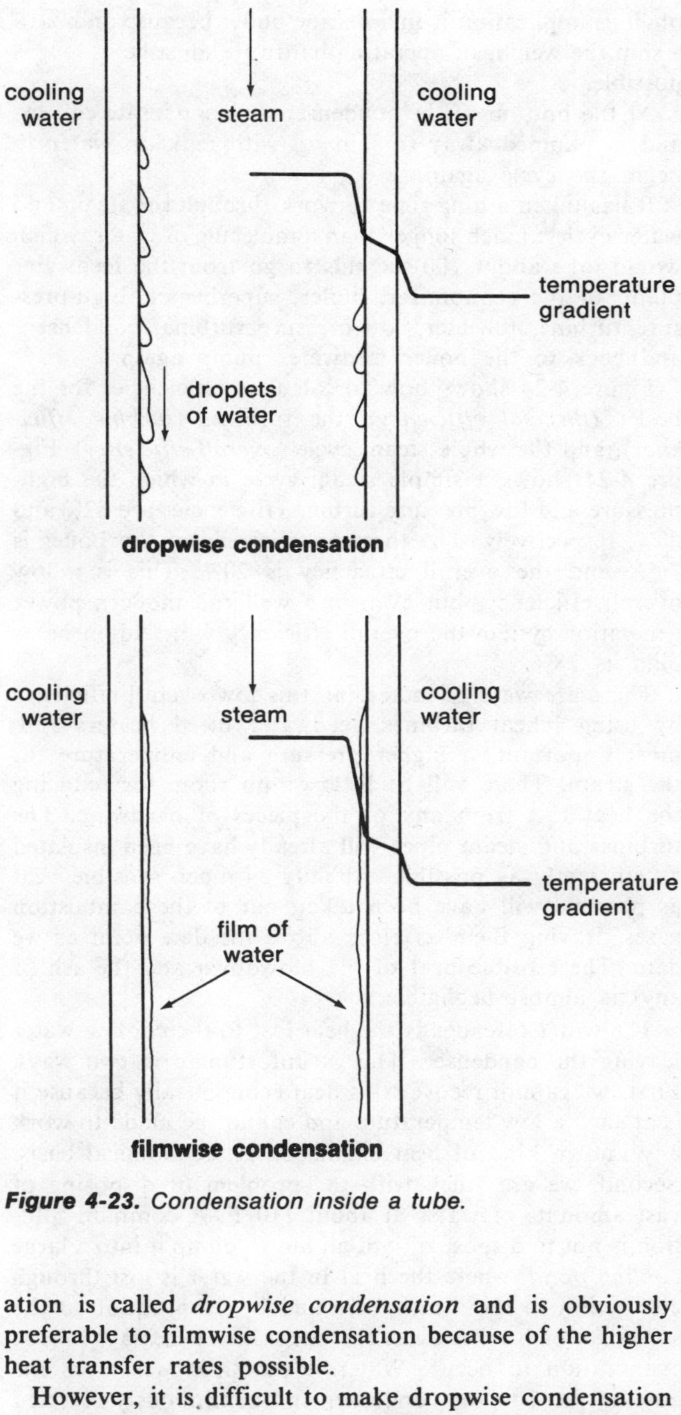

Figure 4-23. *Condensation inside a tube.*

However, it is difficult to make dropwise condensation occur. Promoters, which are usually heavy organic fats with the appearance of wax, are sprayed onto the condenser tubes. Although they promote dropwise condensation for thousands of hours, eventually the harsh conditions in which they must work cause the waxes to wear off the steel surface (and be carried away by the steam), and filmwise condensation sets in again. The use of dropwise condensation promoters is restricted to installations where it is essential to minimize the size of the condenser. An obvious application is in a marine boiler because on board a ship the weight of operating hardware must be as low as possible.

At the bottom of the condenser the condensate collects and is pumped away to join up with makeup water to begin the cycle again.

It has taken a long time to work through the steam and water cycles: much longer than a molecule of water, which would take about 100 seconds to go from the feedwater pump to the economizer, boiler, superheater, high-pressure turbine, reheater, low-pressure turbine, condenser, and back to the boiler feedwater pump again.

Figure 4-24 shows how to calculate efficiencies for the boiler (*thermal efficiency*), the turbines (*turbine efficiency*), and the whole steam cycle (*overall efficiency*). Figure 4-24 shows a simple steam cycle in which the high-pressure and low-pressure turbine efficiencies are 82% and 80%, respectively. The thermal efficiency of the boiler is 79%, and the overall efficiency is 20%. This is a low overall efficiency, but even in a well run modern power generation system the overall efficiency will seldom be as high as 38%.

There are ways of increasing this low overall efficiency by using reheat turbines, feedwater bleed heaters and, most important, a higher pressure and temperature for the steam. There will be little or no room for reducing the heat lost from any of the pieces of hardware. The turbines and steam pipes will already have been insulated as efficiently as possible. Already as much sensible heat as possible will have been taken out of the combustion gases, leaving them as close above the dew point as we dare. The sensible heat of the blowdown and the ash (if any) is almost negligible.

The worst offender is the heat lost to the cooling water leaving the condenser. This is unfortunate in two ways. First, we cannot recover this heat economically because it is at such a low temperature and cannot be made to work any known kind of heat engine on an economical basis. Second, we are stuck with the problem of disposing of vast amounts of water at about 110°F. A common solution is not to dispose of it at all but to pump it into a large *cooling pond*, where the heat in the water is lost through convection to the atmosphere and by evaporation at the surface, which takes away the latent heat necessary for the evaporation to occur. Water is pumped back from the cooling pond to the condenser and continuously recycled with a makeup to allow for the evaporation losses. Virtually all the heat in the water is lost through the surface, so the cooling will be most efficient if the surface is as large as possible.

From your own experience you could have predicted this. For example, hot soup in a tall cup takes much longer to cool than hot soup in a shallow dish. Incidentally, when you blow across the surface of the soup to cool it, you are exploiting the same basic laws as the cooling pond designers. At first sight then, it would appear preferable to have shallow ponds with large surface areas. However, since these ponds often approach 100 acres in area, land costs must be considered. Once again, the designer must balance the virtues of improving efficiency by increasing pond size versus the initial land cost.

The operators of boiler plants producing electricity only are resigned to operating at a maximum possible overall

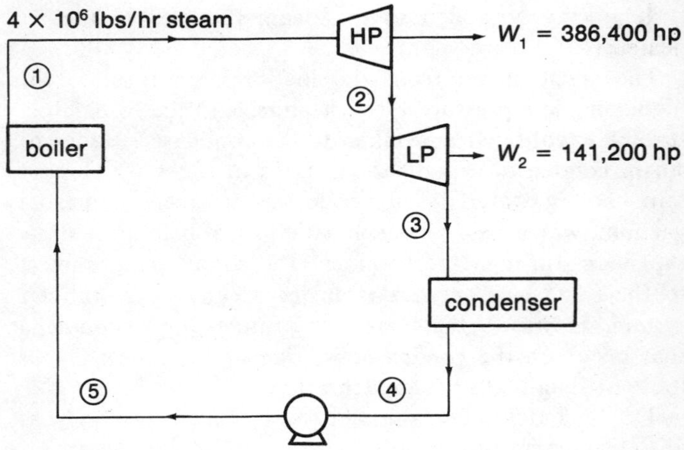

Data:

1. T = 500°F, P = 500 psi, H_1 = 1,492.1 BTU/lb
2. T = 350°F, P = 135 psi, H_2 = 1,192.3 BTU/lb
3. T = 180°F, P = 4.0 psi, H_3 = 1,080.0 BTU/lb
4. T = 180°F, P = 7.5 psi, H_4 = 148.0 BTU/lb
5. T = 180°F, P = 500 psi, H_5 = 148.0 BTU/lb

boiler fuel input = 1,700 BTU/lb of steam produced
high-pressure turbine output = 386,400 hp
low-pressure turbine output = 141,200 hp

Basis: 1 lb steam

boiler efficiency

$$= \frac{\text{heat out}}{\text{heat in}} = \frac{(1{,}492.1 - 148)}{1{,}700} \times 100 = 79\%$$

high-pressure turbine efficiency

$$= \frac{\text{work out}}{\text{heat in}} = \frac{386{,}400 \times 33{,}000 \times 60 \times 100}{(1{,}492.1 - 1{,}192.3) \times 4 \times 10^6 \times 778} = 82\%$$

low-pressure turbine efficiency

$$= \frac{141{,}200 \times 33{,}000 \times 60 \times 100}{(1{,}192.3 - 1{,}080.0) \times 4 \times 10^6 \times 788} = 80\%$$

overall efficiency

$$= \frac{(386{,}400 + 141{,}200) \times 33{,}000 \times 60 \times 100}{1{,}700 \times 778 \times 4 \times 10^6} = 20\%$$

Figure 4-24. *Calculating efficiencies.*

efficiency of less than 40%. If the boiler plant is to be integrated with a chemical plant, however, the situation is different. There are hundreds of uses in a chemical plant for what is called *process steam.*

For example, take a pump somewhere in the plant. There are several methods of driving it. We could buy and store gasoline or diesel fuel and have an internal combustion engine to drive the pump, but this is very unusual. The main choices are (1) electricity to drive an electric motor attached to the pump, and (2) steam expanding through a turbine to provide the pump's drive. It is more efficient to draw steam from the high-pressure turbine bleed to drive this pump. Otherwise, that same steam would go on through the rest of the high-pressure turbine, the pipes, the low-pressure turbine, and the condenser, producing (because of each of *their* inefficiencies) less electrical energy than would be required to drive the pump. In other words, we have used the steam before some of its energy is lost in inefficiencies.

A common use for process steam is as a heating medium. For instance, in the bottom of a distillation column the liquids that collect there need to be reboiled as described in Chapter 5. A simple way of doing this is to arrange steel tubes with steam flowing through them into the reboiler section, i.e., immersed in the boiling liquid. The pressure of the incoming steam would be carefully chosen such that its boiling point (or condensation temperature) would be well above the required temperature of the distilled liquids. Then there would be a temperature gradient across the tubes, and heat would flow into the liquids. It is unlikely that superheated steam will be of any use under these conditions, and it is common practice to bleed steam from a point in the turbine where the steam is approaching its saturation point.

There is a variety of places from which steam can be bled to meet specific requirements of steam. Usually, the most efficient way to work is to calculate the highest pressure of all the individual process steam requirements. Then find a point in the steam power cycle where the steam is at that pressure. Usually, this will be in a turbine. Then arrange to bleed off the steam at that pressure and take it out to the chemical plant for distribution in a steam main. Where particular requirements call for steam at a lower pressure, the steam can be expanded through a small nozzle contained in the pipe, thus reducing its pressure to the desired level. This is an efficient process and little heat is lost.

Steam is also used in a saturated condition for the comfort heating of buildings around the site, including the control room, the boiler plant, research laboratories, lunchrooms, and offices. Live steam is often available at certain points on the plant for steam cleaning and other purposes, and the boiler itself requires steam for sootblowers, which are devices which aim live steam into the spaces between tubes to blow away any deposits.

The study of steam as it varies its properties at different temperatures and pressures is part of a whole branch of science called thermodynamics. Later, when you have a formal course in thermodynamics, you will learn how the behavior of steam under varying pressure and temperature conditions can be exploited.

In this chapter you have seen something of the amazing properties of water and steam in water tubes (the various boiling regimes), in steam drums (foaming), in turbines (conversion of heat and pressure energy to work), and in condensers (dropwise and filmwise condensation). You have been shown the significance of partial pressure and dew point as applied to a gas mixture that contains a condensable vapor. You have seen how solid, liquid, and gaseous fuels can be burned in different hardware systems to give up their chemical energy and how that energy is captured as the sensible heat of the combustion products. You have learned more about heat transfer by radiation, convection, and conduction through various resistances, including boundary layers, steel tubes, and scaly deposits. In addition, you have seen the use of atomization and comminution to increase the surface area of liquids and solids and how reactants diffuse through stagnant layers to reach the surface and react.

Further Reading

Babcock and Wilcox Co. *Steam: Its Generation and Use.* New York: The Babcock and Wilcox Company, 1972.

Clark, W. *Energy for Survival.* New York: Anchor, 1975.

Combustion Engineering, Inc. *Combustion Engineering.* New York: Combustion Engineering, Inc., 1966.

Dryden, I.G.C. (Ed), *The Efficient Use of Energy.* Guildford, England: IPC Science and Technology Press, 1975.

Field, M.A., D.W. Gill, B.B. Morgan, and P.G.W. Hawksley. *Composition of Pulverized Coal.* Leatherhead, England: BCURA, 1967.

Francis, W. *Fuels and Fuel Technology*, Volumes I and II. Oxford: Pergamon Press, 1965.

Gaydon, A.G., and H.G. Wolfhard. *Flames.* London: Chapman and Hall, 1970.

Hottel, H.C., and J.B. Howard. *New Energy Technology.* Cambridge: MIT Press, 1971.

Lyle, Sir Oliver. *The Efficient Use of Steam.* London: HMSO, 1947.

McAdams, W.H. *Heat Transmission.* New York: McGraw-Hill, 1954.

Perry, R.H., and C.H. Chilton. *Chemical Engineer's Handbook.* New York: McGraw-Hill, 1974.

Thring, M.W. *The Science of Flames and Furnaces.* London: Chapman and Hall, 1962.

Chapter 5

The Petroleum Refinery

This chapter discusses the origin of crude petroleum and how it is pumped through pipelines to refineries where it is separated into a number of fractions, each with a different range of boiling points. Also, you will learn the measures taken at a refinery to minimize the energy consumed by the process.

The Origin of Oil

The story of petroleum begins some 500 million years ago, when the oceans teemed with life, both vegetable and animal, and some organisms that could be classified as either or neither, i.e., plankton. Countless plankton lived, competed for food (sunlight for photosynthesis and inorganic salts washed down from the new land masses), and reproduced. They were either eaten by larger creatures or they entered the food chain by dying and being partially consumed by oceanic bacteria living in the shallow part of the ocean where dissolved oxygen was available. Thus, a kind of oxidation occurred, and all that remained of the plankton were the so-called fatty acids, which fell to the ocean bed to be the food source for the anaerobic bacteria that had evolved to live without oxygen in deep waters. After the aerobic and then the anaerobic bacteria had eaten their fill, all that remained were hydrocarbons.

Over the millions of years during which these processes occurred, other events were taking place that would become significant millions of years later when man began to exploit oil. Silt, mud, and clays washed down from the early continents, fell in the oceans, and became intermingled with the remains of the plankton and other decaying matter. Ocean beds became giant mixing bowls. The ingredients were the oily remains of plankton; carbonaceous remains of algae, bacteria, and vegetable matter; sulfur-bearing molecules remaining from the organic matter in the mixture; sand, silt, clay, and mud.

Back in those times the unsettled and young Earth was still subject to enormous structural changes as its insides cooled. Ocean beds became huge mountain ranges in just a few hundred thousand years—almost "overnight" in geological time. The giant mixing bowls—the plankton graveyards—were folded into sandwiches of hard, impermeable rock structures or in some other way were buried under billions of tons of rock. The pressures and temperatures built up on these structures of oil and rock. Reactions and processes occurred that cannot be performed in a laboratory because, although we have the ingredients and the means of providing the correct pressures and temperatures, we simply do not have the several million years of time to observe the experiment!

Today, the results of all these processes are the pockets of oil-bearing rocks deep in the earth, under great pressure and sandwiched between hard rocks which do not allow the pressurized oil to escape. As these processes progressed, some of the lighter hydrocarbon gases were squeezed out from the heavier ones, and in some cases these escaped and were lost. In other cases they became trapped, either directly above the oil or at some distance from it, under an inverted bowl of hard, impervious rock. When the oil boom began in the late nineteenth century, drillers who struck a "gas-well" were most disappointed if oil did not accompany the gas. Of course, today, natural gas is in great demand and there are thousands of gas wells in production.

In the mid-nineteenth century, oil was used as an illuminant in wick lamps. Its main source was coal which, even then, was being carbonized to produce coke. The by-products of carbonization readily yielded an oil which burned with a bright yellow flame.

People were also aware of underground sources of natural oil because the evidence was apparent: streams bubbled out of mountains with a film of petroleum oil on them; in lakes and ponds, an oily scum formed which was sometimes skimmed and treated to make oil.

In 1859 a group of entrepreneurs decided to try to drill for oil in Pennsylvania using the drilling technology that had been developed by companies looking for salt. Their problems were enormous. Equipment was primitive and the rock above the oil seams was very hard. Some days, downward progress was limited to less than 3 feet. However, the men's determination won and, with the drilling stock extended down to 69½ feet, a dark green oil flowed out of the hole. Though a mere dribble compared to the spectacular oil gushers Hollywood would later dramatize for us, those early drillers in 1859 were jubilant. Their determination and engineering skill had triumphed and a new industry had been born.

Immediately an oil boom began. Those Pennsylvania hills were alive with drilling engineers excitedly standing around as their steel drill bits inched downwards. Technologies expanded and developed at breakneck speed. Only two years after this first strike, in 1861, an oil refinery was built to make kerosene and some oils and greases for lubrication. Less than 20 years later the engineering triumph of the age was opened: a 100-mile pipeline across the Allegheny Mountains.

As with any rapidly developing technology, there were spinoffs. In particular, the development of small powerful engines led to the possibility of personalized transport in the form of the automobile. Soon the availability of a source of compact chemical energy in liquid form further fired this technology and the oil and automobile industries became giants in America. Before long it was discovered that crude oil contained literally thousands of chemical species which were regarded as by-products, since in early days the oil was viewed as a source of kerosene and gasoline only. Entrepreneurs researched and gradually built up a wider fuels industry which resulted in today's state of affairs (described in Chapter 6).

Further research showed that "refining" could liberate other useful species of chemicals from the fuels prior to burning them, and before long petroleum became regarded as the best source of feedstocks for chemical processing ever discovered and used. Another industry was born—the petrochemical industry. It is described in Chapter 7. Figure 5-1 shows how these three mighty industries depend upon and work with each other. The dotted lines indicate the boundary between Chapters 5 and 6. Feedstocks for the petrochemical industry come from all parts of Figure 5-1.

After the geologists and petroleum engineers and many other professionals have done their jobs, petroleum oil and natural gas are available at the wellhead. From now on in the story of petroleum, professional chemical engineers take an increasing role. Oil is taken to the great industrial areas of the world by tanker, pipeline, barge, train, or truck. At some stage, however, the oil will find itself in a pipeline which, long or short, has to be designed by using chemical engineering principles.

The Flow of Fluids

To flow through a pipe, oil (or any other fluid) must be pumped into the pipeline at a sufficiently high pressure to overcome friction caused by its flow within the pipe. Otherwise, the oil would never reach the destination chosen for it. That is not to say we cannot, in very long pipelines, have booster pumping stations along the way. For a small-diameter pipe (say, 6 inches) carrying around three-quarters of a million tons of oil a year, it is usual to space the booster stations about 30 miles apart. But for a huge 48-inch diameter pipeline carrying, say, 100 million tons of oil per year, the pumping stations have far more powerful pumps, and the distance between stations may be as great as 200 miles. Since the friction of the flow is so important to a pumping operation (because of the cost of overcoming it), let us consider in detail the processes that cause friction.

Reynolds Number

The best way to understand the peculiar phenomena which occur in flowing fluids is to consider the experiments performed by Osborne Reynolds in 1883. Figure 5-2 indicates the equipment he used. The central glass tube had a flared end, facing the direction flow, into which a small nozzle was pointed. When the water valve was opened, clear water left the outer glass tube and entered and passed through the inner glass tube. Above the tubes was suspended a bottle of ink which was connected through an ink valve to the nozzle pointing into the flared end of the inner tube. Starting with low water flow rates, Reynolds opened the valve and a thin stream of colored ink entered the stream of water. Through the glass, Reynolds noted that the line of ink left the nozzle and flowed down the inner tube intact, and from this he deduced that the flow proceeded along the tube in straight lines. He called it *laminar flow*.

If these experiments are reproduced, another surprising feature can be illustrated: no matter where the tip of the ink nozzle is placed in the cross-sectional area of the tube, under conditions of laminar flow the stream of ink will remain intact because the "streamlines" of the flow are parallel: an observation which has led to laminar flow sometimes being called *streamlined flow*.

In his experiments, Reynolds increased the flow rate of the water gradually and watched the ink tracer line remain steady and straight until at some velocity it began to wiggle around and a small further increase in flow rate caused the stream of ink to disappear completely. He surmised, correctly, that the parallel streamlines had broken down into internal "eddies" or turbulence, and that these were so intense as to act throughout the flowing volume of the liquid to disperse the ink immediately after it entered the bulk fluid flow. Reynolds called this kind of flow *turbulent flow*. After a long series of experiments in which he changed and recorded the flow velocity, the diameter of the tube, and the viscosity and density of the liquid (by using different liquids), he found that these four factors could be put into a formula which would enable him to predict the kind of flow that would occur. He

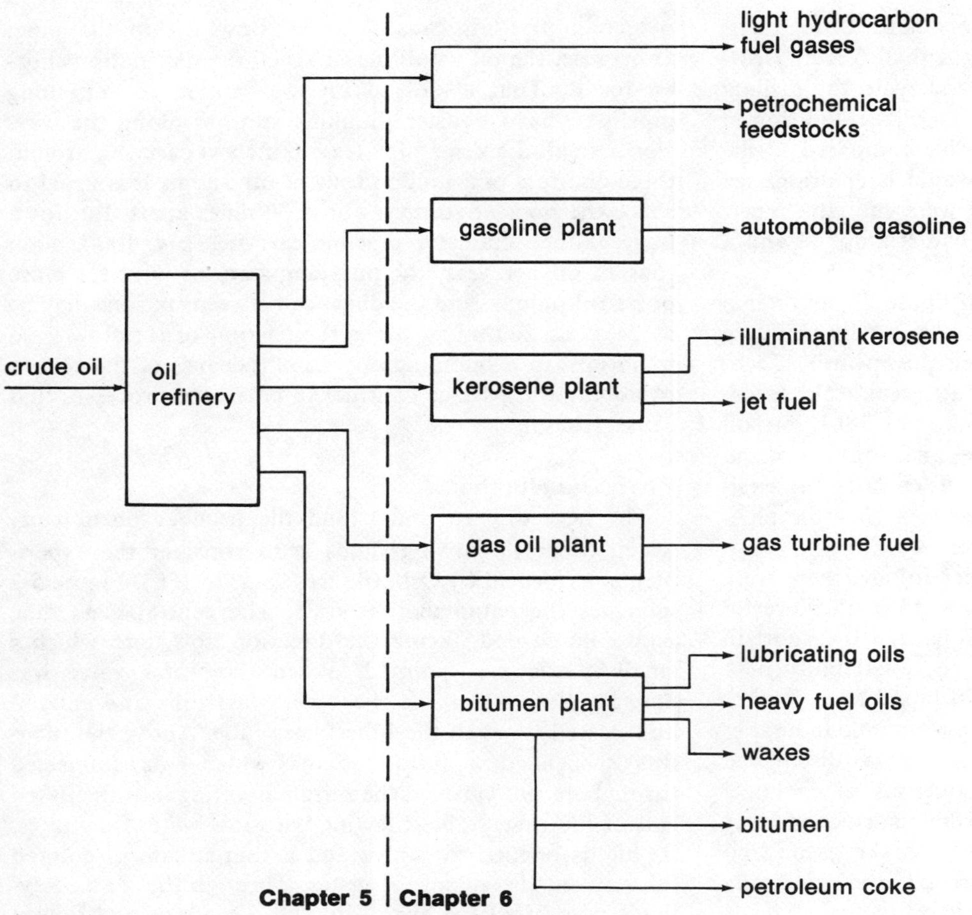

Figure 5-1. The uses of oil.

found the four factors fell into a group which was "dimensionless." This means that the units of each of the factors cancelled out (see Figure 5-3). Since Reynolds' time, and in his honor, the dimensionless group has been called the Reynolds number.

If the Reynolds number of a flowing fluid (gas or liquid) is below 2,100, the flow will certainly be laminar except under extraordinary conditions. If it is greater than 4,000, the flow will generally be turbulent. Between these two limits there exists a transitional regime in which the flow is unstable, and although in a long straight pipe the flow will settle down to be laminar, the slightest disturbance (such as a valve, a bend in the pipe, or even some tiny surface roughness) may break up the streamlined quality of the flow and cause turbulence.

Since the pumping and transport of fluids in pipes is so basic to just about every industry, these phenomena have been studied intensively. Figure 5-3 shows the results of an experiment to measure the velocity profile across a circular tube of a flowing fluid. The graph, in which velocity is plotted against its distance from the axis of the pipe, is superimposed over a sketch of the pipe itself: a common illustrative device in chemical engineering. Notice that for laminar flow the profile is a long, pointed parabola whereas the turbulent flow profile fits a flattened curve.

To understand the difference, consider the mechanism of flow itself. Laminar flow is best compared to a deck of cards. Imagine that you place a deck of cards onto a card table and then, pressing down slightly with the palm of your hand on the top card, you move it several inches out of place. The card beneath it will have moved almost as many inches (in fact, about a sixteenth of an inch less than the top one). The third card down will have moved perhaps a sixteenth less than the second, and so on. The bottom card

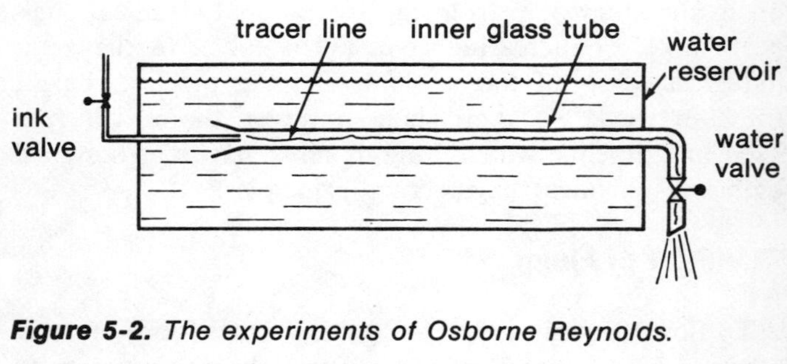

Figure 5-2. The experiments of Osborne Reynolds.

250 gals/hr of water at 75°F

ρ = density
= 62.26 ft³/lb lb/ft³

μ = viscosity
= 6.122×10^{-4} lbs/ft – sec

V = velocity, ft/sec

P = pipe diameter, ft

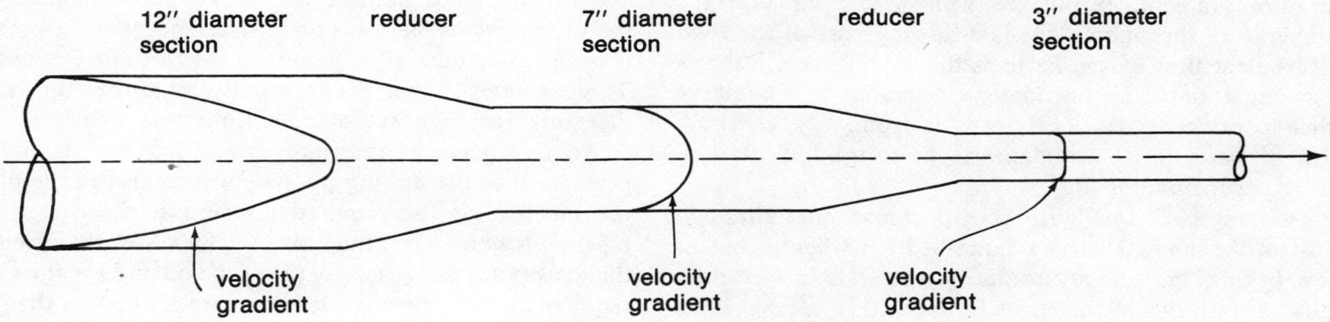

$$V = \frac{\frac{250\ \text{gals}}{\text{hr}} \times \frac{1}{3{,}600}\frac{\text{hr}}{\text{sec}} \times \frac{1}{7.48}\frac{\text{ft}^3}{\text{gal}}}{\frac{\pi\,(12/12)^2\ \text{ft}^2}{4}}$$

$$= 0.0118\ \text{ft/sec}$$

$$\therefore \text{Re} = \frac{\frac{0.0118\ \text{ft}}{\text{sec}} \times \frac{12\ \text{ft}}{12} \times \frac{62.26\ \text{lb}}{\text{ft}^3}}{6.122 \times 10^{-4}\ \frac{\text{lb}}{\text{ft-sec}}}$$

$$= 1{,}202\ \text{(dimensionless)}$$

$$V = \frac{250 \times \frac{1}{3{,}600} \times \frac{1}{7.48}}{\frac{\pi\,(7/12)^2}{4}}$$

$$= 0.0347\ \text{ft/sec}$$

$$\therefore \text{Re} = \frac{0.0347 \times (7/12) \times 62.26}{6.122 \times 10^{-4}}$$

$$= 2060$$

$$V = \frac{250 \times \frac{1}{3{,}600} \times \frac{1}{7.48}}{\frac{\pi\,(3/12)^2}{4}}$$

$$= 0.0473\ \text{ft/sec}$$

$$\therefore \text{Re} = \frac{0.0473 \times (3/12) \times 62.26}{6.122 \times 10^{-4}}$$

$$= 4{,}806$$

Flow is: **laminar** **transitional** **turbulent**

Figure 5-3. *Use of Reynolds Number.*

will not have moved. Friction between it and the card table will have kept it stationary. The shape of the curve that the edges of the cards will now make is similar in shape to the laminar flow profile in Figure 5-3. This is because a similar process is occurring in the tube.

The high-velocity central core of the flowing fluid is analogous to the top card in the demonstration deck. The bottom card is represented by the stationary layer of fluid next to the tube. Momentum, or the driving force for movement, is transferred down the deck of cards by a frictional effect. Similarly, momentum in the flowing fluid is transferred through the layers of flowing fluid. Since the velocity is faster in the middle than at the tube wall, we could say that the momentum is transferred down a velocity gradient. The situation compares well with the process of heat transfer. Just as heat was transported down the temperature gradient, i.e., the driving force for transport was temperature difference, in fluid flow the momentum is transferred down a velocity gradient, and accordingly we can say its driving force is velocity.

As the Reynolds number increases, the streamlines break up and form large eddies that have almost the same diameters as that of the tube. These break up to form smaller eddies which in turn break into yet smaller eddies: a process that goes on until the smallest eddies, about 1 millimeter in diameter, are absorbed by the viscosity of the fluid. Since the eddies are circular, parts of the fluid must occasionally be flowing at right angles to the main flow direction such that the momentum of flow is transferred by means other than the frictional effect which occurred when the flow was still laminar.

This particular effect of turbulence causes other transportation also. If, for instance, the concentration of something dissolved in a flowing liquid initially has a variation throughout the cross-sectional area of the flowing liquid, turbulence will cause a rapid equalization of the concentration throughout the bulk flow, just as you saw in Reynolds' initial experiments. Similarly, if there is a temperature gradient, turbulence will rapidly equalize the temperature throughout the fast-flowing part of the fluid if turbulent flow exists. Perhaps this revelation will throw some light onto the phenomena discussed in Chapter 4 when considering the heat transfer from, say, the bulk flow of water in an economizer tube to the bulk flow of the gas surrounding it.

Referring back to Figure 5-3 you can see that although most of the laminar curve is squashed flat when turbulent flow begins, the velocity profile at the wall surface does show a fall down to zero. In fact, this is the same boundary layer that causes the so-called "film resistance" to heat flow that we encountered in Chapter 4.

Calculating Pressure Requirements

Now, armed with this new knowledge, let us go back to the consideration of the friction which must be overcome to pump fluids along pipelines. At low flow rates, i.e., with low Reynolds numbers, the energy of pumping is lost in the frictional effect of transferring the momentum from the center of the pipe down to those layers close to the wall surface. But surprisingly, as the flow rate increases, the frictional effect *decreases*. The basis for this rather astonishing fact is similar to the basis of the equally astonishing fact that friction between two solids also reduces after the solids are set in motion relative to each other. You know this from your own experience. Imagine a book placed on a drawing table. As the drawing table is tilted the book obstinately stays in position until suddenly the static friction holding it in place is overcome by the portion of the book's weight which acts down the inclined plane. At this point the book begins to move. Of course, there is still some friction between the book and the drawing table, but if you tilt the table back toward the horizontal position, the book will still be moving after you have passed through the angle where it originally began to move. This is because dynamic friction of motion is less than the static friction which held it in place originally. In a flowing fluid we do not have two simple surfaces rubbing against each other. There are many, many millions of tiny friction processes going on within the fluids. As laminar flow gets faster and faster, more of the frictional effects are reduced.

The straight line part of the graph shown in Figure 5-4 displays this. The ordinate in that graph is called the *friction factor*, which can be thought of as the ratio between the pressure drop caused by friction and the energy required to pump the fluid. Figure 5-4 also shows the *Fanning equation* (named after its discoverer) reshuffled to show the important pressure drop on the left-hand side. In the laminar flow region, you can see that the friction factor is very simply related to the Reynolds number. Things get somewhat more complicated when the flow is turbulent, as can be seen by inspecting the curve in Figure 5-4. The relationship between the friction factor and the Reynolds number is complex and depends upon the roughness of the walls and the geometry of the pipe. It is usual to assume an equation of the form in Figure 5-4. The constants a and b can usually be looked up in the literature for each set of conditions and values.

We now have a means of calculating the amount of pressure that the driving pumps must impart to the oil in the pipeline. If the required pressure at the end of the pipeline is known (p_2), and also the length of the pipe (L), the density (ρ) and viscosity (μ) of the oil and the flow rate required to pass through the pipe are all known then we can find p_1, which is the pressure at which the oil must enter the pipeline. This is done as follows: Knowing the volumetric flow rate (gallons per hour) required, we can translate it into cubic feet per hour by dividing the gallons per hour by 7.481 (found in chemical engineering con-

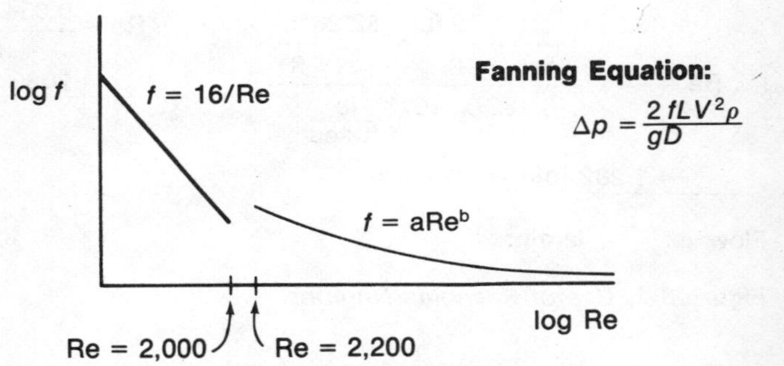

Using data from Figure 5-3: 250 gals/hr of water at 75°F through 12″, 7″ and 3″ pipes

12″ pipe $\quad f = \frac{16}{1{,}202} = 0.0133$

$\therefore$ *Δp for 100 ft of pipe = 1.03 psi*

7″ pipe $\quad f = \frac{16}{2{,}060} = .00777$

$\therefore$ *Δp (100 ft) = 0.893 psi*

3″ pipe $\quad f = 0.079\ Re^{-0.25}$

$= 0.00948$

$\therefore$ *Δp (100 ft) = 4.72 psi*

Figure 5-4. *The friction factor and pressure drop.*

version tables). Dividing by 3,600 brings the flow rate into cubic feet per second. Next, dividing the volume flow rate by the cross-sectional area through which it flows gives the linear flow rate or bulk velocity of flow. Figure 5-4 shows three examples of calculations of velocity for low-, medium-, and high-velocity flows in various pipe diameters. Incidentally, notice how the units cancel to give units of feet per second: units which we know belong to velocity.

A diameter is chosen for the oil pipeline such that the velocity will fall in a certain range depending upon the fluid being pumped. The fluid's density and viscosity may be found in standard data tables such as *The Chemical Engineers' Handbook*. Hence, the calculations of Reynolds number and friction factor can be made as shown in Figure 5-4. Finally, we can calculate the pressure drop, $p_2 - p_1$, using the Fanning equation which takes account of the length of the pipeline.

By the way, it is essential to modify the pipeline's length by adding "equivalent lengths" for pipeline features which cause extra resistance to flow. For instance, it has been found that a right-angle bend in the pipeline causes the same pressure drop as would occur in an extra length equal to nine times the pipe diameter. Similarly, a gate valve (open, of course) causes a pressure drop equivalent to seven pipe diameters. Figure 5-5 shows an example. It is this equivalent length that is used in the Fanning equation. Tables of equivalent lengths for all manner of pipe fittings and other obstructions and events in a pipeline are given in *The Chemical Engineers' Handbook*.

Although these remarks on pumping fluids have been aimed at the operation of major transportation pipelines, they also apply to the flow through any other tubes and pipes, including the economizer tubes described earlier in Chapter 4.

Pumping the Oil

The energy required to cause a fluid to flow is a combination of the energy required to bring the fluid up to its flow velocity (this is called *kinetic energy*), the energy required to raise the elevation of the fluid or pump it uphill (*potential energy*), and the energy required to overcome the combined frictional effects of the pipe itself (and obstructions in the way of flow such as valves, filters, bends and so on). The total amount of energy is put into the fluid by a machine called a pump, blower, compressor or fan. For the sake of this short description I will refer to machines which move liquids as "pumps" and machines which move gases as "compressors." Fans and blowers are kinds of compressors.

All pumps and compressors are rated or specified by four characteristics. The first of these is called *capacity*, and this is the quantity of fluid which is discharged from the machine in a given time. For instance, a water pump could be rated in gallons per minute or an air compressor in cubic feet per minute. The second characteristic is the *increase in pressure*, which is self-explanatory. It is measured in pressure units, and an air compressor with a specified increase in pressure of 20 psi will cause the air entering it at 15 psi to leave at 35 psi. For pumps, the increase in pressure is sometimes called the *head*, which is given in units of length rather than pressure. The head provided by a pump is equivalent to the pressure that

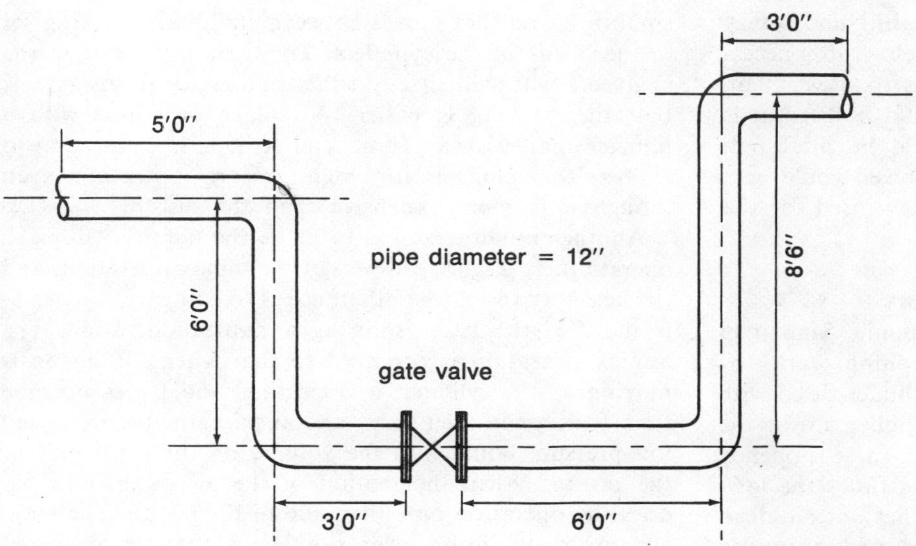

equivalent length of this system = 5 + (9 x 1) + 6 + (9 x 1) + 3 + (7 x 1)
+ 6 + (9 x 1) + 8.5 + (9 x 1) + 3
= 74.5 ft

Figure 5-5. *Calculation of equivalent lengths of pipe.*

would be produced at that point in the pipeline if there were a column of the liquid itself standing up vertically where the pump is. Thus, a water pump providing water at a 120 feet head is equivalent to having a water tower 120 feet high. The head is calculated by dividing the pressure by the density of the liquid.

The third characteristic of a pump or compressor is the *power* it requires. This is the energy consumed by the machine in a given time. Its units are horsepower, an old-fashioned means of rating power still in common use today. One horsepower is equivalent to 550 foot-pounds of work performed each second. It can also be translated into other units; for example, 1 horsepower is equivalent to 2,545 Btu's per hour or 0.746 kilowatts.

The final characteristic of a pump or compressor is its *efficiency*. This is the measure of the machine's effectiveness at transferring the power it consumes into energy of flow of the pumped or compressed fluid. Pumps and compressors in the process industries usually operate at an efficiency between 70 and 80%. This means that for every 100 units of work purchased and provided to the pump or compressor, approximately 75 of them actually are transferred into the fluid's motion.

Types of Pumps and Compressors

There are three kinds of pumps and compressors: reciprocating, rotary, and centrifugal. Sometimes reciprocating and rotary pumps are called "positive-displacement pumps" because they do not have a free passage for flow through the pump unless it is operating; that is, something will be in the way if the pump is switched off. Flow in these is made to occur by positive displacement of the volume of the fluid, and you will see this in some detail by looking at the different kinds of reciprocating and rotary pumps. For all the machines described below, it is necessary to provide motive power: the means to drive them. Although a diesel or gasoline engine could be used, it is more likely that an electric motor would be preferred. Even more probable is that the motive power would be a steam engine or steam turbine as you saw used for the boiler feedwater pump in Chapter 4.

Reciprocating Pumps and Compressors

The simplest one of all is the piston pump. Similar in operation to a car engine, the piston pump works by having a piston move up and down a cylinder fitted with valves as in Figure 5-6. The piston is pulled down the cylinder by the motive power and the inlet valve is opened by a linkage to the main drive shaft. In flows the low-pressure liquid, and when the piston reaches bottom dead center (BDC), the inlet valve is closed and a similar linkage opens up the outlet valve in time for the piston's return up the cylinder to force the liquid out at higher pressure through the outlet valve. At top dead center

Figure 5-6. *Sectional view of a horizontal steam-driven single-piston pump. (Courtesy of Ingersoll-Rand Co.)*

(TDC) the outlet valve closes, the inlet valve opens, and the cycle begins again. Now you can see the meaning of the words positive displacement: the liquid in the cylinder when the piston is at BDC then has its volume displaced by the returning piston.

The reciprocating pump is so called because the piston has a reciprocating action up and down, and I must point out that a single-piston pump is unusual; it is more common to have several pistons pumping away in a sequence chosen to level out the vibrational effects of each single piston.

From this basic description it is convenient to tell you of two modifications, both of which improve the performance of the machine. First of all, the piston can be replaced by a plunger (which is shown diagrammatically in Figure 5-7). Whereas the piston carries piston rings set in grooves around the "skirt" of the piston to act as "packing" to prevent leakage of the pump fluid, the plunger is a smooth barrel that moves between stationary packing set in the heads of the cylinders. The main difference to the engineers who will specify either plunger or piston type is that the packing is easier to replace or adjust with a plunger than with a piston, and so it is more natural to choose the plunger for high-pressure operation even though it is more expensive than the piston.

Another modification is to make the pump a "duplex" operation. In Figure 5-6 we saw a simplex operation, and the best way to tell the difference is to compare Figures 5-6 and 5-8, the latter showing a duplex operation. The duplex operation is arranged so that when the piston is moving up the cylinder, it is ejecting, under pressure, the fluid from one outlet valve and at the same time drawing low-pressure liquid into the volume on the other side of the piston. When the motion of the piston changes, so does the operation on either side of the piston. The main advantage of duplex over simplex is that the delivered flow is somewhat less pulsed, although pulses caused by pumping can be removed by passing the output high-pressure liquid through a reservoir vessel which simply

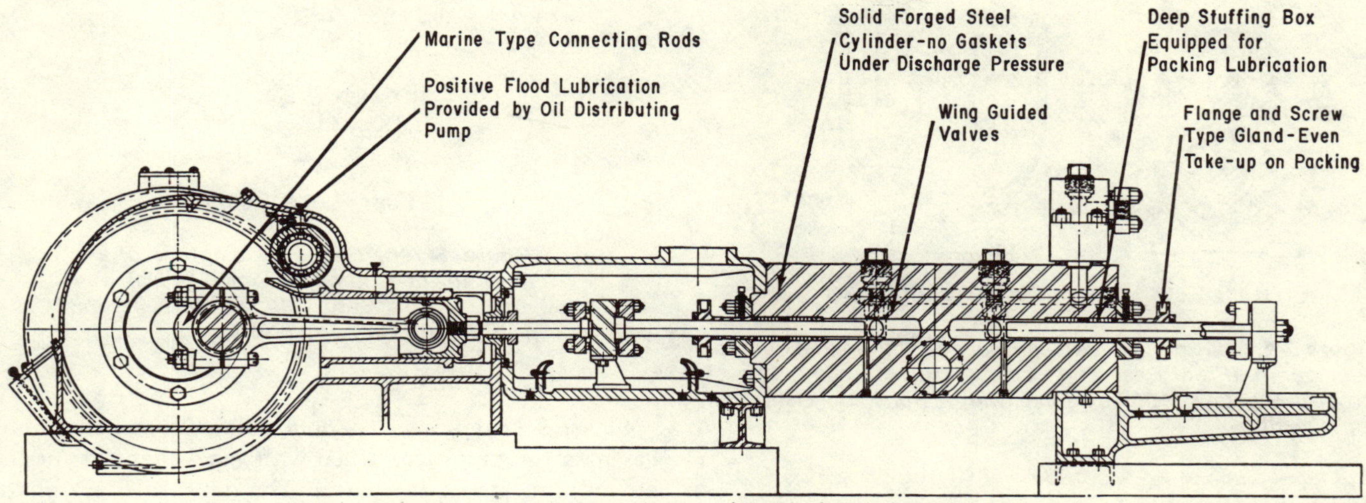

Figure 5-7. *Steam-driven duplex plunger pump. (From Ludwig,* Applied Process Design for Chemical and Petrochemical Plants, *Vol. 1, 2nd edition. © 1978 by Gulf Publishing Co.)*

allows these pulses to die away in its large volume. Another advantage is that duplex pumps tend to transmit less vibration to their foundations.

Rotary Pumps and Compressors

Rotary pumps are so called when they are used to pump liquids, but they are often called *blowers* when used to move gas. Like reciprocating equipment, these machines work by positive displacement. The rotating parts inside the casing move in such a way that they create a space or volume which first enlarges, drawing the fluid in from the low-pressure line, and then, after sealing, is reduced in volume. This forces the fluid through the outlet port or discharge port at the higher pressure. Valves are not required in rotary pumps. Since they are most ingenious in their design, let us examine all five kinds of rotary pumps.

The first is called the *gear pump*, and Figure 5-9 shows an internal and an external gear pump. Notice that in the external gear pump the flow passes around the periphery of the pump, between the space separating the gears and the casing. The gears meshing in the center of the pump act as a seal. The internal gear pump is somewhat more difficult to visualize. In this design the input liquid can follow two paths; either it goes around with the teeth of the inner gear (on the inside of the crescent) or it goes with the external ring gear on the outside of the crescent. If you have any trouble understanding how the internal gear pump works, using Figure 5-9 as a guide, trace the flow of

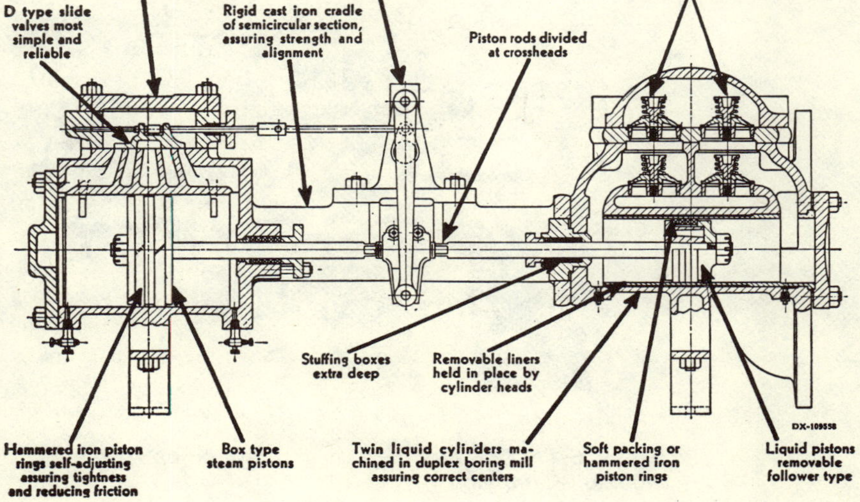

Figure 5-8. *Horizontal duplex piston-pump. (From Rase and Barrow,* Project Engineering of Process Plants. *© 1957 by John Wiley & Sons, Inc. Used with permission.)*

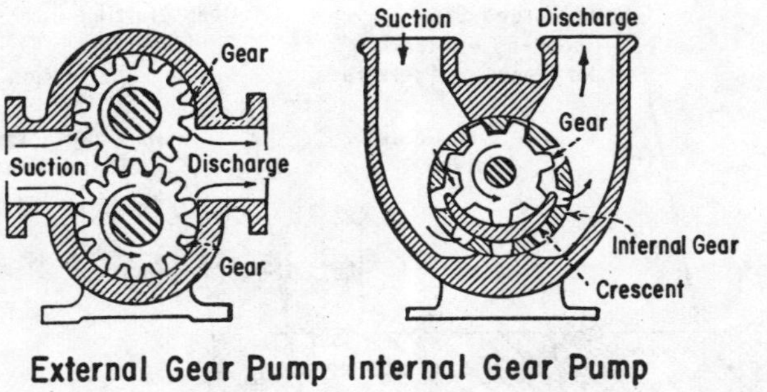

Figure 5-9. *Gear pumps. (A) external gear pump; (B) internal gear pump. (From Rase and Barrow,* Project Engineering of Process Plants. *© 1957 by John Wiley & Sons, Inc. Used with permission.)*

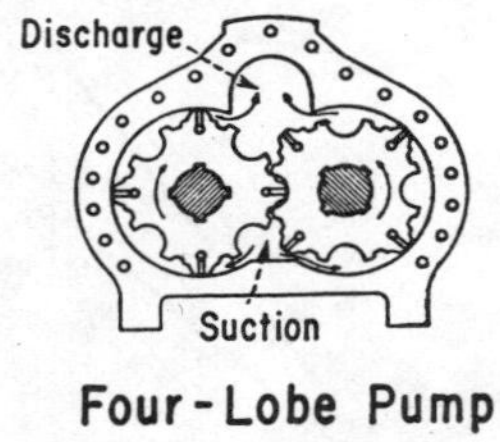

Figure 5-10b. *Sectional diagram of the four-lobe pump. (From R.E. Dolman,* Chem. Eng., *March 1952, p. 159. Used with permission.)*

the liquid around as the inner and outer gears rotate counterclockwise.

Figures 5-10a and 5-10b, show two variations of *lobe pumps*. The first is called a three-lobe blower or pump and the second is a four-lobe version. The lobe pump can be thought of as a gear pump with fewer teeth (either two three or four). However, whereas the gear pump only had drive to one of the gears, the lobe pump has both *impellers* driven separately through external gearing. Thus, a finer design is possible, one that avoids actual contact of the impellers with each other and with the casing. Tolerances are kept to a few thousandths of an inch, which is sufficient to prevent any major leakage from the high-pressure to the low-pressure side while drastically reducing friction and wear in the moving parts. In this way maintenance costs are lower generally for a lobe pump.

The *screw pump* is a common kind of blower for air (see Figure 5-11). A central driven rotor meshes with two "idler" rotors that form fluid-tight seals in the housings. The fluid enters at either end of the rotors, and because it is sealed between the housing and the rotors, it is forced to the central discharge.

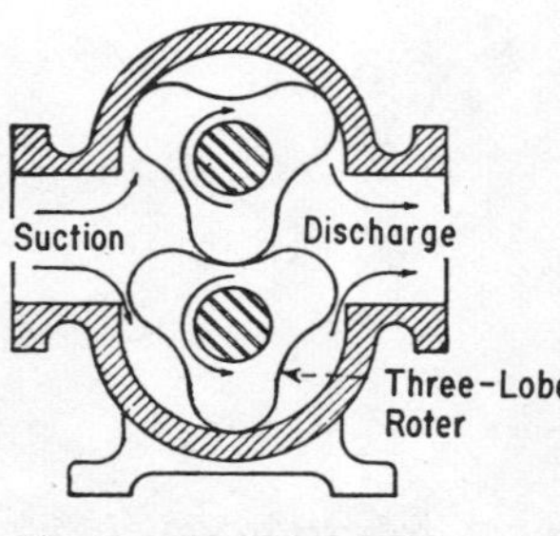

Figure 5-10a. *Cutaway view of a three-lobe pump. (From R.E. Dolman,* Chem. Eng., *March 1952, p. 159. Used with permission.)*

Figure 5-12 shows a *vane pump*, which is constructed by having a rotating barrel fitted with vanes that slide in and out of slots cut in the barrel. The outward edges of the vanes maintain a sliding contact on the elliptical casing and thus provide the means once again of expanding a volume to draw in fluid, sealing it, and then decreasing the volume to compress and discharge the fluid.

Finally, the most difficult to visualize—the *cam pump*—is illustrated in Figure 5-13. A circular rotor is mounted eccentrically. As it rotates it causes the circular plunger or cam to vary its position around the compression space.

Centrifugal Pumps and Compressors

In the early days of our profession, rotary and reciprocating pumps were the most common. In recent years, though there has been a tremendous increase in the use of *centrifugal* pumps and compressors. The reason for this is the improvements made in manufacturing technology. Not only is it possible these days to have finely engineered turbine blades made to stringent tolerances, but their costs have been reduced as well by using computer-controlled cutting methods. In addition, the materials of construction

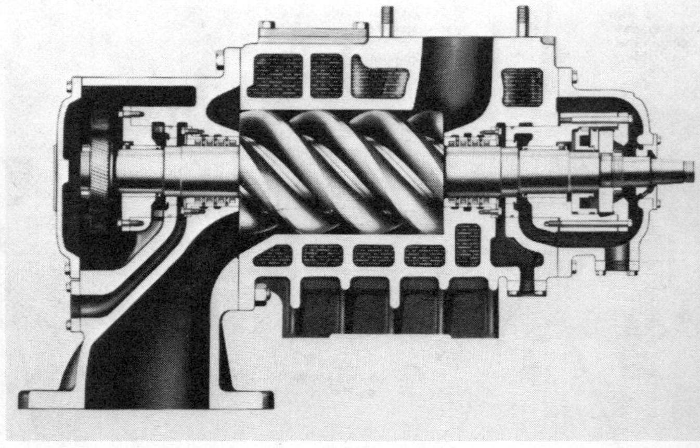

Figure 5-11. *Cutaway view of a screw pump. (Courtesy of Ingersoll-Rand Co.)*

Figure 5-12. *Diagram of a sliding-vane pump. (Courtesy of Ingersoll-Rand Co.)*

have been improved dramatically by persistent metallurgical research.

A centrifugal pump or compressor works exactly in the reverse mode of the turbines described in Chapter 4. Instead of a fast-moving fluid being allowed to expand through a small gap containing a turbine blade which is deflected by the motion of the fluid to give power along its shaft, a pump or compressor works by receiving power through the shaft, from a motor that forces the blades around at an angle to the fluid, causing it to be ejected at a higher pressure. The best common analogy is the fan mounted between the engine and radiator in your automobile. The rotating motion of the fan and the angle of the blades causes a flow of air at right angles to the plane of the fan. A centrifugal compressor can be thought of as a fan enclosed in a casing and with many, many more blades.

There are other kinds of centrifugal pumps and compressors. Some rely on the propeller effect (like the automobile fan) and some rely on a centrifugal effect using an impeller. It is best to classify centrifugal machines according to whether they are *axial flow* or not. Axial flow is the automobile fan type, where the fluid leaves the low-pressure side, travels through the propeller to the high-pressure side, and whose path does not vary much from a straight line. The *non-axial flow* type exploits the centrifugal effect: the fluid flows into the machine and is flung outward by the impeller into an outer casing from which the outlet port projects. Figure 5-14 shows some axial flow impellers. Figure 5-15 shows several kinds of non-axial flow impellers.

There are all sorts of variations on all three kinds of pumps and compressors. All I have tried to do here is give you an idea of the ingenuity that has gone into the design of these most essential parts of our plants. In a later class you will learn the mathematical descriptions of the processes that occur inside pumps and compressors, and such knowledge will enable you to "size" or "rate" a pump or compressor for a particular duty. However, pumping and compressing technology changes rapidly; whenever you need to provide one of these machines in your plant, the best first move is to contact several pump or compressor manufacturers and discuss the particular application with them.

The Basis for Separation of Crude Petroleum

Crude oil contains many different chemical species, as many as 25,000 or more. The simplest of these is hydrogen gas which, along with methane, ethylene, ethane and increasingly heavier hydrocarbon gases, is dissolved in the oil. Obviously, to sell the oil components for different purposes, some means must be found of separating them. Starting with hydrogen, a list of the 25,000 components in ascending order of molecular complexity (an impossible task), would show that the boiling points of the components increases with molecular complexity.

Now, it is obvious that to separate two or more substances from each other there must be some means by which the substances can be told apart. If we were simply

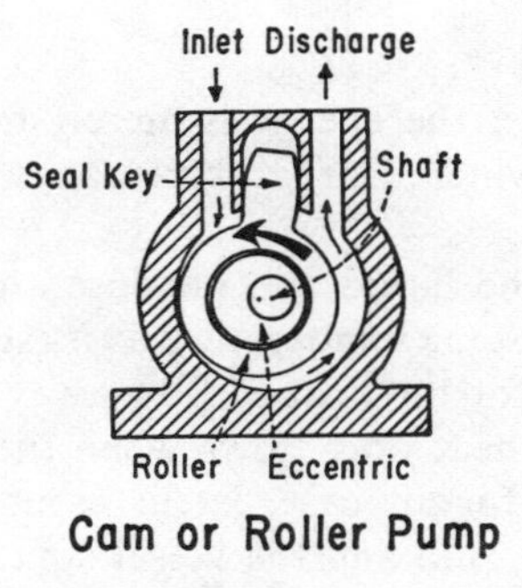

Figure 5-13. *Cutaway view of a cam pump. (From Ludwig,* Applied Process Design for Chemical and Petrochemical Plants, *Vol. 1, 2nd edition. © 1978 by Gulf Publishing Co.)*

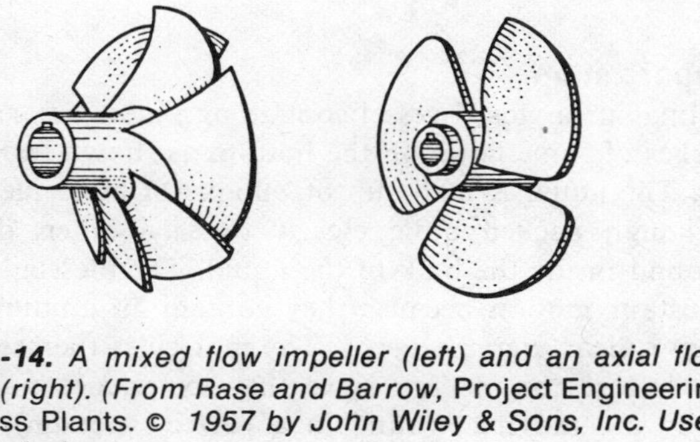

Figure 5-14. *A mixed flow impeller (left) and an axial flow impeller (right). (From Rase and Barrow,* Project Engineering of Process Plants. *© 1957 by John Wiley & Sons, Inc. Used with permission.)*

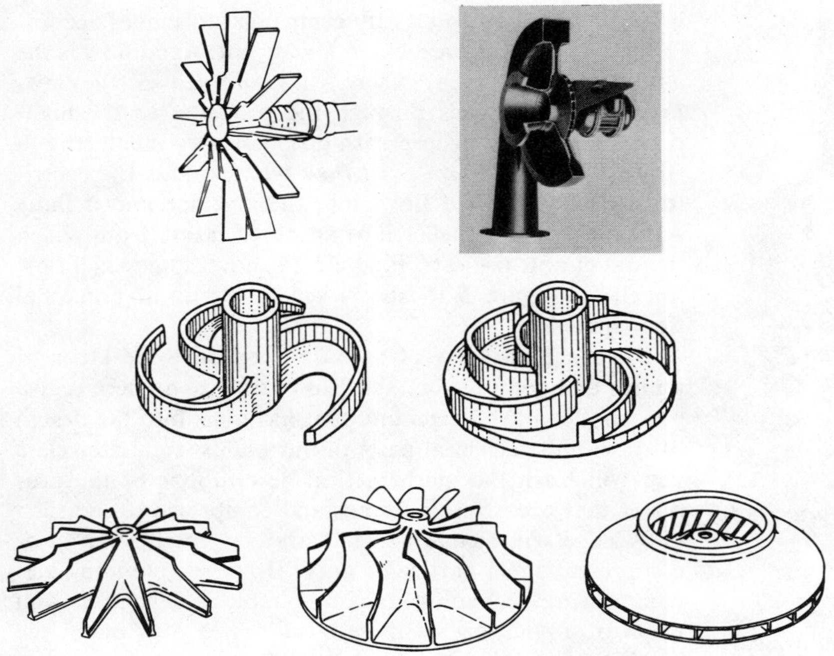

Figure 5-15. *Non-axial impellers. (From Rase and Barrow,* Project Engineering of Process Plants. *©1957 by John Wiley & Sons, Inc. Used with permission.)*

separating ripe and unripe tomatoes, for example, the physical property which would form the basis of each separation would be their color. Similarly, in order to separate clean coal from dirty coal by removing shale and rock, we can float the dirty coal into a bath of liquid that has a specific gravity of 1.5. Coal (specific gravity of 1.3) will float on the liquid and can be skimmed off and sold as clean coal. On the other hand, rock and shale (specific gravity of over 1.7) will sink in the liquid and periodically can be scraped from the bottom of the bath and taken away for disposal. This means of separation exploits the density or specific gravity difference of the materials.

With crude oil, we have already noticed that the physical property that seems to vary the most between the materials in the oil is the *normal* boiling point at atmospheric pressure (14.7 psi). You will recall from Chapter 4, that boiling point is in fact a function of pressure. This topic needs more discussion before we consider the operation known as distillation.

Vaporization

Imagine a closed vessel containing nothing except a few inches of some liquid in the bottom as shown in Figure 5-16. The liquid is made up of billions of molecules which, although packed quite closely to each other, do move around inside the bulk of the liquid. The molecules are in constant motion because they contain an amount of energy called *internal energy*. The molecules themselves will never lose this energy since they experience no friction (they are moving around in a perfect vacuum), and although they lose some of their energy on each impact with another molecule, the energy is merely transferred to the new molecule which, for a while anyway, will have a greater velocity.

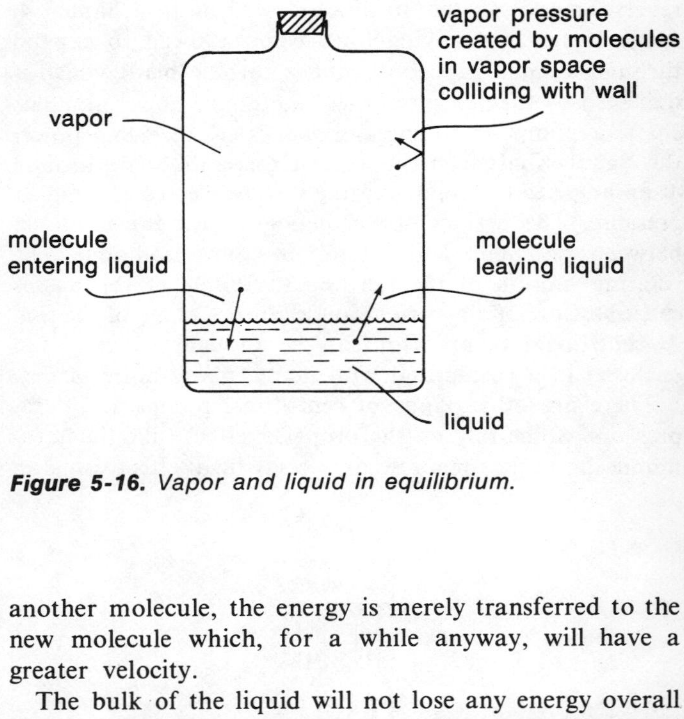

Figure 5-16. *Vapor and liquid in equilibrium.*

The bulk of the liquid will not lose any energy overall since it is at the same temperature as its surroundings, and there is no way for the heat to flow away from the liquid. Remember that heat only flows when there is a temperature difference. In our case, let us assume that the temperature of the liquid and the vessel containing it remains the same as the temperature of the general surroundings, which means the internal energy of the liquid will be constant. Consider the following possibility: A molecule is moving around, bouncing off its fellows, in an absolutely random pattern. Then, our molecule is suddenly hit from below by a fast-moving molecule which sets ours in motion vertically upward. Imagine further that the molecule, rather than being stopped by collisions with other molecules in its path, is given further boosts by faster molecules from below, causing our molecule to gain speed until it actually reaches the surface of the liquid.

Consider, before we go on, the circumstances of the molecule in question. Since all the molecules are tiny bits of mass, they attract each other with a weak force similar to the gravitational force that causes the moon and the earth to be attracted towards each other. Normally, when our molecule has been in the center of the bulk of the liquid, it has felt the attractions of all the molecules around it. Thus the attractions together have no effect on the individual molecule because they cancel each other out coming equally from all directions (see Figure 5-17).

Now, when our molecule is actually at the surface, it can only feel attractions from molecules beneath it. There are no attractions above because there are no molecules there (see Figure 5-17). Our molecule is pulled back by the attractive forces of its fellows. Imagine our molecule is again in a series of collisions that also happen to propel it toward the surface with an even greater velocity. In spite of the attractions of its fellows, the molecule breaks through and enters the space above the surface of the liquid. It is able to maintain its higher velocity because few of its fellows have actually made the jump out through the surface. Consider the situation now: There are slow-moving, jostling molecules in the liquid and a few fast escapees bouncing around freely in the volume above the liquid. The cumulative effect of their bouncing against the wall of the container is how the vapor exerts its pressure on the wall.

Molecules that bounce off the wall or other molecules

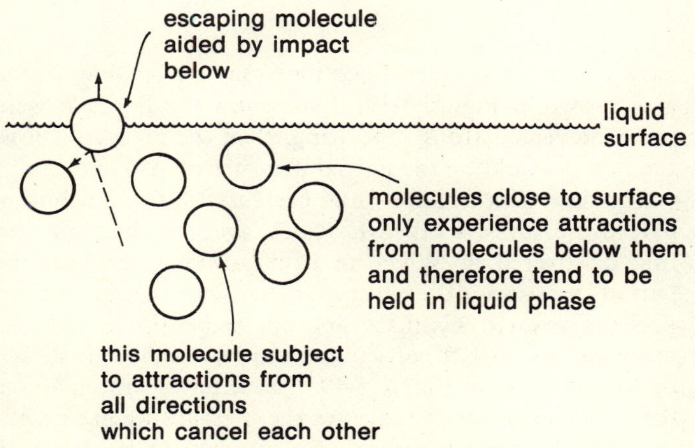

Figure 5-17. *Attractive forces acting on molecules near the surface.*

and find themselves heading downward may pass through the surface and be slowed down again by the crowded conditions in the liquid. Now, since the speed of the molecules depends upon the internal energy of the liquid, and since an increase in the temperature of the liquid increases its internal energy, you can see that at higher temperatures, more molecules will escape. However, if the temperature is held constant, in a very short period of time the faster-moving molecules in the space above the liquid reenter the liquid at the same rate that others are leaving. In other words, the vapor space has a net gain of zero molecules in any given time after a steady state has been reached at any given temperature. Put another way, at constant temperature an equilibrium will be reached where the number of molecules in the vapor space remains essentially constant; thus the pressure they exert on the walls of the vessel also will be constant. It is called the *vapor pressure*.

Figure 5-18 shows a modification to the vessel we have been using for this hypothetical experiment. Consider now a smooth-walled cylinder with its open end upwards. Assume that the piston has no weight, makes no friction between itself and the walls of the cylinder, and provides an absolutely molecule-tight seal. (These are ideal, nonexistent assumptions, but they serve to illustrate a very real principle.) The piston is then free to ride up and down, and initially it sits on the two lugs that stick out in the cylinder wall a few inches above the surface of the water. We know from our earlier discussion that the space between the liquid and the piston contains molecules that exert a vapor pressure on the piston. However, the piston does not rise even though its underside is exposed to the vapor pressure, because its upper side is exposed to the atmospheric pressure bearing down upon it (the weight of the air above the piston), which happens to be 14.7 psi.

Let's suppose we start this experiment at 70°F. Increasing the temperature increases the pressure. We can look up in the vapor pressure columns of Steam Tables the vapor pressure of water at 70°F and find the value is 0.363 psi. With each square inch of the piston subjected to 0.363 pounds force from beneath and 14.7 pounds force from above, it will remain firmly and snugly against the lugs that hold it in place. Now, imagine increasing the temperature. At 120°F the vapor pressure (from Steam Tables) is 1.69 psi. Still no move. At 190°F the vapor pressure is 9.34 psi. Still no action. However, we find at 210°F the vapor pressure, at 14.12 psi, is close to atmospheric, but still the piston remains in place. Finally, we can see in Steam Tables that at 212°F the vapor pressure of water is equal to 14.7 psi and the piston now has an equal pressure in either side.

Meanwhile, back in the liquid, all sorts of action is occurring. The increased internal energy has speeded up the molecules and an increasing number have been escaping. The vapor pressure has been rising until, at 212°F, so

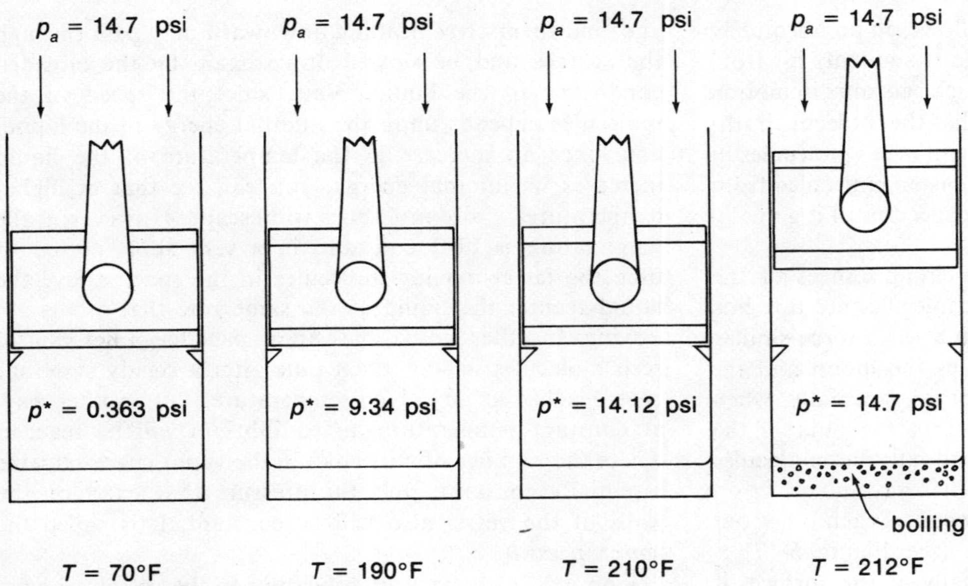

Figure 5-18. The effect of temperature on vapor pressure.

many molecules have the necessary escape velocity that instead of just one or two molecules passing through the surface, there is a mass migration. In fact, the molecules in the liquid are moving so fast that they begin to make space for themselves within the liquid and form pockets of vapor (bubbles) which rise to the surface since they have a lower density than the surrounding liquid. Have you guessed what this phenomenon is? We call it boiling.

A British explorer being interviewed by the press upon his return from the summit of Mount Everest remarked that they had celebrated their achievement while on the summit by making tea. He lamented that the tea had tasted awful. This is hardly surprising. The rarefied atmosphere on the top of Everest at 29,000 feet above sea level exerts a much lower pressue (4.7 psi) than that experienced at sea level (14.7 psi). The pressure of 4.7 psi meant that the water boiled at 159° F, and thus the tea-brewing process was not complete and the result was awful. In Chapter 4 we saw that the boiling point of the water in the boiler was 467° F, and that is because the vapor pressure of water does not reach 500 psi (which was the pressure inside the boiler) until that temperature; hence 467° F is the boiling point at that pressure.

Distillation

Now consider the mixture of two liquids, A and B, in the piston and cylinder arrangement repeated in Figure 5-19. Starting off with a vacuum above the liquids as before, and assuming a constant temperature, molecules of A and molecules of B escape through the surface of the liquid and exert a pressure on the piston. The total pressure exerted on the piston is made up of the *partial pressure* of vapor A plus the partial pressure of vapor B. The partial pressure of any gas or vapor is the pressure that it *would* have exerted if it alone filled the volume available. In the case shown in Figure 5-19, if somehow the liquid B were to be prevented from vaporizing, then the pressure above the liquid would be the partial pressure of A alone.

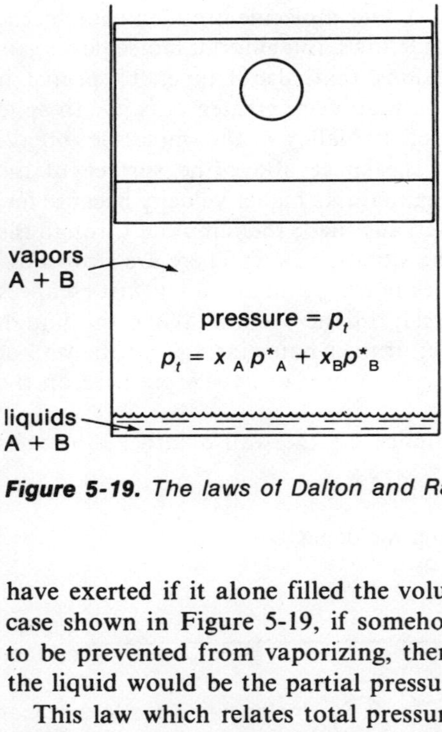

Dalton's Law:

$p_A = y_A p_t$

$p_t = p_A + p_B$

or, at equilibrium

$p_t = p^V_A + p^V_B$

Raoult's Law:

$p^V_A = x_A p^*_A$

$p^V_B = x_B p^*_B$

$= x_A + x_B$

Figure 5-19. The laws of Dalton and Raoult.

This law which relates total pressure to its contributing partial pressures is called Dalton's law after the man who first discovered it. When an equilibrium is reached, the partial pressure of each vapor will be an equilibrium pressure exerted by that vapor in competition with the other vapors at that particular temperature. This is shown by the equation in Figure 5-19. The letter V written above the p as a superscript indicates the pressure we are speaking of is a vapor pressure. (An asterisk (*) written as a superscript indicates an equilibrium vapor pressure exerted by the pure vapor, i.e., with no other vapors present.)

Another law we should consider is Raoult's law, which states that in a mixture of liquids the vapor pressure exerted by each component present is equal to its equilibrium vapor pressure multiplied by its mole fraction in the liquid. This is shown by the third equation in Figure 5-19. Unfortunately, Raoult's law is not correct for all liquid solutions, although most found in practice come so close to it that it is often reasonable to assume it is true. The implications of Raoult's and Dalton's laws together are enormous for us. Dalton's law, when fully stated, is that the mole fraction of a component in a gaseous mixture is equal to the partial pressure of the component expressed as a fraction of the total pressure. Notice the deduced equation which says that since all the mole fractions add up to one, then all the partial pressures add up to the total pressure, which is the initial way I stated Dalton's law.

Returning to the example of the mixture of A and B, it will become clear to you by considering Dalton's and Raoult's laws together that unless the vapor pressures of both A and B happen to be the same at the given temperature, the mole fraction of each in the vapor space will be different from the mole fraction of each in the liquid space. To demonstrate the importance of this conclusion, consider the worked example in Figure 5-20. You can see that in the vapor space we have 67.4% benzene and 32.6% toluene, and yet in the liquid phase we had 40% benzene and 60% toluene. Now look at Figure 5-21. In this picture the piston has been replaced by a sealing canopy to the vessel and there is a tube leading the vapor away from the vapor space through a condensing apparatus. Even though the original liquid contained 40% benzene, the condensate that the equipment will produce contains 67.4% benzene and only 32.6% toluene. The hardware design shown in Figure 5-21 is a means of enriching the benzene content in the condensate or the tolune content of the residue.

Imagine that we do the same thing again and again as shown in Figure 5-22. That is, we take the condensate, put it in another vessel with a vacuum above it, hold this new vessel at 100°F and, as soon as equilibrium is established (almost instantaneously), draw off the vapor and condense it. Figure 5-22 shows the hardware and the calculations, again based on the simple concepts of Dalton's and Raoult's laws. After the second separation, the condensate would be 86.4% benzene and 13.6% toulene. The third separation shows how the benzene could be enriched further to 95.2% opposed to the 4.8% toulene. You can imagine that this could go on until we approached a condensate which would be just about pure benzene. Where did the toulene go? At each "stage" of our vaporization and condensation the liquid left behind was somewhat enriched with toulene.

We now have the means of separating two components with different vapor pressure characteristics, which is the same as saying we have two components with different boiling points, for the reasons outlined earlier. Consider Figure 5-23. Instead of the arrangement shown in Figure 5-22, this time the vaporization vessels are stacked on top of each other: the vapors pass upward and the liquids flow downward from each stage to fall to the stage below. Actually, Figure 5-23 shows the workings of a simple

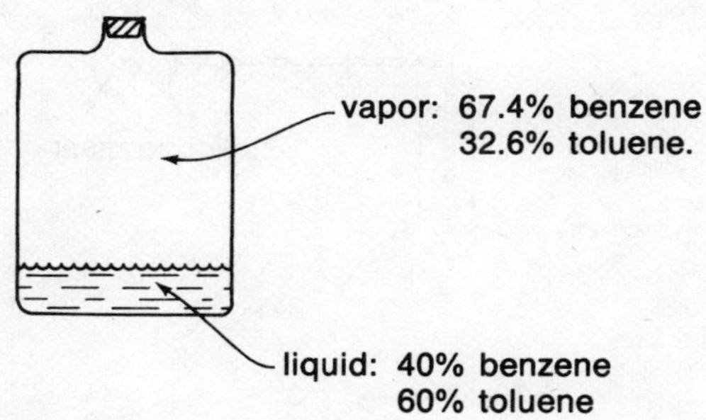

Calculation

Basis: liquid contains 40% benzene, 60% toluene.
temperature = 100°F

$\therefore x^L_{benz} = 0.4$

$x^L_{tol} = 0.6$

(superscript L means mole fraction in liquid)

from thermodynamic tables, $p^*_{benz} = 3.20$ psi, at 100°F

$p^*_{tol} = 1.03$ psi, at 100°F

$\therefore p^V_{benz} = 0.4 \times 3.20 = 1.28$ psi

$p^V_{tol} = 0.6 \times 1.03 = 0.62$ psi

$\therefore p^V_{total} = 1.28 \times 0.62 = 1.90$ psi

$$\therefore y^V_{benz} = \frac{1.28}{1.90} = 0.674\ (67.4\%)$$

$$y^V_{tol} = \frac{0.62}{1.90} = 0.326\ (32.6\%)$$

Figure 5-20. *The vapor above a mixture of benzene and toluene.*

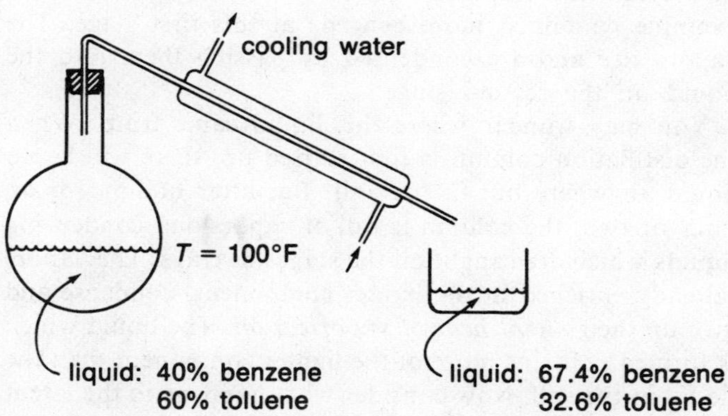

Figure 5-21. *Enriching the benzene-toluene mixture by distillation.*

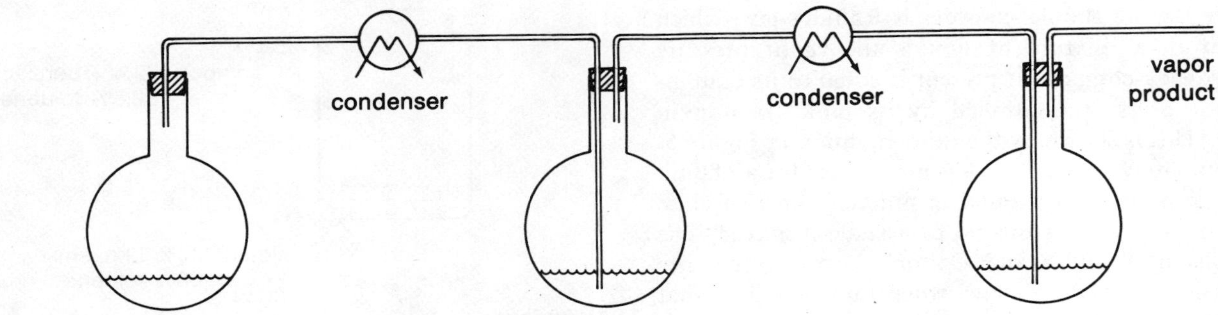

$x^L_{benz} = 0.4$

$x^L_{tol} = 0.6$

$p^v_{benz} = 0.4 \times 3.2 = 1.28$ psi

$p^v_{tol} = 0.6 \times 1.03 = 0.62$ psi

$p^v_{total} = 1.9$ psi

$y^v_{benz} = \frac{1.28}{1.90} = 0.674$

$y^v_{tol} = \frac{0.62}{1.90} = 0.326$

∴ vapor leaving
first stage is
67.4% benzene
32.6% toluene

$x^L_{benz} = 0.674$

$x^L_{tol} = 0.326$

$p^v_{benz} = 0.674 \times 3.2 = 2.16$ psi

$p^v_{tol} = 0.326 \times 1.03 = 0.34$ psi

$p^v_{total} = 2.50$ psi

$y^v_{benz} = \frac{2.16}{2.50} = 0.864$

$y^v_{tol} = \frac{0.34}{2.50} = 0.136$

∴ vapor leaving
second stage is
86.4% benzene
13.6% toluene

$x^L_{benz} = 0.864$

$x^L_{tol} = 0.136$

$p^v_{benz} = 0.864 \times 3.2 = 2.76$ psi

$p^v_{tol} = 0.136 \times 1.03 = 0.14$ psi

$p^v_{total} = 2.90$ psi

$y^v_{benz} = \frac{2.76}{2.90} = 0.952$

$y^v_{tol} = \frac{0.14}{2.90} = 0.048$

∴ vapor leaving
third stage is
95.2% benzene
4.8% toluene

Figure 5-22. *A series of distillations.*

distillation column. The bottom vessel is called the *still.* It is constantly supplied with sufficient heat to cause the liquid to boil. The vapors above that liquid contain more of the lighter component of the two, just as our earlier example contained more benzene at this first stage. The vapors rise and are condensed by passing them into the liquid on the second stage.

You may wonder where this liquid came from. When the distillation column is first started up, there will be no liquid anywhere but in the still. But after boiling for an hour or two, the column is full of vapors and condensing liquids which are caught on the stages or trays. The vapors (already enriched in the lighter component) condense and give up their *latent heat of vaporization.* The liquid which is formed contains more of the lighter component than the liquid in the still. Now consider what happens to the latent heat of vaporization given up by the vapors. It causes some of the liquids on the second stage to evaporate. Because of Raoult's law, the new vapors are even richer in the lighter component than the second stage liquid they leave behind as they ascend to the third stage. They condense in the third stage and liberate vapors which are led away from the column and condensed as shown.

The condensed stream or *condensate* is divided into two parts: one is called the *product* in Figure 5-23 and the other is called the *reflux.* Remember that the reflux has the same composition as the product. Now imagine that we close off the outlet marked product and return all the condensate as reflux. The top tray of the column now contains a liquid which is greatly enriched in the lighter component. Consider this top tray only. Because the reflux liquid is condensed from vapors which themselves are richer in the lighter component than the liquid on the top tray (where they came from), the top tray liquid gradually becomes more and more enriched in the lighter component. Furthermore, the effect begins to spread downward to the lower trays in the column. Thus, after several hours of operation an equilibrium is reached where the concentration of the lighter component is greatest in the vapors leaving the top of the column and the concentration of the heavier components is greatest in the bottom still.

You can already speculate on the next step. Instead of simply having three stages in our column, we can increase the number of stages such that the enrichment of the lighter component is sufficient to cause the composition of the stream of vapor leaving the top tray to be almost pure A and the still composition to become almost pure B. At that point, and referring to Figures 5-23 and 5-24, we can now consider making the distillation operation a continuous one. Suppose we have a feed stream that is 40% benzene and 60% toluene. We could imagine that initially the entire column is full of this mixture and that all of the "tops" product (often called the *overhead product*) is returned as reflux to the column. After several hours, the composition of the reflux on a 30-stage column would be,

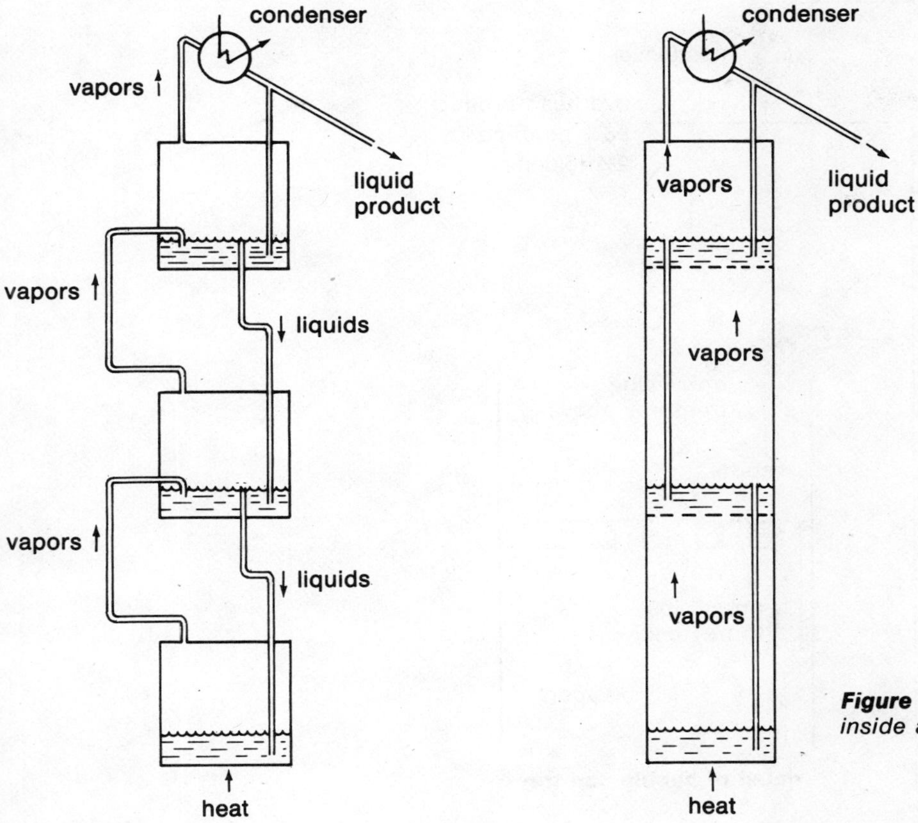

Figure 5-23. *Making the three-stage distillation inside a column.*

say, 98% benzene, whereas the composition in the still would be, say, 99% toluene. At one of the stages in between, the composition of the liquid would be very close to the 40-60% composition of the feed stream. It is onto this stage or tray that we should input the feed stream at a rate low enough not to disturb the overall equilibrium. Incidentally, it is only by calculations such as the one in Figure 5-22 that the composition on the trays can be determined in order to decide before building which tray is to be the *feed-tray*.

After establishing the overall equilibrium, the product line can be opened and some of the overhead product drawn off, in this case with a composition of 98% benzene and 2% toluene. The rest of the overhead stream is returned to the top tray as reflux.

Similarly, some *bottoms* product can be collected by continuously removing a small amount of the liquid from the still. This composition will be 99% toluene and 1% benzene. The only constraint is that the total weight of products removed from the column per minute (in both overhead and bottoms streams) is equal to the weight of feed poured onto the feed-tray each minute.

You have just seen the operating method of a distillation column which will continuously separate two components by virtue of their differences in vapor pressure alone. The distillation column is an extremely important tool in almost every process plant. It is used to separate raw materials from each other before processing them into new, useful products. It is also used to separate the products and by-products from each other, as you will see later in this book. We can now consider how to make use of the distillation operation to process the crude petroleum oil after it has been taken from the ground and transported to the refinery.

Fractionation

Crude oil is not a simple mixture of two components, as was the liquid in the above example. Rather, it is a mixture of around 25,000 components. However, the same principle of distillation can be used in what is called a *fractionating column* in order to separate the thousands of chemical species, not into individual components, but into mixtures that have similar boiling points or similar vapor pressure characteristics. Figure 5-25 shows a schematic representation of a fractionating column. The difference between "binary" distillation (with only two components), and this "multi-component" distillation (or *fractionation*) is that we remove products at many points in the column. Fractionation must be the single most important operation in an oil refinery since it is the first step in the long path from crude oil to consumer products.

The crude oil is first heated up in a *pipe still*, which is similar to the boiler unit described in Chapter 4. The crude oil enters a number of steel tubes which are exposed

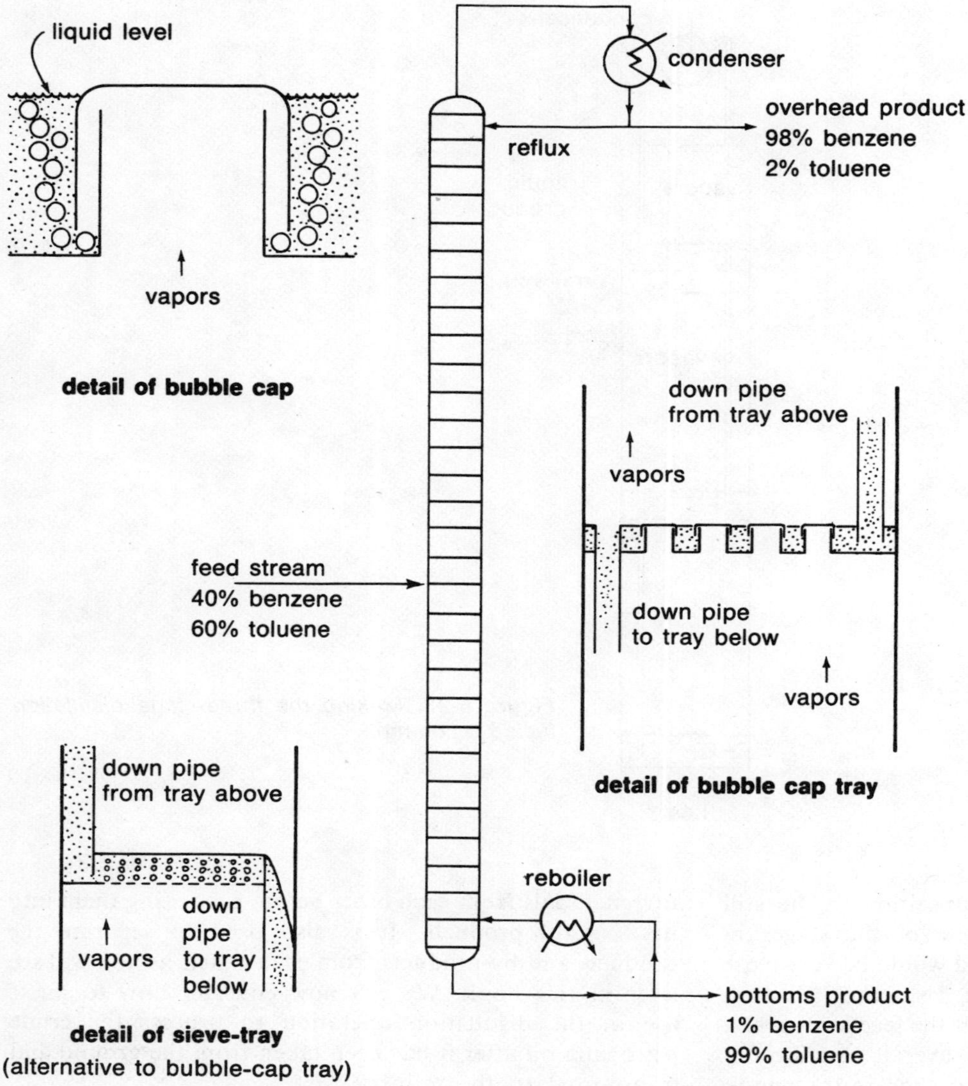

Figure 5-24. *Continuous binary distillation.*

to a flame produced by burning fuel oil or natural gas as described in Chapter 4. Some of the crude oil boils or evaporates, and the mixture of liquid and vapor is fed into the fractionating column at the point *F* (Figure 5-25), which is the feed-tray. Immediately, the vapors begin to rise up the column and the liquids in the mixture fall down the column toward the still. As the vapors rise, because of the vapor pressure laws you have already seen, the liquids on the higher trays become enriched in lighter components. For this reason the section of the distillation column above the feed point is called the *enriching section*. As the liquid falls from the feed-tray downward in the column the lighter components are stripped from it (to ascend the column); hence the portion of the fractionating column beneath the feed point is called the *stripping section.*

The overhead which leaves the top of the column is condensed and part of it returned to the top tray of the column as reflux, as shown in Figure 5-25. The liquid in the still at the bottom of the column can be treated in two ways. The first way is to withdraw the liquid and divide it into two streams; one is the bottoms stream or residue and is taken away to a feed-heater, which I will describe in a moment. The other portion is reboiled in a steam-heated reboiler from which the vapors are fed back into the column beneath the lowest tray.

The second method of treating the bottoms product is to have a constant drawoff from the still but cause the reboiling to occur in the still by injecting hot steam into it.

Before we leave the fractionating column itself, let us take a more detailed look at the trays or stages of the column (often referred to as *plates*). The early fractionating towers had plates called "bubble-cap trays" and the column itself was called a "bubble tower." Figure 5-24 shows

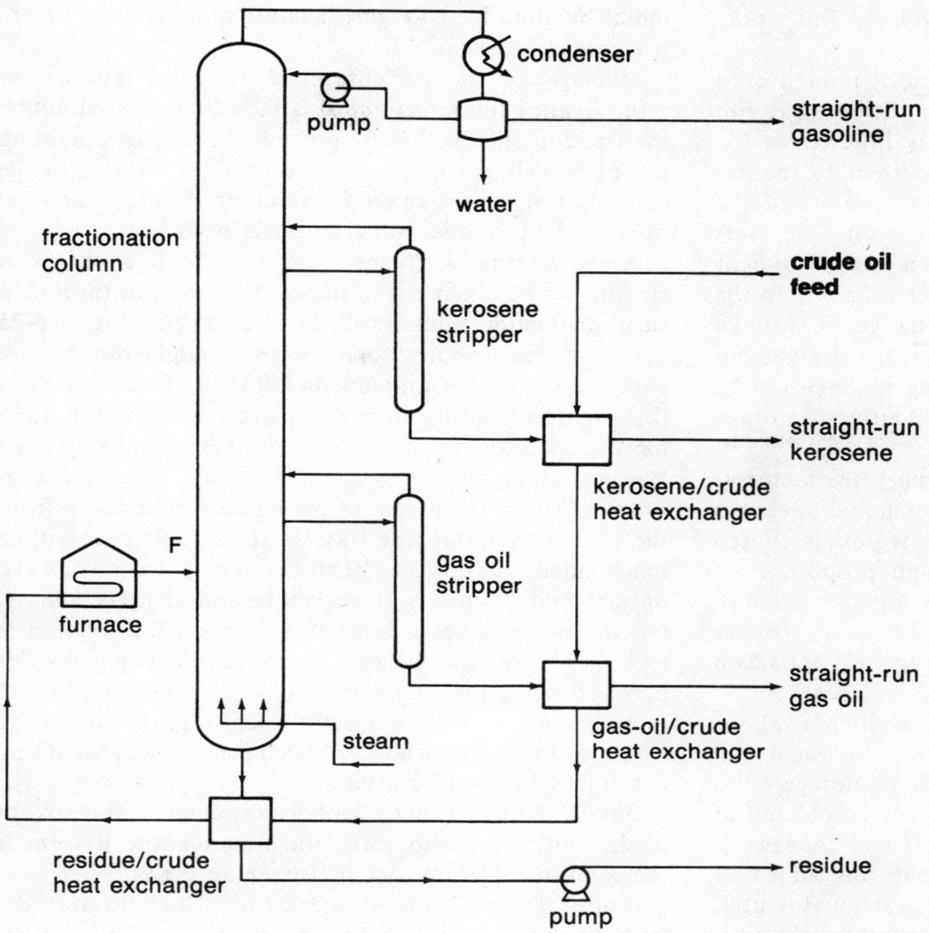

Figure 5-25. *Crude-oil fractionation.*

a detail of a bubble-cap tray with an indication of the action of the tray immediately above and below it. Each bubble-cap causes the rising vapors to be bubbled through the liquid present on that tray. Thus, an intimate contact exists between the rising vapors and the liquid at any point in the column. The liquid condensates build up on a tray until they pass over the overflow and into the down-pipe which takes them to the tray below. Notice the lower end of the downpipe is immersed in the liquid on the tray below in order to prevent vapors rising up it.

Although there are still many bubble-cap towers actually operating today, new designs use a slightly different device called a *sieve-tray*. The sieve-tray is a disk set into position inside the column. Its entire area is drilled with small holes whose diameters have been calculated very carefully so that when the column is in operation, the descending liquid will fall at such a rate that it will maintain a layer of liquid on each tray, held there by the vapors bubbling through each of the holes in the tray. Thus, there is intimate contact as a bubble leaves a hole and rises through the liquid. The operation of the sieve-tray is shown in Figure 5-24. Now, having seen all the individual aspects of a fractionator, let us consider its continuous operation.

The feed leaves the pipe still or furnace at around 575° F and enters the fractionator at the feed-tray. At the steady state which is soon established, the vapors ascend the column through the enriching section and the concentration of the liquids on each tray becomes more and more enriched in the lighter components. Also, the temperature in the column falls gradually since the heated stream of vapors is meeting the falling condensed liquids from the reflux. Just a few trays up from the feed-tray, the temperature is around 480° F, and there is a constant withdrawal of liquid from this tray because at this boiling point the fractional product will contain high proportions of fuel oils, lubricating oils, grease, and other useful products. This fraction is called *straight-run gas oil.* "Straight-run" means that the product has only been through the crude oil fractionator and has not yet been processed any further: "gas oil" is the name given to this product because formerly it was used to make gas. It is worth noting that before leaving the area of the fractionator, the straight-run gas oil is taken through a little distillation column called the *gas oil stripper*. This works as follows.

The gas oil leaves the fractionator and is poured onto the top tray of a small distillation column. It begins to run down through the trays. Hot steam is injected at the bottom of the column to rise counter-current to the gas oil. This has the effect of "stripping out" some of the lighter hydrocarbons remaining in the gas oil. The overhead product from the gas oil stripper is a mixture of light hydrocarbons and steam, which is injected back into the main fractionator a tray or two above the gas oil offtake tray. The bottoms of the gas oil stripper, the straight-run gas oil itself, is cooled in a heat exchanger, which will be described in a moment, and then pumped either to storage or to its next processing unit.

Higher up the fractionator column, when the temperature has fallen to around 330° F, another liquid stream is taken off. This particular temperature region is chosen because the liquid here contains a high proportion of kerosene and other useful products. The liquid is taken to a small distillation column called a *kerosene stripper* which, like the gas oil stripper, is charged with hot steam to strip out some of the lighter dissolved hydrocarbons.

Again, the tops are returned to the fractionator above the kerosene offtake tray and the bottoms (*straight-run kerosene*) are cooled and pumped either to storage or to their next process. Finally, at the top of the column at around 250° F, the overhead is taken off and condensed. The condensate contains the water that was added as steam in the two side strippers and the fractionator itself (see below). This is removed in a vessel where the condensate from the condenser is allowed to form into two layers. The bottom layer is water, which is bled constantly from the unit. The top layer is a mixture of straight-run gasoline and lighter hydrocarbon gases. This stream is split into two, one part is returned to the top of the fractionator as reflux at about 120° F. The other part of the stream (straight-run gasoline) is taken away to storage or for further treatment.

You can see how the fractionator is splitting up the crude oil into mixtures of components with similar boiling points. These are called *fractions*. The degree of separation between fractions depends upon the number of trays in the column and the percentage of overhead product which is returned as reflux. If there are more trays in the column, the cut-off between the boiling points of one fraction and the one below it is "cleaner." Of course, more trays means that the column will be more expensive to build. If a column is built with less trays, it will cost less initially, but the operating costs for the refinery will increase because extra processing will be required to separate the fractions after they leave the column. Perhaps you sense that this is another of those trade-off situations where there is no general correct answer. Each situation has to be considered carefully, and the interest of the initial building loan weighed against running costs befor reaching a solution.

Similarly, if the percentage of overhead returned a: reflux (called the *reflux ratio*) is increased, then althougt the fraction split will be improved, the cost of operating the plant will go up since more heat will be lost in the condenser at the top of the fractionator. Again, a carefully calculated economic compromise is necessary.

As an alternative to more trays in the fractionator, a stream can be taken out of the column and put through a small distillation column called a *side-stripper*. Figure 5-25 has two side-strippers: one on the straight-run gas oil stream and one for straight-run kerosene. Even though at first sight this seems more expensive than simply putting more trays in the main fractionator, it is more likely to be cheaper to design the system this way. The reason is simple: Only a small part of the liquid will be taken from the straight-run kerosene tray, so the side-stripper will be much smaller in diameter than the main column since the upward and downward flows will be considerably smaller. On the other hand, a huge tray in the main column is extremely expensive in itself. Also, not having side-strippers will make the column taller, which means it will cost more. Again a trade-off; and each case is solved by chemical engineers who can calculate the way of doing the fractionation at least cost.

Finally, let us consider the bottoms stream. This stream is the hottest one to leave the fractionator. It is at a temperature of about 540° F (or at an even higher temperature). Steam has been injected into the bottom of the fractionator to assist in the distillation process by reboiling the liquids in the still and by helping to strip out the lighter components from the descending liquids. Since the residue stream is so hot and has a high heat capacity, the sensible heat it takes away from the fractionator is a significant heat loss. If the residue were to be taken directly to further processing requiring it to be hot, the sensible heat would be useful in the next process and not lost. However, the residue will often be stored before being processed further, or it could even be sold and taken away from the plant. In any of these cases, the sensible heat will be lost unless steps are taken to recover it. The crude oil entering as feed, on the other hand, has to be preheated in the furnace or pipe still before being charged into the fractionator.

Heat Exchangers

It would be useful if we could transfer the sensible heat from the bottoms product into the crude oil feed before it enters the pipe still—that way less heat would be required in the pipe still. This is a most important consideration in these days of high oil and energy costs. A device called a *heat exchanger* helps us out. The role of a heat exchanger is to take heat from a stream that no longer needs it and give it to a stream that does. Because this is a common situation in a plant, it often pays to lay a little extra piping to carry a stream with sensible heat (that would otherwise be wasted) to a heat exchanger, where the heat can be transferred to another stream which will shortly have to be heated anyway.

Early heat exchangers were simply two heads on the opposite ends of a cylindrical shell joined by a number of tubes that were sealed from the shell. Figure 5-26 gives a schematic. One fluid flows into the left-hand head, through the tubes, into the right-hand head, and out through the "nozzle." The other fluid flows into the right-hand nozzle on the shell, circulates around the inner tubes, and leaves via the left-hand nozzle on the shell. These flows are shown in Figure 5-26.

A careful choice will have been made as to which fluid to pump through the tubes and which will go through the shell. If one of the fluids had suspended solids in it, it would be troublesome if it were the "shell-side" fluid, since the suspended solids would tend to be deposited in low-velocity areas such as odd corners around the tube bundle and become baked onto the heat exchanger walls. This would have two effects. First, the baked-on solid would cause an extra resistance to heat transfer through the heat exchanger walls and reduce its effectiveness. Second, by obstructing flow to some extent, the pumping power required to push the liquid through the exchanger would be increased, as would the cost of operating the pump. The exchanger also would need a bigger pump, which would mean a bigger capital outlay when the plant was built.

For these reasons, if one of the liquids does contain solid material, then it would be normal to put it through the tubes since, apart from in the heads themselves, there are no stagnant pockets or low-velocity parts of the flow inside the tubes. Any deposits that did form in the tubes could be cleaned off easily by removing a head and quickly scouring out the tubes with ramrods. Whereas if the solids-containing fluid were put through the shell-side, one would be faced with the tedious and time-consuming job of cleaning the space between the tubes inside the shell. If there were no other considerations (such as suspended matter), the decision would be an easy one: the hot fluid would go through the tubes. If the design were to be the other way around, then the hotter fluid would be closer to the outside atmosphere and cause a greater heat loss.

Figure 5-26 also shows the temperature profiles of the hot and cold fluids as they pass through the heat exchanger. Notice how they have been drawn on the same axes in order to show the difference between *counter-current* and *parallel-flow* operation. You know from previous chapters that the heat will flow at a rate corre-

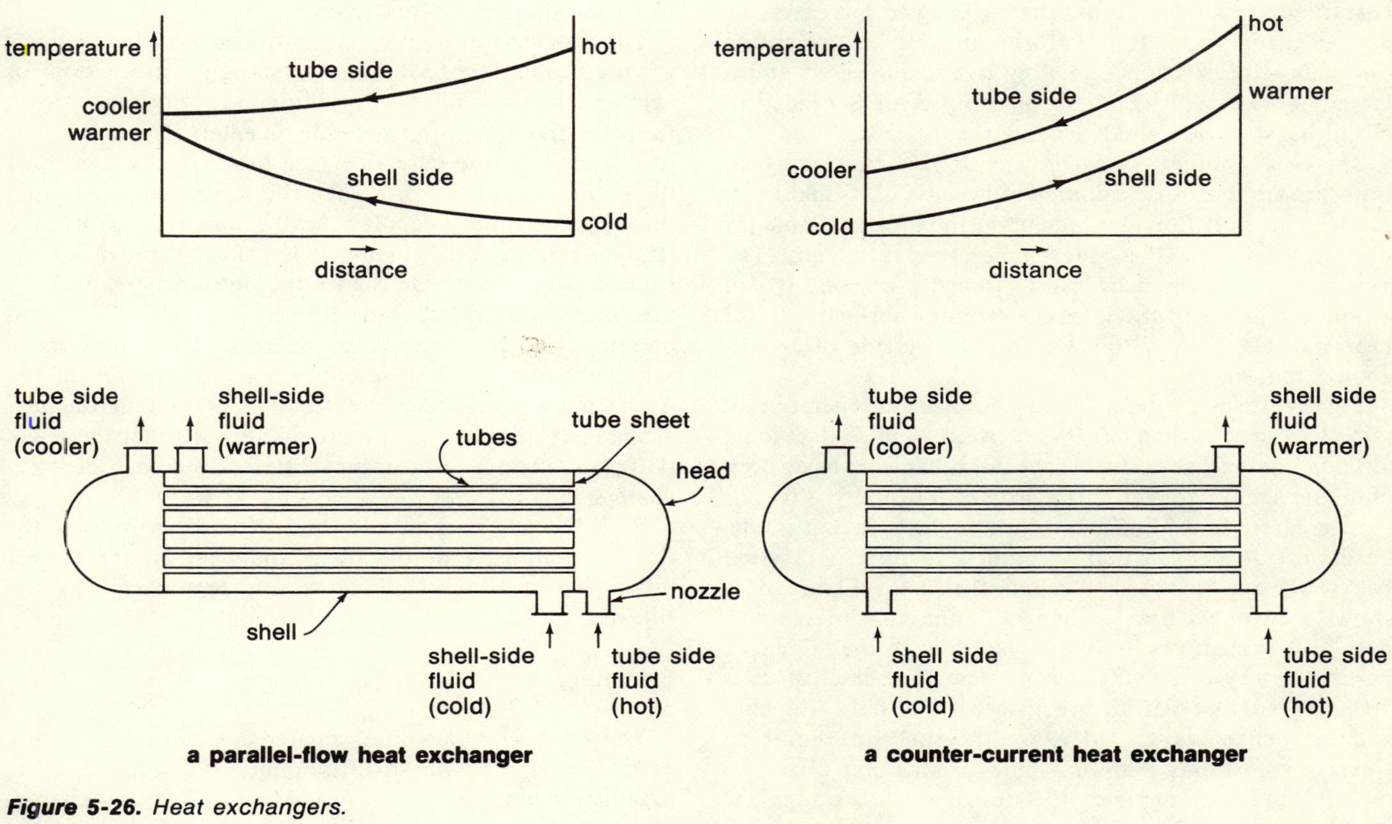

Figure 5-26. *Heat exchangers.*

sponding to the temperature difference between the fluids. Consider the two temperature-distance diagrams in Figure 5-26. In the parallel-flow exchanger, the temperature of the cooler fluid starts off well below the temperature of the hot one and the heat flow rate is rapid. Toward the end of the tubes, however, the temperature gap has narrowed considerably, with a corresponding drop in the heat flow rate. On the other hand, in a counter-current exchanger, the temperature difference changes little along the tubes and the heat flow rate hardly varies.

Which is the better exchanger? Again there is no simple answer. It depends on the particular case, and a design engineer must do the necessary calculations before a decision can be made. However, the one-pass heat exchanger—the type shown in Figure 5-26—is rarely used these days, if at all. And you will see that its replacement is a combination of both types of arrangement: counter-current *and* parallel-flow.

The exchanger shown in Figure 5-26 is called "one-pass" because the tube-side fluid passes from head to head only once. Figure 5-27 shows a two-pass heat exchanger. The tube-side fluid flows from head to head and back again before leaving the exchanger. The temperature profiles show how the flow is first parallel flow and then counter-current. Notice the baffles placed in the flow of the shell-side fluid, causing it to flow a tortuous path. This arrangement increases the turbulence of the shell-side flow and helps to "scrub" away the boundary layer resistance to heat flow. In other words, the turbulence increases the heat exchange rate. It is unfortunate that in promoting turbulence the resistance to flow is also increased and a larger, more expensive pump or compressor is needed to push the shell-side fluid around the baffles.

There is another effective way to increase the heat transfer rate, but it is seldom used because of its high cost. Remember, heat flow depends upon the area available for heat transfer as well as the temperature difference. The area can be increased by using "finned" surfaces. If you are unfamiliar with the idea of extending surface area for heat transfer, look at the fins on the outside of an air-cooled motorcycle engine.

Finally, let us consider the appearance and operation of a modern heat exchanger. It is between 8 and 20 feet long, although larger and smaller heat exchangers have been built. Its shell diameter varies from 1 foot up to 8 feet or even greater. It is almost certainly multipass; that is, the heads are divided so that the tube-side flow is diverted backward and forward through the shell. Figure 5-27 shows a two-pass heat exchanger. Four- and even eight-pass heat exchangers are used according to the designer's particular way of solving the problem. The shell space is arranged so that baffles are placed quite close together and in such a way as to cause the shell-side liquid to traverse a tortuous path as it goes from input nozzle to output nozzle. Sometimes the designer will arrange the

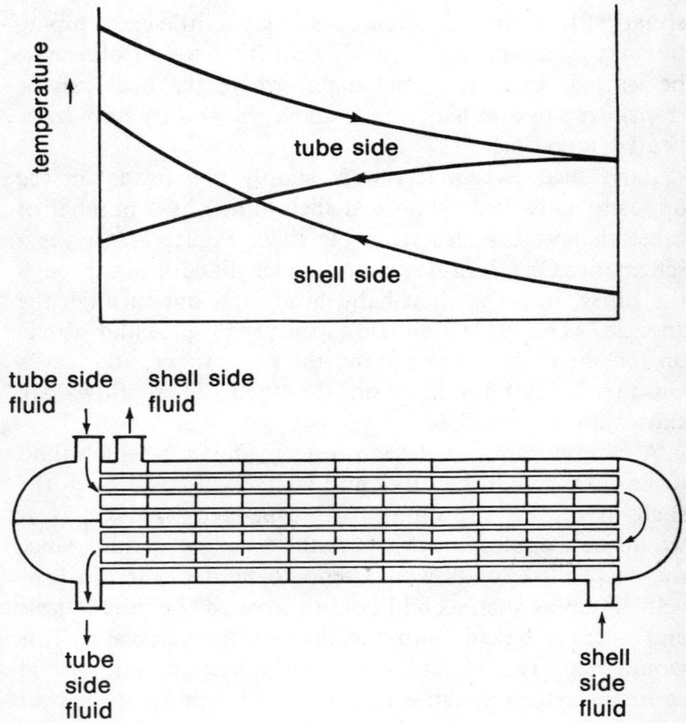

Figure 5-27. *A two-pass heat exchanger.*

baffles and shell arrangement so the the shell-side fluid performs more than one pass.

The flowsheet in Figure 5-25 shows how heat exchangers are used to extract heat from the straight-run fractions of kerosene and gas oil and the bituminous bottoms stream from the fractionator. The crude oil enters the plant and is first heated on the cold side of a heat exchanger against the straight-run kerosene. Then the warm crude is passed through the cold side of a heat exchanger against the straight-run gas oil. Finally, it is passed through an exchanger whose hot side carries the bottoms stream. Why this particular order? The answer is simple: The temperatures in the fractionator decrease from bottom to top. Therefore, the bottoms stream is hotter than the gas oil, which is hotter than the kerosene. Arranging the heat exchangers as shown in Figure 5-25 means that there will be the maximum temperature difference in each heat exchanger.

Both the side streams and the bottoms stream leave the fractionation part of the plant and pass on for further processing into fuels, lubricants, and petrochemicals, the subjects of Chapters 6 and 7.

Summary

You have seen how oil originated as a complex mixture of hydrocarbons with small amounts of sulfur and other organic remains. We have considered the fluid flow characteristics that it shares with other liquids. You saw how the Reynolds number, a dimensionless group, was useful not only in predicting the type of flow in a pipe but also in calculating the friction factor to predict the pressure drop that the friction would cause. You then learned the full significance of the phenomenon known as boiling, and how mixtures of different boiling points can be fractionated into different streams of successively higher boiling ranges working down the column. In the next chapter you are going to see how the four side-streams from a typical fractionator can become the feedstocks for a variety of fuels and lubricating oils.

Further Reading

British Petroleum. *Our Industry Petroleum*. London: British Petroleum Company, Ltd., 1970.

Brown, G.G. *Unit Operations*. New York: Wiley, 1960.

Cook, G.A. *Survey of Modern Industrial Chemistry*. Ann Arbor: Ann Arbor Science, 1975.

Kern, D.Q. *Process Heat Transfer*. New York: McGraw-Hill, 1950.

Nelson, W.L. *Petroleum Refinery Engineering*. New York: McGraw-Hill, 1958.

Peters, M.S. *Elementary Chemical Engineering*. New York: McGraw-Hill, 1954.

Rase, H.F. and M.H. Barrow. *Project Engineering of Process Plants*. New York: Wiley, 1957.

Robinson, C.S. and E.R. Gilliland. *Elements of Fractional Distillation*. New York: McGraw-Hill, 1950.

Royal Dutch Shell. *The Petroleum Handbook*. London: Shell International Petroleum Company, Ltd., 1959.

Thompson, E.V. and W.H. Ackler. *Introduction to Chemical Engineering*. New York: McGraw-Hill, 1977.

Chapter 6

Production of Fuels and Lubricating Oils

In this chapter you will learn how the straight-run products of crude-oil fractionation are further processed to the fuels that power the wheels of industry and the lubricating oils that keep those wheels running smoothly. Figure 6-1 shows a greatly simplified flow diagram of how the four straight-run fractions are further processed. Notice how the final gasoline, which is sold as motor spirit for automobiles, comes from all four of these fractions. Although the automobile fuel obtained by refining the straight-run gasoline is perfectly adequate in composition, there simply is not enough of it to meet the enormous demand. On the other hand, if the world's fuel industry were to process sufficient crude oil to meet the demand from the straight-run gasoline fraction alone, then the surplus products in other fractions would either have to be stored on an enormous scale or burned. Clearly, neither of these options is practical nor desirable, and so means have been developed to convert some of the surplus into gasoline.

Figure 6-1 shows schematically the production of the various fuels and lubricating oils from the straight-run fractions. Though this diagram looks somewhat complicated, it is in fact highly simplified from the flowsheet that would represent a modern refinery complex. This is because the actual flowsheet contains pumps, compressors, heat exchangers, storage tanks, and many other types of equipment which, although they perform vital roles in the plant, are not considered major operations for the purpose of this chapter.

Straight-Run Gasoline

The top fraction from the column, straight-run gasoline, is sometimes referred to as a *light distillate fraction* to separate it from those side streams from the middle part of the fractionator, which are called *middle distillate fractions*. The straight-run gasoline is mixed with light hydrocarbons produced by processes (described later in this chapter) carried out on the middle distillates (kerosene and gas oil fractions) and the residue. The resulting mixture is passed through a distillation column called the *light distillate column*, in which the temperatures are so arranged that the bottoms stream contains mainly kerosene and the overhead product contains hydrocarbons with boiling points lower than the kerosene range. The light distillate column, then, is simply a means of catching any kerosene that had been carried higher in the column than the straight-run kerosene side stream.

The overhead product is taken to another distillation column called a *depropanizer*, where the temperatures are carefully arranged such that the overhead product contains predominantly propane and lighter gases. The bottoms stream from the depropanizer is taken to a similar distillation column called a *debutanizer*, whose overhead is predominantly butane and any remaining light hydrocarbons. The bottoms product from the debutanizer is a mixture of hydrocarbons in the gasoline range. For purposes of identification when we return to discuss the further processing of this stream, let us call it a crude gasoline stream and rejoin its progress later.

The overhead products from the depropanizer and the debutanizer contain butane, butylene, propane, propylene, ethane, ethylene, methane, and some other light gases including hydrogen and, perhaps, hydrogen sulfide. Figure 6-2 shows the chemical structure of some of these compounds. The crude gasoline stream is split into two. The actual percentage of the split depends upon market conditions. In a sales area where the propane and butane have a high value due to great demand for bottled gases, perhaps very little further processing will be carried out.

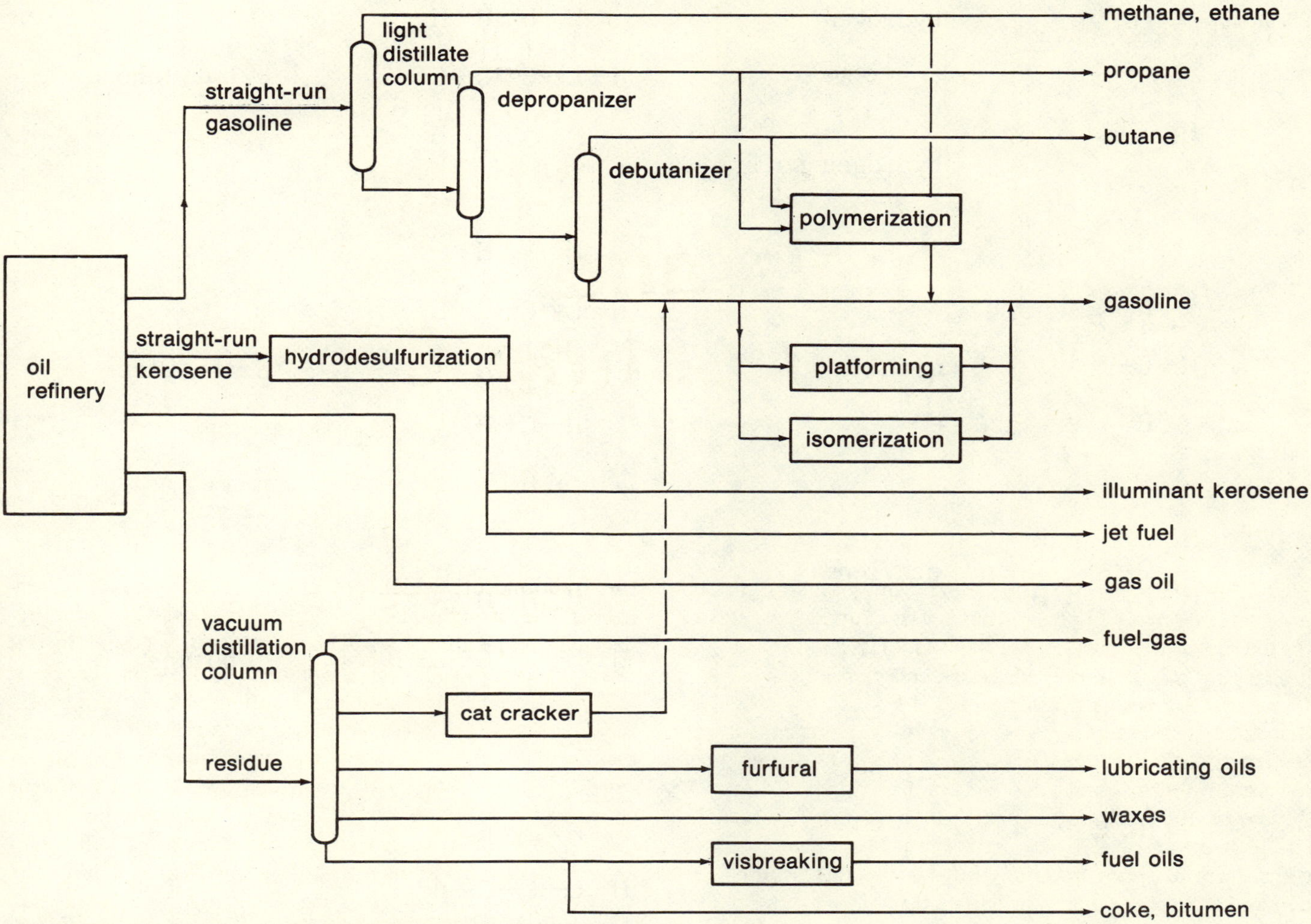

Figure 6-1. *The fuels and lubricating oils industry.*

Conversely, if there is a great demand for gasoline and a low demand for propane and butane, it is good economic practice to convert the low-value propane and butane into higher-value gasoline.

Polymerization

Polymerization is an operation whereby the smaller molecules of butane and propane are joined up with each other to make the larger molecules of hydrocarbons in the gasoline range. The process works by feeding the debutanizer overheads into contact with sulfuric acid (about 70% concentration) at about 105°F. The sulfuric acid behaves as a catalyst. We will be meeting many catalytic reactions in this and the next chapter, so let's pause to consider the action of these agents.

Catalysts puzzled early scientists who noticed that although the catalyst was unchanged at the end of the reaction and none had been used in the reaction, the reaction would not occur without it or, at best, would proceed much more slowly. It is now known that the catalyst does take part in intermediate reactions which take some of the catalyst to form intermediate products. Then, the intermediate products react with the catalyst, each other and the fresh feed to produce more intermediate products. However, the conditions in the reactor vessel (see the discussion on reactor design in Chapter 7) are designed to encourage the intermediate substances to undergo final reactions that produce the original catalyst and the products sought. It is as though nature for some reason cannot go the direct route from reactants to products and so chooses a more complicated route which temporarily makes use of the catalytic agent.

Even though the intermediate reactions have been studied and are understood in some cases, it is conventional to set down the chemical equations to relate only the reactants and products. Figure 6-3 shows the polymerization reaction as though the catalyst had no part to play. In the sulfuric acid polymerization process the molecules of iso-

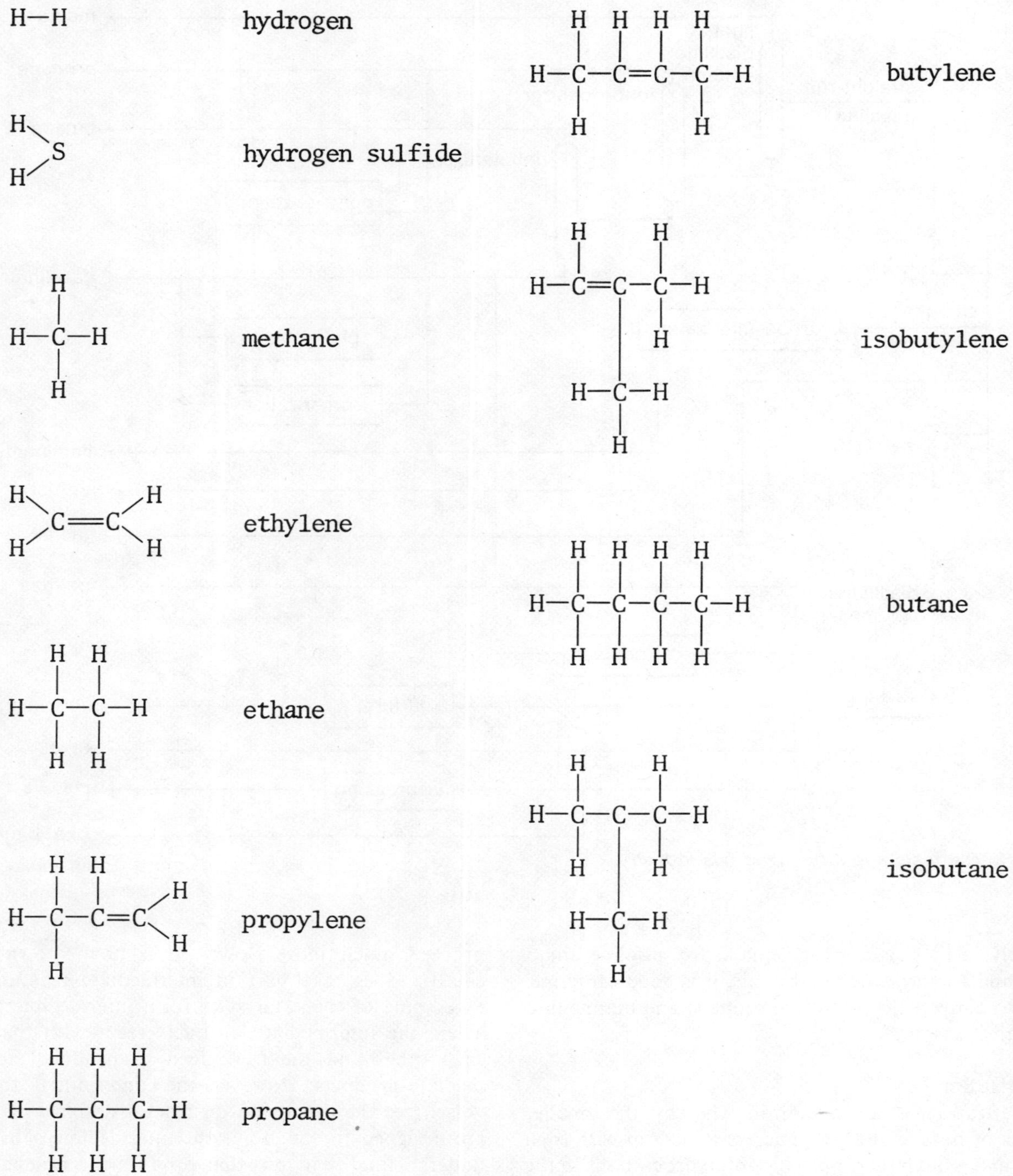

Figure 6-2. *The molecular structure of some overhead products from the debutanizer.*

butylene in the feed react with the acid to form complex molecules that remain dissolved in the acid. The unreacted hydrocarbons are then separated from the acid, which is heated to about 185° F, and the complex molecules break down into isobutylene polymers (the products shown in Figure 6-3) and the original sulfuric acid, which is returned to the reactor.

The polymers are passed through a tower where they are sprayed with sodium hydroxide to neutralize any sulfuric acid remaining in them. Finally, they are taken to a small fractionator where the polymers are separated. That compound which is made up of just two isobutylene molecules is called the *di*mer, and it is separated from the *tri*mer, which is made from three isobutylene molecules.

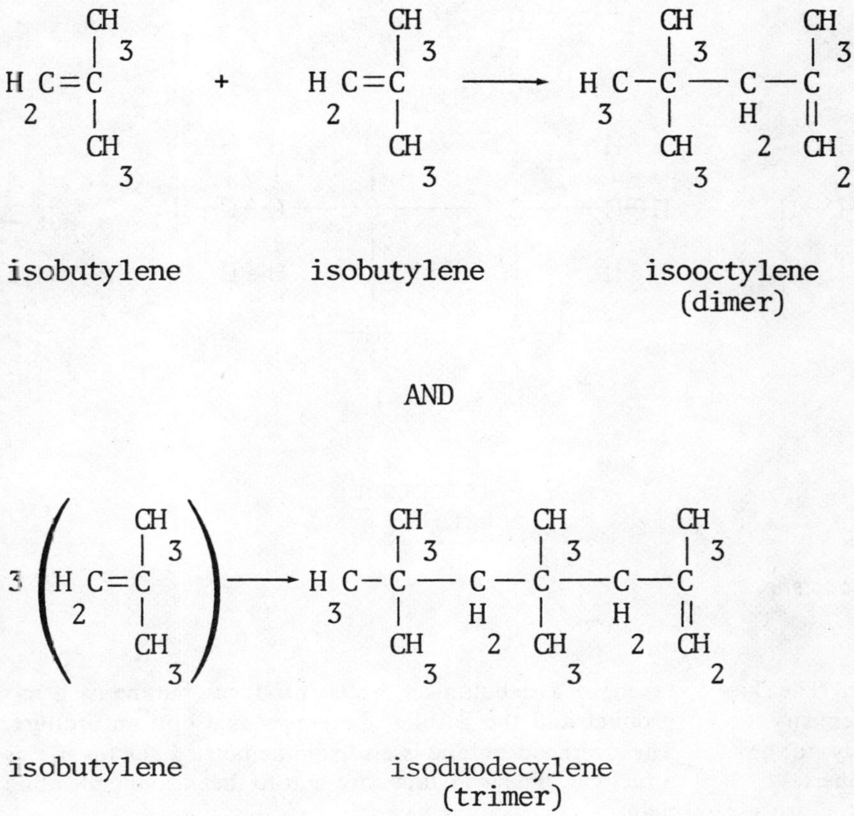

Figure 6-3. *The polymerization of isobutylene.*

The dimer contains hydrocarbons having eight carbon atoms that are not in a straight chain like ordinary octane but rather are in branched chains as seen in Figure 6-4, in which the two types are compared. These are called octylenes, and they can be *hydrogenated* (see Chapter 7) by mixing them with hydrogen and passing them over a catalyst to form a product rich in isooctane.

Why bother with all this? The reason is the very ready market for gasoline in all countries in the world. But gasolines vary, and the most important property as far as the consumer is concerned is called the *octane number*. A gasoline with a high octane number can be used in high-performance internal combustion engines since it burns very quickly and very smoothly and is not subject to the bad running phenomenon called "knocking." Lower-octane-number gasolines, if used in high-performance engines, tend to ignite spontaneously before the ignition spark occurs at the spark plug. When this happens, a flame front begins at the point of spontaneous ignition and, similarly, an ignition front is initiated by the spark. The two or more flame fronts flash through the volume of air and gasoline, and when they hit each other they cause a loud "ping" or knocking noise; hence the name of this phenomenon. The knocking explosions, which may occur on every revolution of the engine, are often violent and can even cause the piston or spark plug to shatter in extreme cases.

Isooctane and other substances, predominantly those called *aromatics*, which I will describe later, have high octane numbers, and amounts of these substances are blended into gasoline according to its purpose. In your local service station, the premium grade gasoline has a high octane number because of the addition of some of these materials. Regular gasoline has hardly any of the additives and instead relies upon the addition of a compound called tetraethyl lead, which improves octane number without the addition of the more expensive hydrocarbons. Lead-free gasoline must, by federal law, not contain lead, and it has been given some of the additives which increase octane number (although not so many as in premium grade). Since the isooctane and aromatic additives are costly compared to the tetraethyl lead, the price of lead-free gasoline is higher than regular grade.

The sulfuric acid polymerization process, then, is a means of converting one of the products of fractionation into another judged to be more valuable in the geographical area where the refinery is situated. Before we leave this operation known as polymerization, it is worth mentioning the "phosphoric acid polymerization process," which also produces high-octane additives for gasoline. The catalyst used in this process is phosphoric acid. To promote the maximum area of contact between the phosphoric acid and the feed stream (the overhead from the debutanizer), the phosphoric acid is soaked into pellets of

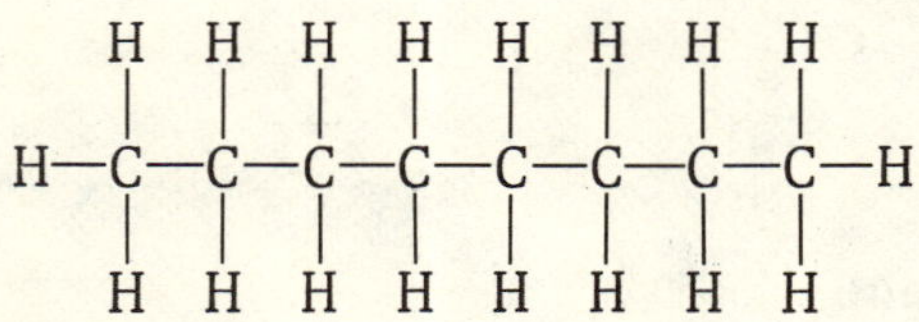

normal octane
C_8H_{18}

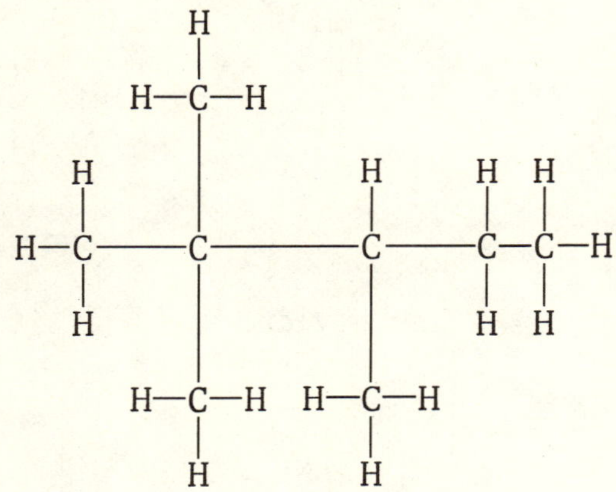

isooctane
C_8H_{18}

Figure 6-4. *Normal octane and isooctane.*

a naturally occurring clay called kieselguhr. The kieselguhr plays no part whatsoever in the chemistry occurring in the reactor: its function is merely to be a support for the acid and to be chemically inert.

Figure 6-5 shows a schematic flowsheet of the process. The feed is passed through the cold side of a first heat exchanger and then preheated in a second one that has hot steam running through its tubes. The hot feed then passes through a packed bed of the catalyst described above. The butanes and propanes react with the phosphoric acid and each other to form intermediate products which then break down, while still in the reactor, into the seven- and eight-carbon-atom compounds with the complex side chains we call the isopentanes and isooctanes. The products are now even hotter than when they entered the reactor because the polymerization reactions have an effect of releasing stored chemical energy which ends up as heat. Incidentally, reactions that do this are called *exothermic*, as opposed to *endothermic* reactions, which need to be supplied with heat that they convert into chemical energy.

The product stream from the reactor is passed through the hot side of the feed heat exchanger and then into a depropanizer, which removes unconverted propane as overhead product to be mixed with other propane on the site and probably sold in steel bottles as a domestic fuel. Consumers buy propane for a wide variety of purposes, perhaps the most familiar of which is its role as a bottled gas fuel in farms, ranches, and other outlying places that do not have a source of city gas.

The bottom stream from the depropanizer is split and some is heated in a steam reboiler and injected back into the bottom of the column. The bottoms product stream is taken to a debutanizer which produces butane as a top product and the gasoline polymers as a bottom product. The overhead butane is also sold as bottled gas for use as a fuel gas. The isooctanes are sent to the gasoline blending plant.

The overhead from the original debutanizer was split into two. The part that is *not* polymerized as described enters a depropanizer with conditions arranged to give a butane bottoms product and lighter gases as overhead. The overhead product is taken to a final distillation column where the bottoms product is propane and propylene. These two are separated later. The propylene is sold as a useful feedstock in the petrochemical industry (Chapter 7) and the propane joins the propane stream for sale as bottles gas. The overhead product from the final column is a fuel gas which comprises methane, ethane, ethylene, and some hydrogen. The ethane and ethylene may be separated and used as feedstocks to make polyethylene and the remaining gases used on the refinery site as a fuel gas in pipe stills, furnaces, and steam boilers.

Returning to the crude gasoline stream that we left earlier, we find that there is little processing left to do. All that needs to be done is to pass this stream, together with any other surplus products in the same boiling range from other parts of the plant, to a *gasoline splitter*, where the overhead product is a light gasoline and the bottoms product is a heavy one. These are both sent directly to the gasoline blending plant.

Straight-Run Kerosene

Now let's see what happens to the kerosene straight-run fraction—the lighter of the two middle distillates from the

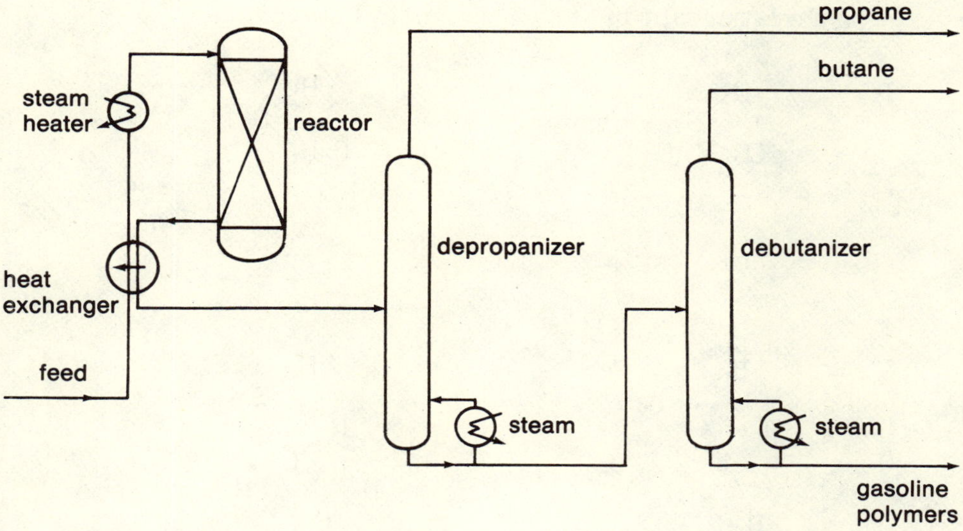

Figure 6-5. *The phosphoric acid polymerization process.*

fractionator. The required end-products from the kerosene fraction are called *illuminant kerosene* and *vaporizing kerosene*. The first type is sold for use in lamps. The bulk of the kerosene, however, is made into the vaporizing kerosene used as jet fuel and for which there is an ever-increasing demand.

Desulfurization

It is a simple matter to split the kerosene into two fractions, their relative sizes being determined by the demand for each, and processing the illuminant fraction by removing the aromatic components to prevent a smoky flame. But there is another undesirable substance in the kerosene—sulfur—which must be removed from both kinds of end-product kerosene before the separation process. The undesirable sulfur is combined in lower-boiling materials called mercaptans, which are chemically similar to hydrogen sulfide as shown in Figure 6-6, and most of the similarities they share are undesirable. First, these substances have an odor reminiscent of rotten eggs, *very* rotten eggs. Another unpleasant characteristic of these compounds is that they are very corrosive to metals. Since the bulk of the kerosene will be used in finely engineered jet engines in aircraft where engine failure due to corrosion or anything else is most unwelcome, the sulfur-bearing compounds must be either removed or broken down so that the sulfur can be removed.

The process developed to carry out the desulfurization of middle distillates such as this straight-run kerosene is called the *hydrodesulfurization refining process*, also referred to as *hydrofining*. Figure 6-7 shows a flowsheet. The principle is to react hydrogen with the "sour" oil (an adjective commonly used to signify that the oil contains more than just a tiny trace of sulfur). The sour oil and its dissolved hydrogen are heated on the cold side of a heat exchanger and passed through a furnace that heats the stream up to about 700° F. The feed pump which moves the stream produces a pressure of 800 psi. The highly compressed oil and hydrogen pass through the reactor, which is a vertical steel tower containing steel racks bearing a packing of small blue pellets. The pellets are made by mixing cobalt and molybdenum oxides with alumina (aluminum oxide) and water to form a paste which is then "extruded" down a tube, the end of which is perforated with small holes about ⅛ inch in diameter. Each of the small holes behaves like a toothpaste tube, and the paste comes out in a long "snake" that is immediately dried and broken up to make the small pellets.

The cobalt and molybdenum oxides contained in the pellets behave as a catalyst for reactions in which the sulfur leaves the mercaptan molecules and reacts with the hydrogen to make hydrogen sulfide. It is astonishing to me that in spite of the very high temperature and pressure, and in spite of the presence of three metal oxides, any of which could behave as a catalyst, very little else occurs in the reactor. Hardly any of the hydrocarbons are affected. Hydrofining is an example of how extremely specific catalytic reactors can be. And I am equally astonished that anyone, by accident or guided research, should have stumbled upon the particular temperature and pressure conditions which, with this specific catalyst, will do the required trick.

The products leaving the reactor are a mixture of sweet oil and hydrogen sulfide. The stream is very hot and so it passes through the hot side of the feed heat exchanger. Incidentally, the reaction is exothermic and the extra heat is usually sufficient to heat up the feed stream all the way to 700° F without use of the furnace after the initial "start-up." Under these circumstances after start-up, the furnace is taken out of the feed line by an arrangement of valves.

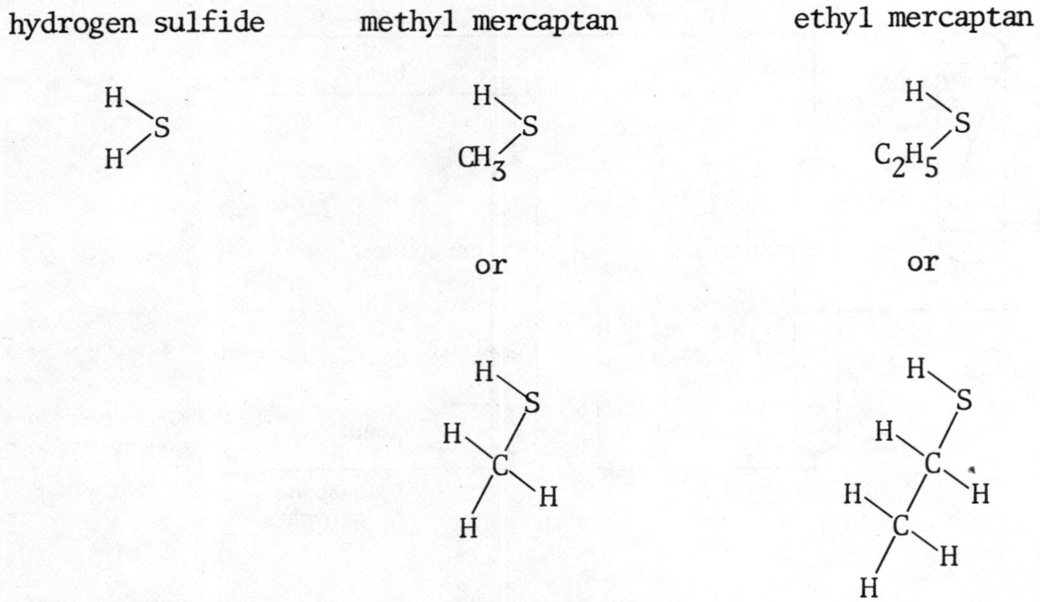

Figure 6-6. *Some sulfur compounds in petroleum products.*

The products leave the feed heat exchanger and are cooled against water in another heat exchanger before falling into a horizontal drum called a *separator*. The sweet oil remains in the drum, but the hydrogen and hydrogen sulfide go overhead, and after separation (described below), the hydrogen is compressed and mixed with "makeup hydrogen" before being mixed with the fresh sour oil feed. A constant drain of sweet oil from the separator is taken to a stripper which is analogous to the lower half of the fractionator described in Chapter 5. In the stripper the feed stream enters at the top of the column and, together with the reflux, descends as a liquid. Intimate contact occurs with the vapors from the reboiler, resulting in removal from the sweet oil of the low-boiling hydrocarbons that were formed by the decomposition of the mercaptans and thiophens. Dissolved in the sweet oil, there may be some lighter hydrocarbons resulting from decomposition of some of the remaining hydrocarbons.

Rectification

The overhead product is cooled and passed through a *reflux drum*, where the liquid is separated out into two streams: a reflux and a product stream of low-boiling hydrocarbons. The material that leaves the reflux drum as a gas is used on site as fuel gas. The low-boiling hydrocarbons stream must be separated further before the individual products can be taken from it, and so this stream is taken to the straight-run gasoline stream, where it is processed to give propanes and butanes as you saw earlier.

The reason for passing the sweet kerosene through this stripper is that if the low-boiling hydrocarbons are left in the kerosene, they would lower the temperature at which the kerosene begins to boil. There is often a need to remove low-boiling hydrocarbons from a product stream for this purpose. The operation, which uses a stripper as you have just seen, is called *rectification* or, sometimes, *stabilization*. Fuel oils, including gasoline, would be very impractical products if they began to boil at low temperatures—gasoline might boil in the tubes leading it from the gasoline tank to the carburator and cause "vapor lock": an annoying and unnecessary cause of breakdown for the automobile driver. The stabilization process removes the low-boiling hydrocarbons and the product quality is improved accordingly.

Absorption

The sweet oil leaves the refining plant after being cooled as shown in Figure 6-7. Going back through the process you will find that there is a mysterious box in the flowsheet in Figure 6-7 which is called "Girbotol" and which has the function of separating the hydrogen sulfide from the stream of hydrogen to be recycled.

The Girbotol process is shown in flowsheet form in Figure 6-8. The gaseous feed is blown in the base of an absorber and it rises through a number of bubble cap trays in exactly the same way as those used in the fractionating column for crude oil. However, since the feed gas is well above its boiling point, it does not condense, and most of it leaves the top of the column as a purified gaseous product as shown in Figure 6-8. The liquid descending the column is a *solvent*. It is chosen to have the property of dissolving out one specific type of product (in our case, hydrogen sulfide) without reacting in any way with the remainder of the stream. As the solvent cascades down from tray to tray, the "sour" hydrogen bubbles up through it. Again, the adjective sour indicates the presence

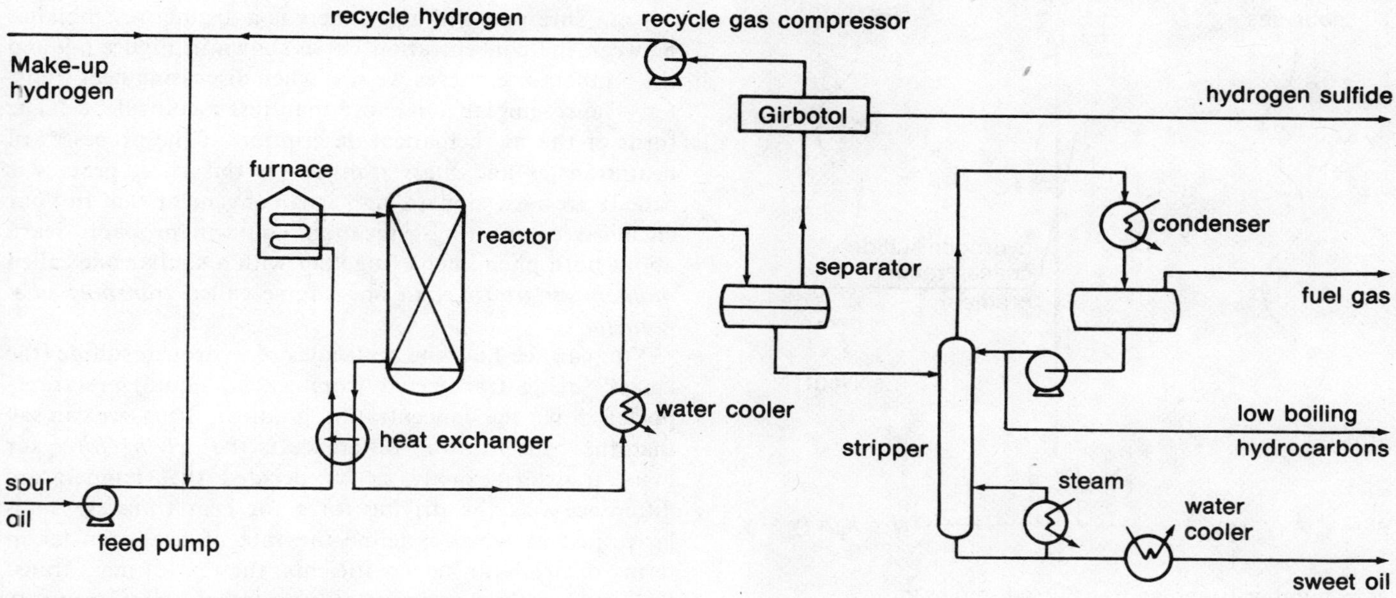

Figure 6-7. *The hydrodesulfurization process.*

of a sulfur compound. The hydrogen sulfide is transferred from the sour gas into the solvent stream (diethanolamine, which meets the above specification for a solvent in this case). Let us examine in detail how the hydrogen sulfide is transferred from the gas medium to the liquid medium. In many ways the phenomenon is similar to heat transfer described in earlier chapters.

Consider the diagram in Figure 6-9. It shows an imaginary point in the absorber where the liquid solvent is flowing down alongside the sour gas, which is shown ascending on the left of the diagram. The thick line down the center represents the interface between the two phases. In the gaseous phase the molecules of hydrogen sulfide and hydrogen are moving around at great speed, as you have seen before in a gaseous medium. In the liquid medium the diethanolamine molecules and any dissolved hydrogen sulfide molecules are moving much more slowly, as you know to be the case in liquids. The fast-moving hydrogen sulfide molecules, by sheer random probability, occasionally gain sufficient speed to find themselves hitting the gas side of the interface. This probability obviously increases the closer and closer we get to the interface.

When a hydrogen sulfide molecule hits the interface, it passes straight through—notice the similarity to the earlier discussion on vapor pressure in Chapter 5. The difference

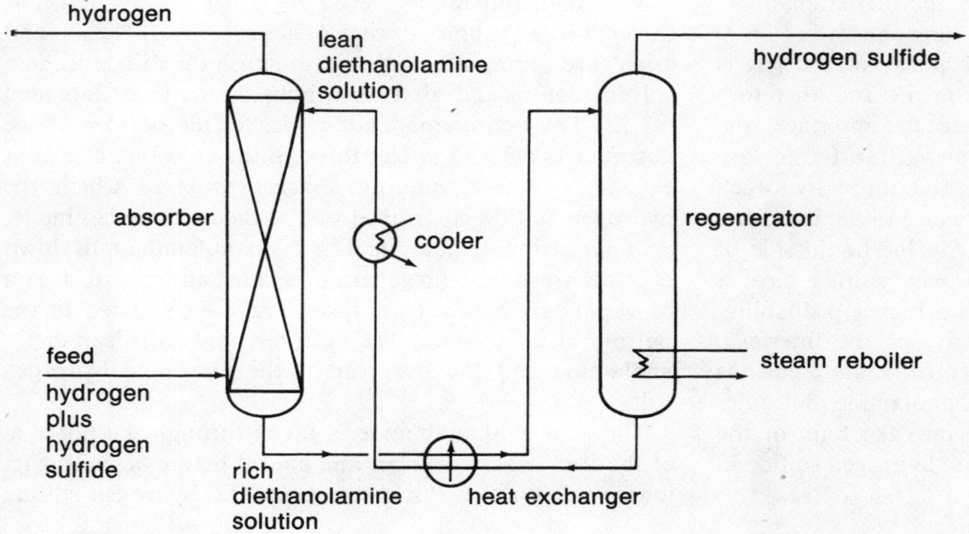

Figure 6-8. *The Girbotol process.*

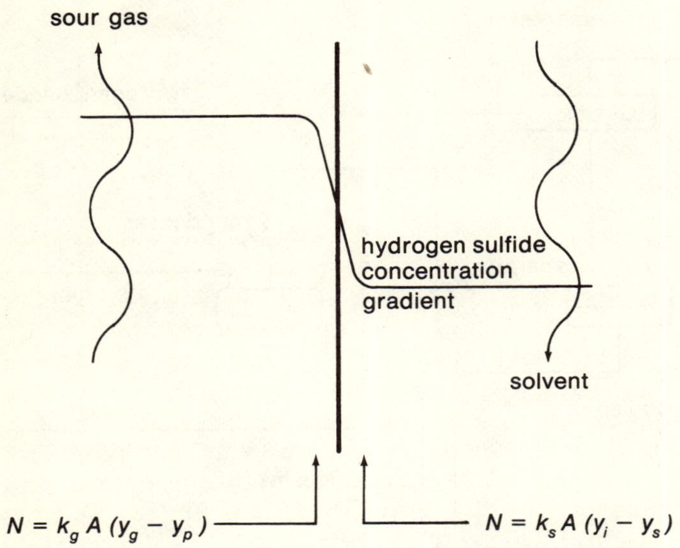

N = number of moles transferred, moles/hr

k_g = mass transfer coefficient in the gas side of the interface, moles/ft^2 hr

k_s = mass transfer coefficient on the solvent side, moles/ft^2 hr

A = area available for mass transfer, ft^2

y_g = mole fraction of H_2S in gas

y_i = mole fraction of the H_2S at interface

y_s = mole fraction of H_2S in solvent

Figure 6-9. *Mass transfer.*

here, though, is that the hydrogen sulfide molecules prefer being in the liquid phase. The reason for this is that the attractive forces between the hydrogen sulfide molecule and the molecules of diethanolamine are stronger than the ones that existed between the hydrogen sulfide molecule and the hydrogen, which is another way of saying that the hydrogen sulfide is more soluble in the diethanolamine.

Now look at the concentration gradient shown in Figure 6-9. The concentration in the bulk phase of the gas is almost flat because its flow is turbulent and therefore mixed efficiently. However, closer to the interface, the probability of the hydrogen sulfide being transferred into the liquid stream becomes higher, and more hydrogen sulfide molecules leave the thin layer close to the interface, which is called the *boundary layer*. On the liquid side of the interface the molecules of hydrogen sulfide have a much lower velocity, and so there is a higher probability that they will stay within the vicinity of the interface. Gradually, however, they will undergo sufficient collisions with other hydrogen sulfide and diethanolamine molecules to migrate away from the interface into the bulk of the liquid stream and leave room for more hydrogen sulfide in the liquid boundary layer.

I am sure you will agree there is a distinct resemblance between the concentration curve shown in Figure 6-9 and the temperature curves we saw when discussing heat transfer. The resemblance is more than just a coincidence. The form of the mathematical description of the processes of heat transfer and *mass transfer* (as this latest process is called) are also similar, and it is convenient that in your chemical engineering education you will probably learn about both phenomena (together with a similar one called *momentum transfer*) in one course called *transport phenomena*.

You can see how the molecules of hydrogen sulfide (the "mass" being transferred from gas to liquid) are transported down the concentration gradient. Thus, we can say that the concentration difference is the *driving force* for mass transfer, exactly as we decided that temperature difference was the driving force for heat transfer. Similarly, just as we can define the rate of heat transfer in terms of *heat* transfer coefficients, the rate of mass transfer can be defined in terms of *mass* transfer coefficients. A major difference here is that although a heat transfer coefficient on the gas side of a heat exchanger is much lower than that on the liquid side, the reverse is true of mass transfer coefficients. In the Girbotol process, the mass transfer is slower on the liquid side of the gas-liquid interface.

One other point to note here is the area term in the mass transfer equation shown in Figure 6-9. As with heat transfer, the rate of mass transfer is directly proportional to the area of contact between the gas and liquid phases. Obviously, it is desirable to make the mass transfer rate as high as possible. This is done by promoting turbulence to cause a scrubbing effect on the two boundary layers and by increasing the area of contact. Absorbers can be made to have a high area of contact by using bubble-cap trays or by filling an empty column with packing over which the liquid spreads. Porcelain shapes are usually chosen for the packing because they are impervious to chemical attack.

The diethanolamine becomes richer as it descends through the bubble tower. (The adjectives "lean" and "rich" are commonly used to distinguish the difference in a solvent before and after absorbing the material intended for it.) The rich diethanolamine leaves the bottom of the absorber tower and passes through the cold side of a heat exchanger before entering a *regenerator* in which the hydrogen sulfide is stripped out of the diethanolamine by a simple stripping column. The diethanolamine falls down over the trays in the regenerator against an upward stream of vapors from the steam-heated reboiler situated in the bottom of the column. The bottoms product is lean diethanolamine and the overhead is the unwanted hydrogen sulfide.

The lean diethanolamine is taken through the hot side of the feed heat exchanger and cooled before beginning its journey down the absorber again. The hydrogen sulfide joins hydrogen sulfide streams from other hydrofiners which are busy removing sulfur from gasoline, fuel oils, and various other products. The hydrogen sulfide could be burned to sulfur dioxide in a small furnace and then vented from a very tall chimney into the atmosphere. This is most undesirable since the sulfur dioxide would contribute to atmospheric pollution, and today all companies are most concerned with environmental protection. Industry is determined not to repeat the terrible damage inflicted upon our environment during and after the Industrial Revolution.

Sulfur dioxide is one of the worst pollutants, and for this reason alone it would not be vented to the atmosphere. But in a refinery where hydrodesulfurization is used extensively, the total production of hydrogen sulfide from the entire plant could easily be 60 tons per day! With these amounts, it is economic to turn the hydrogen sulfide into yellow, elemental sulfur using a simple process which works as follows. The hydrogen sulfide is burned in a closed vessel to sulfur dioxide with a limited oxygen supply. Instead of supplying sufficient oxygen to cause all the hydrogen sulfide to burn to sulfur dioxide, exactly one-third of that amount of oxygen is supplied. The product of this combustion is a mixture of two-thirds hydrogen sulfide and one-third sulfur dioxide. These react with each other to give elemental sulfur and water (Figure 6-10). The sulfur is dried and sold to chemical companies who use it to make sulfuric acid.

After hydrodesulfurization, the sweet kerosene is ready to be sent to the jet fuel mixing plant where it will have some lower grades of gasoline added to it and additives to keep it from oxidizing, or attacking the aircraft's metal fuel tanks, and prevent it from freezing in high-altitude aircraft. A small amount of the sweet kerosene is taken through a process that removes the aromatics (which smoke when burned) to make the kerosene suitable as an illuminant. This process is known as the Edeleanu process. Figure 6-11 shows a flowsheet. It exploits a basic principle of chemical engineering that we have not encountered before: liquid-liquid extraction, sometimes called solvent extraction.

Solvent Extraction

Solvent extraction is a mass transfer process similar in most ways to absorption except that the extract is taken from one liquid by another. A new word you need to learn here is *raffinate*, which is the name given to the product stream after the unwanted ingredients have been extracted away by the solvent. The feed of sweet kerosene is pumped into the base of the extraction tower. The lean solvent is pumped to the top of the tower and the two liquids begin their journey through the tower countercurrently; that is, one upward, the other downward. Since the tower is packed with mesh-like material, both liquid streams are broken up into tiny channels within the other liquid. Thus, there is a large area of interface surface between the two liquids. As you saw in the absorber, this will cause the migrating molecules of the extract to be offered every opportunity of passing from one phase to the other, in this case to the solvent from the sweet kerosene.

$$2H_2S + 3O_2 \rightarrow 2SO_2 + 2H_2O$$
$$\downarrow$$
$$4H_2S + 2SO_2 \rightarrow 6S + 4H_2O$$

Figure 6-10. *Making sulfur from hydrogen sulfide.*

The solvent is liquid sulfur dioxide made elsewhere in the plant by burning some hydrogen sulfide and cooling and compressing the gaseous sulfur dioxide product into a liquid. The concentration gradients away from the surface of the two liquids are more shallow than they would be if either of the phases were a gas. For this reason, mass transfer from liquid to liquid is a slow process, and therefore the column must be high and the contact between the liquid must be as intimate as possible hence the mesh packing which causes very tiny channels of droplets of one liquid in the other. As with absorption, mass transfer coefficients can be measured by experiment, and a mathematical description of the solvent extraction process gives us a means of designing the necessary equipment. Solvent extraction is not a particularly common operation and no further description will be given in this book.

The kerosene leaving the top of the extraction tower (sometimes called the solvent extraction unit) is taken to a stripper called the *raffinate solvent evaporator*. This is necessary because some of the sulfur dioxide solvent dissolved in the kerosene. The stripper removes it by allowing the raffinate to cascade down a bubble tower against the rising vapors from a steam reboiler set in the bottom of the stripper. The bottoms product is the raffinate which is an illuminant-quality kerosene; the overhead product is sulfur dioxide, which joins the bulk of the sulfur dioxide solvent to be recompressed and condensed into liquid in a heat exchanger on the cold side of which is liquid propane. The propane has been brought from some other part of the plant to do the cooling *and* because it needs to be heated before entering the particular process for which it is destined! Large plant integration means that this kind of advantage can be taken to save energy and materials—two birds with one stone.

The bottoms stream from the solvent extraction tower is the rich sulfur dioxide which now contains aromatics. The rich solvent is fed into an *extract solvent evaporator* which again is a stripper that works by trickling the rich solvent down the bubble column against the rising vapors from a steam reboiler in its bottom. The overhead solvent joins up with solvent from the raffinate stripper and is reused. The bottoms stream from the column is the extract—a range of aromatic compounds which can be added to gasoline to improve its octane number.

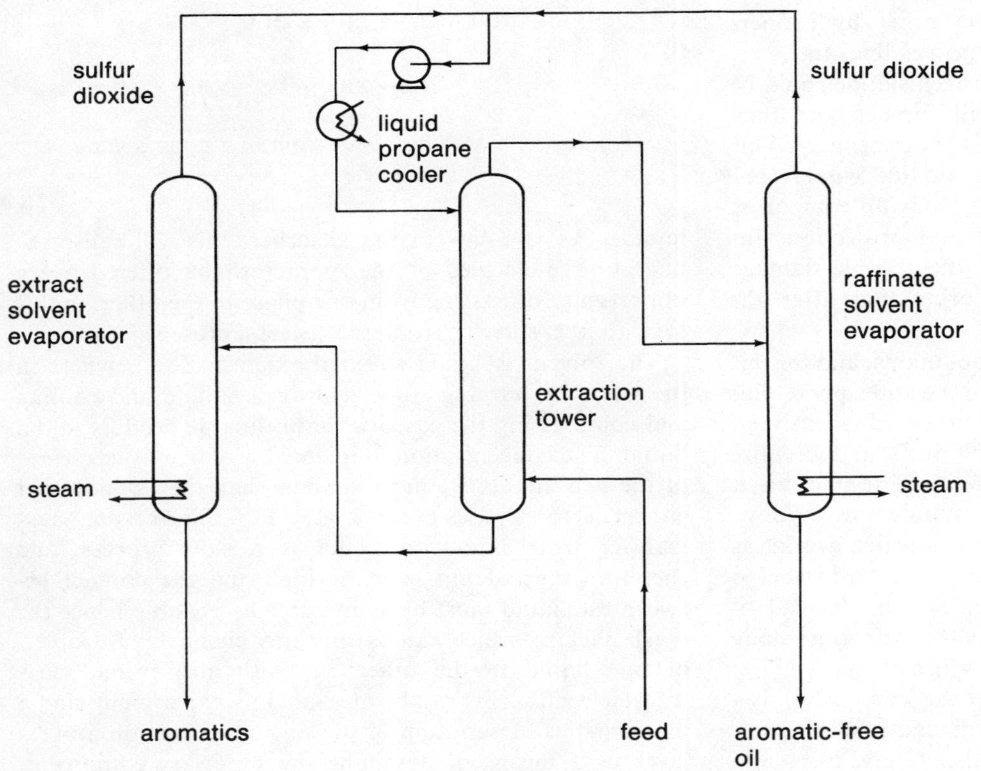

Figure 6-11. *The Edeleanu process.*

Straight-Run Gas Oil

Returning to the crude oil fractionator, we find that the other middle distillate side stream is the straight-run gas oil fraction. In the early days of oil refining, this oil was used to make gas for distribution to domestic users in cities, and this explains the origin of its name. These days, use of this stream depends upon the geographical location of the refinery. In some parts of the world the gas oil is *cracked* (see below) into extra gasoline and fuel gases. A more common use for this stream, however, is as a fuel for engines that operate on the *gas turbine* principle.

Figure 6-12 shows such an engine, and you can see how the gas oil is burned in a small combustion chamber using compressed air as the *oxidant*. As the oil burns it gives off its chemical energy as heat which heats up the combustion gases and causes them to expand rapidly. They are channeled through the "expansion zone" of the gas turbine, where they pass through rings of blades, alternately stator and rotor blades as in the steam turbine, and as they do they give up their pressure energy and some of their sensible heat to the turbine shaft in the form of power. The exhausted combustion gases still contain a lot of sensible heat, which is removed in a device called a *waste heat boiler*. This is best thought of as a conventional boiler in which the flame has been replaced by the exhaust gases from the gas turbine.

As the shaft of the turbine turns around, some of its work is used up in sucking in air from the atmosphere and compressing it to use as the oxidant within the turbine.

In some refineries the gas oil is further separated by distillation into heavy gas oil and a light gas oil which is blended with some of the heavier gasoline fractions to make diesel fuel for automobile, truck, and locomotive use. The heavier gas oils are used in gas turbines in ships and sometimes in stationary gas turbines to generate electricity.

The Fractionator Bottoms Stream

The final straight-run fraction from the crude oil fractionator is the residue or bottoms stream. This product once was used in its entirety to make lubricating oils and greases, heavy fuel oils for marine boilers, and fuel to be burned in steam boilers to raise steam. The remainder was a tarry residue called bitumen. Today the market for these products is less demanding than the market for automobile gasoline, so some of this bottoms fraction is converted into motor gasoline.

The Vacuum Distillation Column

After the stream leaves the bottom of the fractionator, it is further distilled in a *vacuum distillation column*. This is an interesting application of the distillation and vapor

Figure 6-12. *The gas turbine. (Courtesy of General Electric Co.)*

pressure principles you learned in Chapter 5. This black, tarry stream contains hydrocarbons with very high boiling points, the lowest of which is around 660° F. Any attempt to separate the thick oil using distillation at atmospheric pressure (14.7 psi) would require heating it to temperatures so high that the heavy, long molecules would decompose. The alternative is to lower their boiling points by distilling them in a column which itself is maintained under a "partial vacuum"; i.e., the pressure inside the column is chosen to lower the boiling point range of the hydrocarbons in the feed to around 650-700° F. The partial vacuum is maintained by means of a vacuum pump on the overhead stream.

A vacuum distillation unit has several interesting features, not least of which is the careful use of heat exchangers to heat up the feed stream using the sensible heat of three of the products (see Figure 6-13). The residue from the crude oil fractionator first is heated up against the light lubricating oil that comes from the upper, cooler regions of the vacuum distillation column; then it is heated against the heavy lubricating oils that come from the middle of the vacuum column; and finally it is heated against the hottest product, the vacuum distillation column bottoms—bitumen.

Before entering the column, the feed is further heated up to the chosen operating temperature (this depends upon the nature of the feed) and then it is allowed to enter the vacuum atmosphere of the column. Most of it boils immediately because of the reduced pressure, and the vapors rise quickly up the column against a descending cascade of reflux from the condensed overhead. As in the crude oil fractionator, a steady state develops in which the composition of the liquids on each tray in the column remains fairly constant. The overhead product is taken and condensed in a heat exchanger against cooling water. A pipeline from the condenser to the vacuum pump maintains the chosen vacuum pressure in the column. The condensate from the overhead product is split into a light gas oil fraction and a reflux which is returned to the column. We shall return in just a moment to the fate of this light gas oil.

There are two middle distillate streams taken from the vacuum distillation column in normal practice. The upper one of these is chosen such that the product is a light lubricating oil suitable for use in automobile and aircraft engines. The stream is taken from a tray in the vacuum column, steam-stripped of any remaining lower-boiling hydrocarbons in a side stripper, and after passing through the hot side of the first feed heat exchanger, it is sent to storage and subsequent sale. At a lower point in the column another side stream is taken off, stripped of lower-boiling hydrocarbons in a side stripper and cooled on the hot side of the second feed preheater. This stream contains heavy lubricating oils that can be separated further in a distillation column to give a heavy lubricating oil suitable for locomotives and a range of greases for a wide variety of lubrication jobs. The quality of lubricating oils is impaired if they contain aromatics, those compounds characterized by having a benzene ring in their molecular structures (which will be discussed shortly). To improve the lubricating oils leaving the vacuum distillation unit, the aromatics are extracted. The Edeleanu process could be used but it is more common to employ the Furfural process.

The Furfural Process

Furfural is a curious material. It is made by the action of dilute acids on bran and other waste fibrous material from the cereal industry. It is a liquid which boils at 324° F and has a high solubility for aromatics. Figure 6-14 shows the flowsheet. Essentially, the Furfural process is a solvent extraction operation. The lubricating oil feed ascends the solvent extractor tower against the descending furfural. Mass transfer of the aromatics occurs and they dissolve in

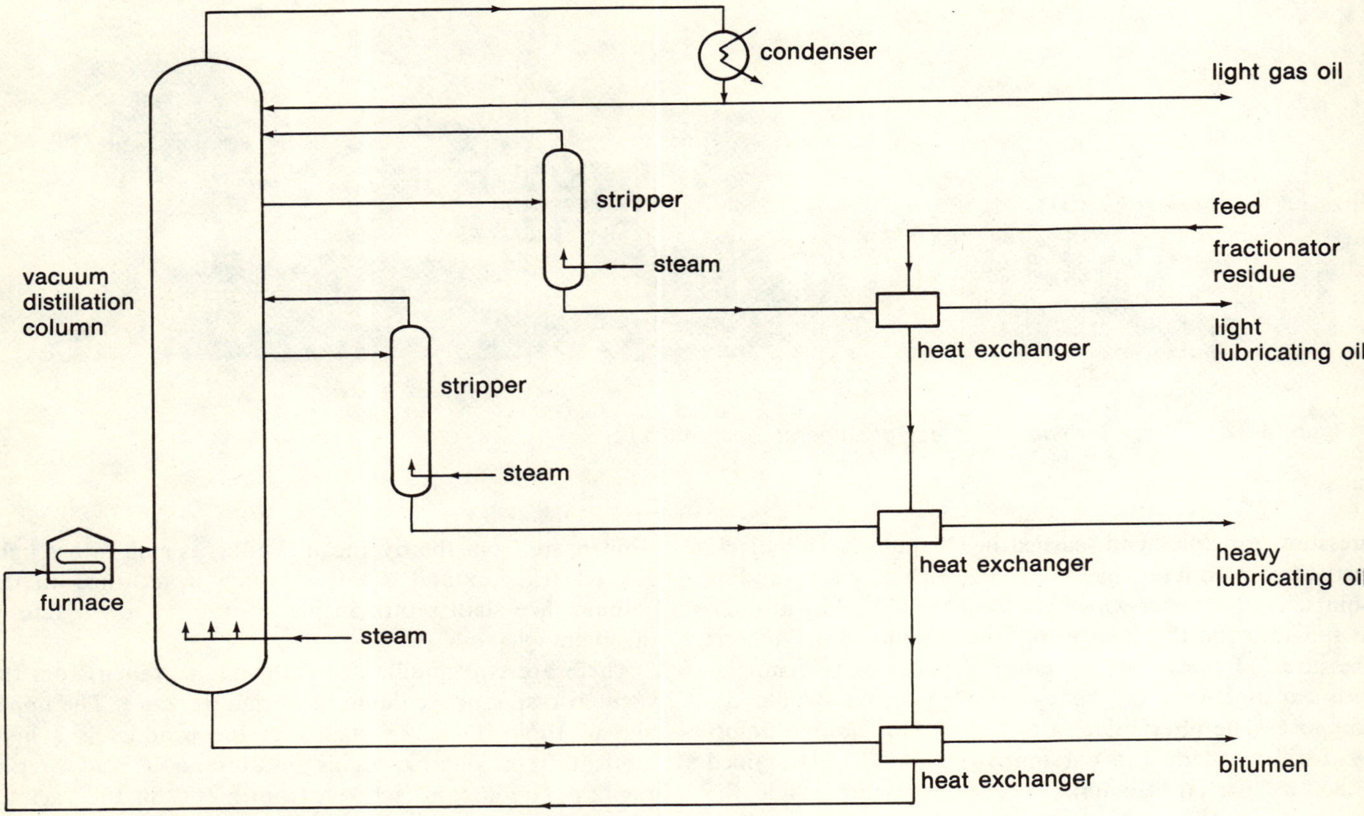

Figure 6-13. *A vacuum distillation plant.*

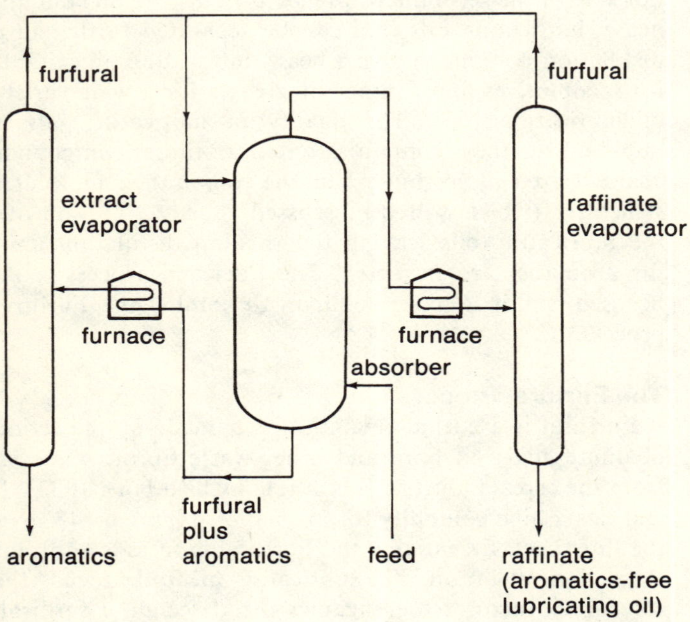

Figure 6-14. *The Furfural process.*

the furfural which leaves the bottom of the extractor, is heated in a furnace and then passed into a distillation column where the furfural goes overhead for recycling and the bottoms stream is the extracted aromatics. The raffinate (aromatic-free lubricating oils) has furfural in it and it is necessary to heat it and feed it to a raffinate evaporator (which is a distillation column). The overhead is furfural for recycling and the bottoms stream is the lubricating oil.

The Vacuum Distillation Bottoms Stream

At the bottom of the vacuum distillation column hot steam is injected. This has two purposes: first, it causes a reboiling action, sending vapors rising up the column; second, the hot steam itself joins with those vapors and helps to strip the lighter components from the descending liquid. Some of the bottoms liquid is drawn off as a thick, black bitumen which may be treated in several ways. Most commonly, the bitumen is divided into three separate streams, the proportions also depending upon local market conditions. The first stream is bitumen, and this is left unchanged and sold as “road tar”, the agent used to bind loose gravel together in the blacktop on our highways. A second stream is used to make "petroleum coke."

Coking

Refineries situated in areas where there is little demand for fuel oil will use a *coking process* to turn the heavy bitumen into coke. The bitumen is heated to about 950° F and passed into the bottom of a steel drum where the high temperature causes the large hydrocarbon molecules to break down or *crack* into smaller ones. As these complicated reactions occur some of the carbon atoms are stripped of all their hydrogen and are deposited as carbon on the inside of the drum. Over a period of time the carbon deposit builds up into a thick coke material, while the light hydrocarbons are pumped out of the drum and taken to a distillation unit that separates them into suitable boiling fractions which can then be added to appropriate streams in the refinery, such as the straight-run gasoline, kerosene, and gas oil fractions. Eventually, the coke drum is full and the feed is switched to a second drum. The first drum is allowed to cool, is opened up, and the coke removed.

If the original crude oil contains very low amounts of sulfur and ash (the remains of sand and silt laid down millions of years ago with the organic material), then the coke is mixed with a heavy oil binder and extruded like black, one-foot diameter toothpaste into long rods which are baked to set them into hard, graphitic electrodes suitable for use in the aluminum and steel industries in electric arc furnaces. If the petroleum coke cannot be used for electrode production, either because it contains too much sulfur or ash or because there is no local demand for electrodes, the coke is used as a fuel to raise steam for power on the refinery site.

Fuel Oil Thermal Cracking

The third stream from the bottom of the crude-oil fractionator is used to make more heavy fuel oil which will be sold to burn in boiler furnaces to raise steam for power. Since it contains high quantities of wax and is solid at atmospheric pressures and normal temperatures, it cannot be used just as it is. However, by using a mild cracking treatment, the waxy materials are broken down into lighter hydrocarbons that give the fuel oil product much better handling characteristics: the product will be liquid at atmospheric temperature. Although this could be done by mixing the fuel oil with some of the lighter gas oil, this is undesirable since the gas oil has a higher value.

A more common means of reducing this viscosity is to use a mild cracking action to break down enough of the heavy viscous compounds into lower-boiling, less viscous compounds. This process is known as *viscosity breaking* or more commonly as "visbreaking." The fuel oil product of visbreaking requires far less gas oil for dilution to a suitable viscosity. Visbreaking and coking are examples of an operation called "thermal cracking." This process is so called because large molecules are cracked or broken into smaller ones by the use of heat alone. The name thermal cracking also serves to distinguish this kind of cracking operation from another, called *catalytic cracking*, or more commonly, "catcracking."

Hydrocracking Light Gas Oil

Now let us return to the light gas oil which was the overhead product of the vacuum distillation column. The light gas oil stream is divided into two parts; and a cracking operation is performed on both. In both cases the primary objective is to crack the light gas oil into gasoline fractions that will be sent to the gasoline blending plant I will describe shortly. The first of these streams will be subjected to a cracking process known as *hydrocracking*. The hydrocarbons known as paraffins have a straight-chained molecular structure, which is typical of a *homologous* series. Figure 6-15 shows the first few members of four homologous series known as paraffins, olefins, acids, and alcohols. There are many more, but their structures are similarly arranged and each successively higher entry in each series has one more CH_2 group, as you can see in Figure 6-15.

Shown alongside each of the molecular structures of the paraffin series in Figure 6-15 is the atomic ratio of hydrogen to carbon. You can see that methane, CH_4, has a hydrogen-to-carbon ratio of 4. Pentane has a ratio of 12/5 or 2.4; decane has 22/10 or 2.2, and terdecane has a ratio of 62/30 or 2.067. The point here is that the hydrogen-atom-to-carbon-atom ratio decreases as hydrocarbons contain more and more carbon atoms. Imagine now that if we were to crack down the larger hydrocarbons (with a low hydrogen-to-carbon ratio), there would be a shortage of hydrogen atoms to make the lower ones. We saw what happened in a thermal cracking reaction in the coking process, where the scarcity of hydrogen atoms led to the deposition of solid carbon.

There is an alternative approach. If hydrogen is injected into the hydrocarbons before cracking, then high yields of either gasoline or gas oil (or both) will be obtained from the light gas oil and wax distillate from the lubricating oil part of the refinery. This process is called *hydrocracking*. It can also be used to make gasoline from gas oils. However, it requires extreme conditions. The reactors must be made of fine steel up to 8 inches thick in order to contain a pressure of 2,500 psi and a temperature of around 800° F.

Figure 6-16 shows a typical flowsheet for hydrocracking. The feed is a mixture of light gas oil and wax distillate. It is pumped to a pressure of 2,500 psi and about 12,000 cubic feet of hydrogen per ton of feed is injected into the stream, which then passes through the cold side of a heat exchanger and on through a furnace to be heated up to about 800° F. The hydrogen used for hydrocracking

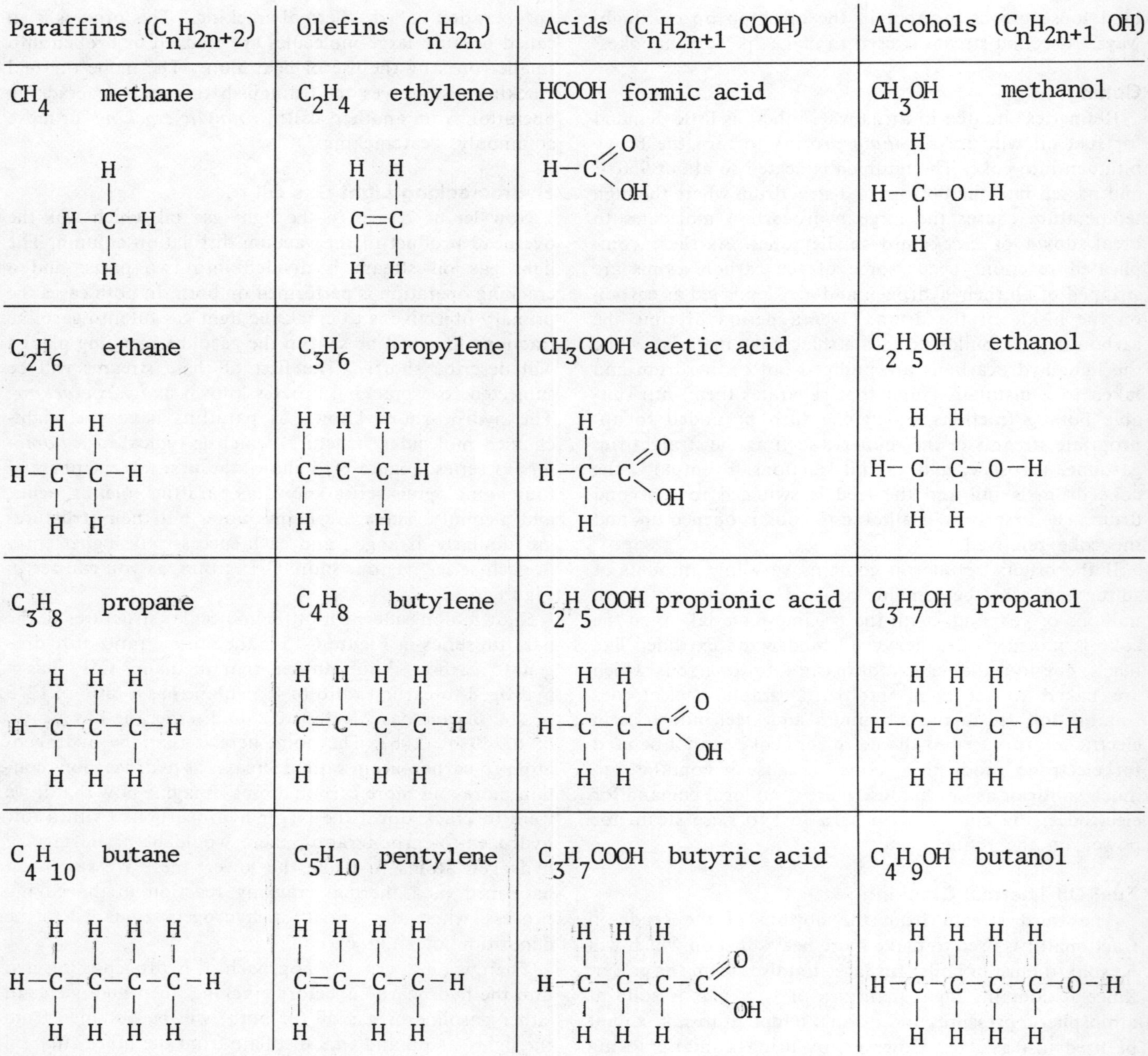

Figure 6-15. *Some homologous series.*

comes from other refinery operations such as reforming (see later). But there is seldom enough hydrogen from other refinery operations and in most cases a special hydrogen manufacturing plant has to be built to provide it.

The gas oil and wax distillate feed passes into the first reactor as shown in Figure 6-16. There a catalyst causes the desired cracking. The hot products pass through the hot side of the feed heater and are further cooled in a water-cooled heat exchanger before entering a drum in which the surplus hydrogen bubbles out of solution and is collected by a recycle compressor for recycling into the feed stream. The liquid product is sent to a stripper where a light fuel gas is taken off as overhead. The bottoms of the stripper are taken through a similar process which involves a secondary reactor containing a catalyst different from the first, so that slightly different cracking reactions occur. Again, the excess hydrogen is separated out for

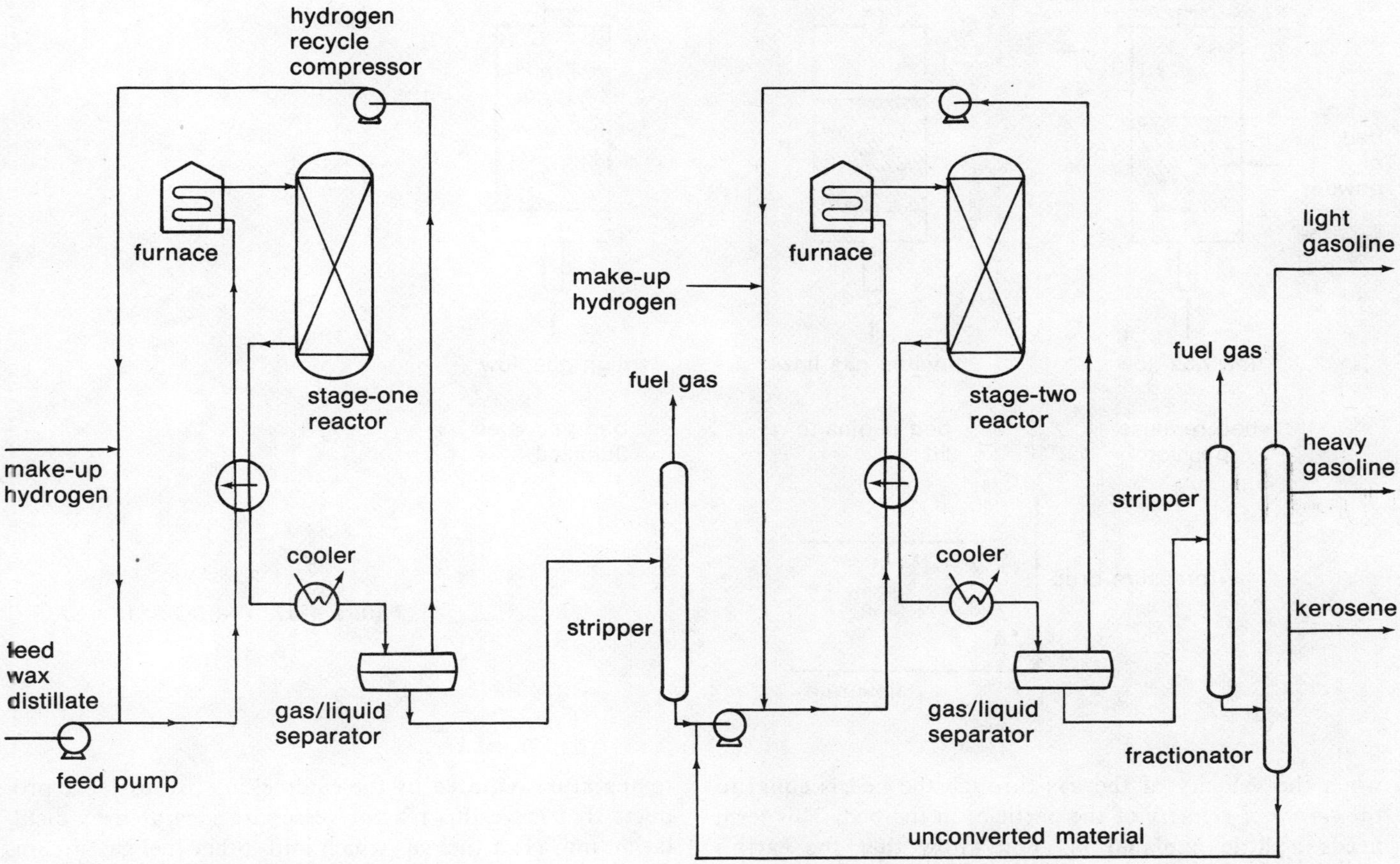

Figure 6-16. Hydrocracking.

recycling and the liquid product is stripped of any light fuel gases before being fractionated into an overhead light gasoline and two side streams, one of heavy gasoline and another of either kerosene or gas oil, depending upon the conditions in the second reactor.

The bottoms stream from the fractionator is returned to the feed pump to be sent back through the second reactor. The fractionation products are sent to an appropriate place in the overall refinery flowsheet, and you can see in Figure 6-1 that the light and heavy gasolines are sent to the gasoline blending plant and the other stream to the straight-run gas oil stream.

Catalytic Cracking of Light Gas Oil

The other light gas oil stream is subjected to a different cracking process—catalytic cracking or "catcracking." Research and development teams just a few years ago showed that if heavy oils were passed through beds of Fuller's earth (a natural material containing aluminum silicate) or a similar material, they would crack and give a good yield of high-octane-number gasoline. The only snag was that the Fuller's earth powder became coated with carbon (for reasons stated above) and its catalytic action stopped. It was found, though, that if the carbon was burned off the powder, the catalyst became as active as before.

Early catcrackers worked by having two reactor vessels. The material to be cracked was passed through a bed of Fuller's earth until the catalyst was "spent." The feed was then switched to a second reactor vessel of Fuller's earth and air was blown through the first in order to burn off the carbon. More recently, a more efficient means of regenerating the catalyst was found—the *riser cracker*. It relies for its operation upon a chemical engineering principle called *fluidized bed operation*. Figure 6-17 shows a vertical cylinder in which a bed of powder rests upon a plate perforated with many tiny holes. Beneath the plate is a semicircular head which acts as a means of distributing the flow of gas coming through the nozzle. The gas passes through the perforated plate and makes its way through the tortuous channels between the particles in the bed until it reaches the top and escapes through the top nozzle in the vessel. Now imagine that we slowly increase the gas flow rate. Those tiny channels through the bed will become wider and wider, and the net effect on the bed will be to expand it slightly as shown in Figure 6-17. If we further increase the gas flow rate, there comes a point

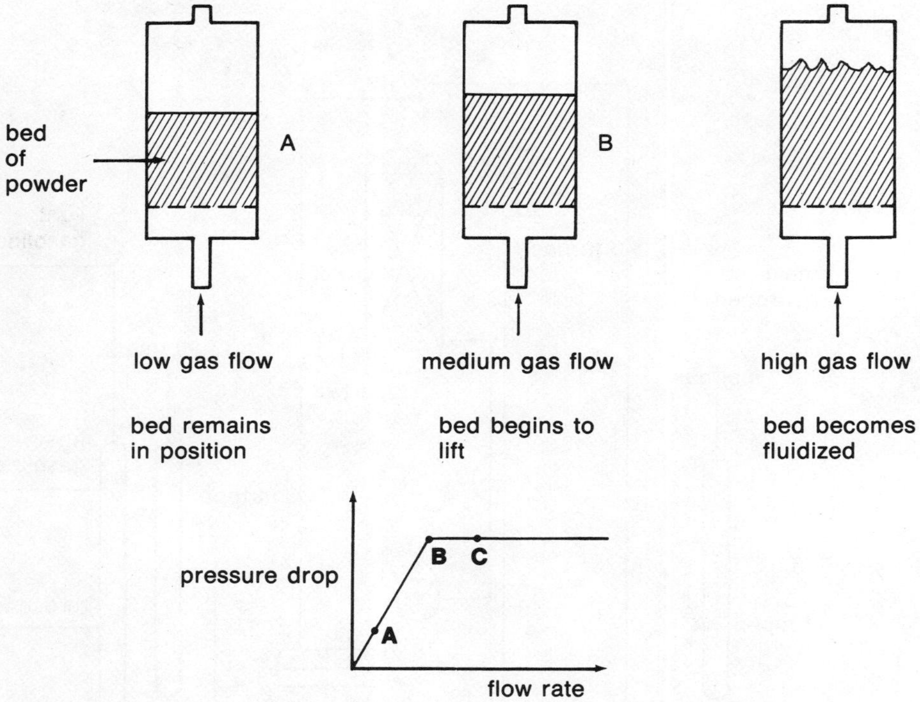

Figure 6-17. *The fluidized bed.*

when the velocity of the gas through the bed is equal to the *terminal velocity* of the particles in the bed. This term needs a little explanation. You know that the earth's gravity causes objects to fall toward it. The gravity effect is to accelerate the falling object so that its speed increases all the time. The falling object falls faster and faster. However, here on earth the object has to contend with the friction of the air in its path. Frictional effects increase with velocity and a point is reached when the force of gravity is balanced out by the frictional force on the particle and from that moment the downward velocity of the particle remains constant. This velocity is called the terminal velocity, and it depends upon the weight and diameter of the falling particle.

In the packed bed the particles are not falling; the gas stream is moving up through them. The effect is the same. At a certain velocity—the terminal velocity—the particles suddenly are suspended on the rising gas flow. Then a most surprising phenomenon occurs: the bed of solid particles behaves exactly like a fluid! If the velocity of the gas flow increases still further, "splashing" occurs and streamers of solid particles are blown clear from the bed. This can be avoided by maintaining the upward velocity at the terminal velocity of the particles.

Because of the fluid-like behavior of the bed under these conditions, the device is known as a fluidized bed. The riser cracker employs two fluidized beds as shown in Figure 6-18. In the first the fluidizing medium is a mixture of feed gas oil and steam flowing at the terminal velocity of the catalyst particles in the fluidized bed and at the temperature required by the catcracking process. The products that leave this reactor vessel are separated by distillation into (1) a fuel gas which joins other fuel gas streams we have made in various parts of the plant, (2) propane and butane which also join similar streams, (3) a high-octane gasoline that is sent to the gasoline blending plant and (4) a gas oil fraction that is recycled through the reactor.

Notice that in the lower part of the fluidized bed there is a takeoff stream of solid, spent catalyst which descends under the action of gravity into a second fluidized bed. Here the fluidizing medium is air, which burns off the carbon from the catalyst. The regenerated catalyst leaves the second fluidized bed down a chute that causes it to enter the hot feed stream to the first fluidized bed.

There are some advantages of the fluidized bed which are particularly useful in the riser cracker. First, like all fluidized beds, the pressure drop across it no longer rises at the same rate as it would in a packed bed (see Figure 6-17). Second, once fluidized, the bed remains stable. If there is a slight variation in flow rate of the feed, the response of the bed will restabilize the fluidization. The reason for this is not difficult to see: If the flow rate increases slightly, then the particles begin to lift, and that action widens the spaces between particles. This means that the gas flow has a larger free cross-sectional area through which it can flow, and that in turn means the velocity falls and so the particles settle back a little. If, on the other hand, the gas flow is reduced for some reason, the bed settles a little and the channels between the

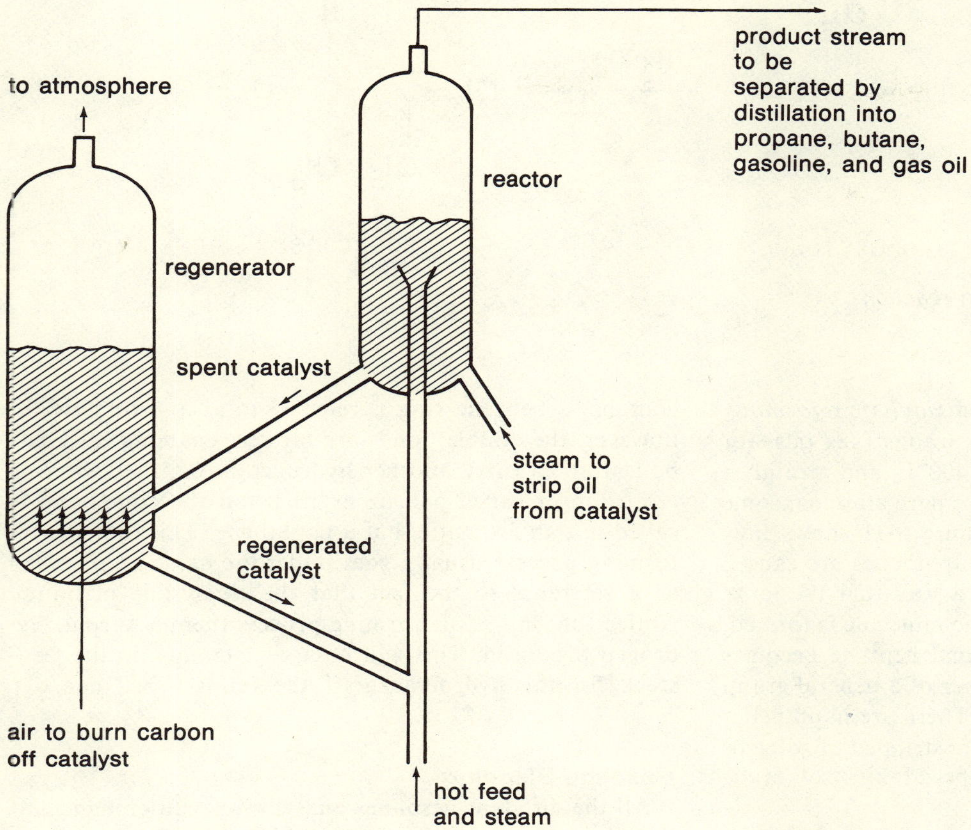

Figure 6-18. *The riser cracker.*

particles become that much narrower, so the gas velocity rises and the status quo is maintained.

The most important characteristic of the fluidized bed exploited in the riser cracker is that, because of the rapid random movement of particles around the bed, the temperature of the particles is uniform throughout. This means that the reaction occurs equally throughout the bed and thus maximizes the use of space. Furthermore, for the same reasons of particle mobility, all the catalyst is used in the reaction. There are no low-velocity corners where particles can accumulate and become coated with carbon so quickly that they are no longer of any use. Quite the contrary, the particles move around so quickly that they all become equally spent and each has an equal probability of being drawn down for regeneration in the second fluidized bed, where similar advantages exist. More recently the Fuller's earth catalyst has now been replaced by a synthetic catalyst that is a mixture of silica (silicon dioxide) and alumina (aluminum oxide).

Alkylation, Isomerization, Catalytic Reforming

There are two more ways in which gasoline's octane number can be increased. Just as catalytic cracking breaks large hydrocarbon molecules down to gasoline size, there is a process called *alkylation*, which takes molecules of a smaller size and builds them up to molecules in the gasoline range. Using a feed of butane, isobutane, isobutylene and a catalyst of strong sulfuric acid, the result is isooctane as shown in Figure 6-19. Isooctane is a valuable component for high-performance gasoline. However, because fewer automobiles are made with high-performance engines (except in Europe) and piston-engine aircraft have been replaced by jets, the importance of the alkylation reaction has declined, although a few units are still operating.

There are other ways of making these high-octane-yielding, branched-chain hydrocarbons such as isooctane. One is *isomerisation* and it is most effective with a feed of hydrocarbons with a boiling range of 85-160°F. It was found that by passing the feed of straight-chain hydrocarbons over a platinum catalyst at a temperature of 300°F then some of the straight-chains in the feed were reformed into branched-chains (see Figure 6-20). Fortunately, this reaction is quite successful and a high reforming rate can be obtained in just one step. I say "fortunately" because the reforming rate is so high that it is unnecessary to have a distillation unit in which to separate the product isomers from unreacted feed. Instead, the product stream can be sent straight to the gasoline blending plant.

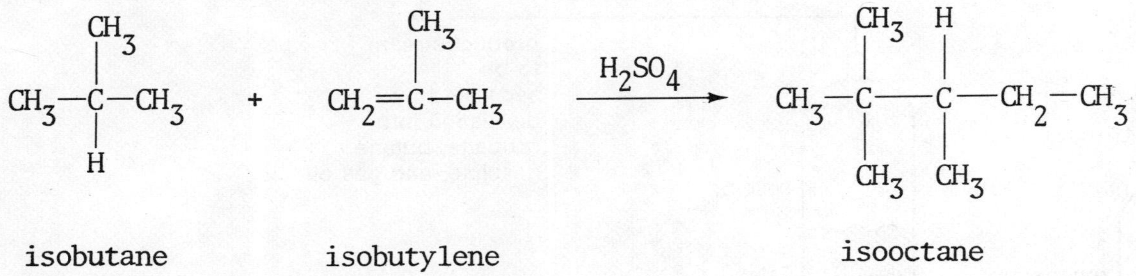

Figure 6-19. *The alkylation reaction.*

Finally, there is an operation called *catalytic reforming.* Using a feed of high-boiling hydrocarbons (gas oils, for instance) with a boiling range up to 400° F, and a catalyst containing ½% platinum deposited on pure alumina, some most interesting reactions occur. Figure 6-21 shows how straight-chain hydrocarbons called naphthenes are caused to wind around themselves in such a way that the loose ends react with each other and a cyclic molecule is formed. This figure also illustrates how normal heptane becomes methylcyclohexane, which is a member of a general group of hydrocarbons called naphthenes. There are naphthenes in the feed to begin with and these are stripped of some of their hydrogen atoms to form a special kind of cyclic molecule called an aromatic.

Notice in Figure 6-21 the conversion of methylcyclohexane to toluene. The ring of six carbon atoms bonded by alternate single and double bonds, as shown, is called a *benzene ring.* As a general rule, any compound that contains a benzene ring is referred to as an aromatic. If, however, the double bonds are broken, enabling each carbon atom to carry another hydrogen atom, then the ring is no longer called a benzene ring and the compound is called not an aromatic, but a naphthene. This catalytic reforming process usually goes under the name *platforming* as a reference to the fact that the catalyst is platinum. Notice that in the platforming process there is surplus hydrogen produced. This will be used, perhaps, in the feedstock for the hydrocracker if the refinery has one.

Gasoline Blending

All the different gasolines and gasoline-enriching agents that have been made are stored in tanks around the gasoline blending unit (Figure 6-22). A pipeline runs from each tank to a pump, a meter, and an automatically controlled valve. A master control unit is "set" to produce

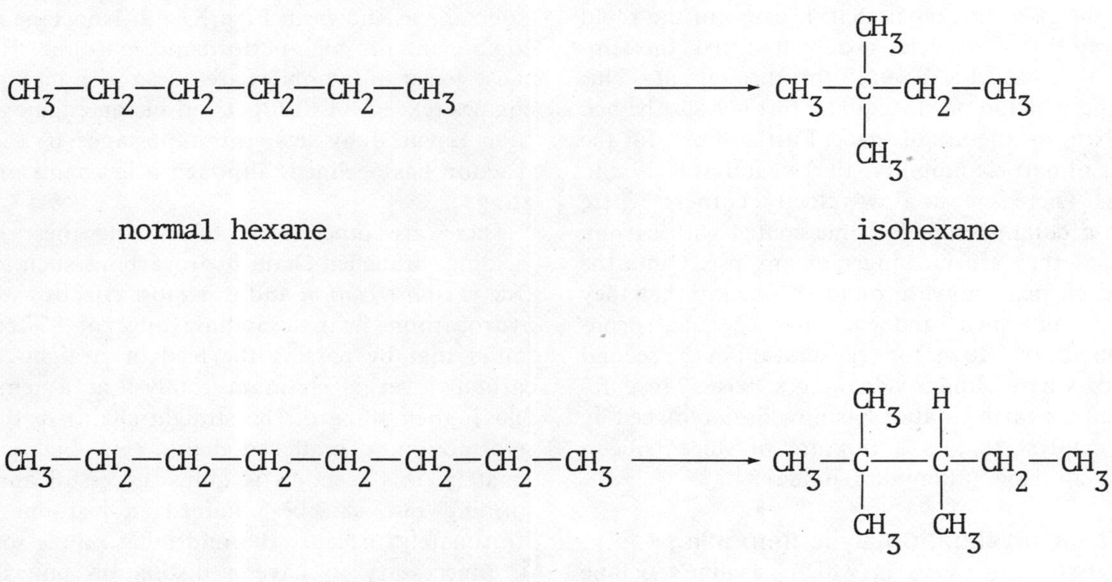

Figure 6-20. *Isomerization reactions.*

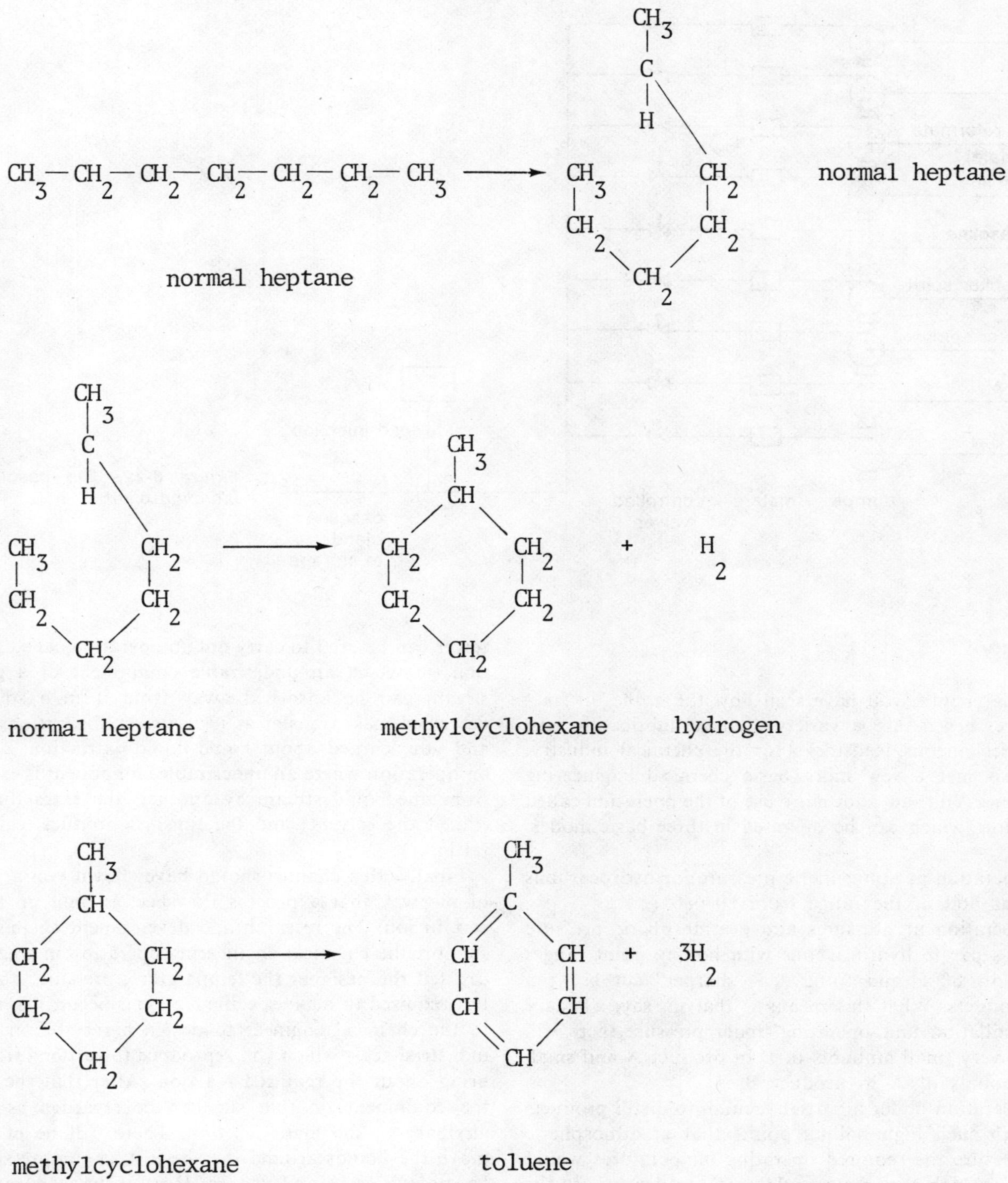

Figure 6-21. Catalytic reforming.

the desired blend, whether it be for aviation fuel or premium, regular or unleaded automobile fuel. The master controller is a tiny computer programmed to open the valves to the correct setting to produce the required final blend. The master controller receives a signal from each fuel line meter and uses that piece of information to adjust the flow value whenever necessary. The end-product of the gas blending plant is the final blend of fuel, which is run straight into tankers, railway wagons, and tank trucks for distribution to the market.

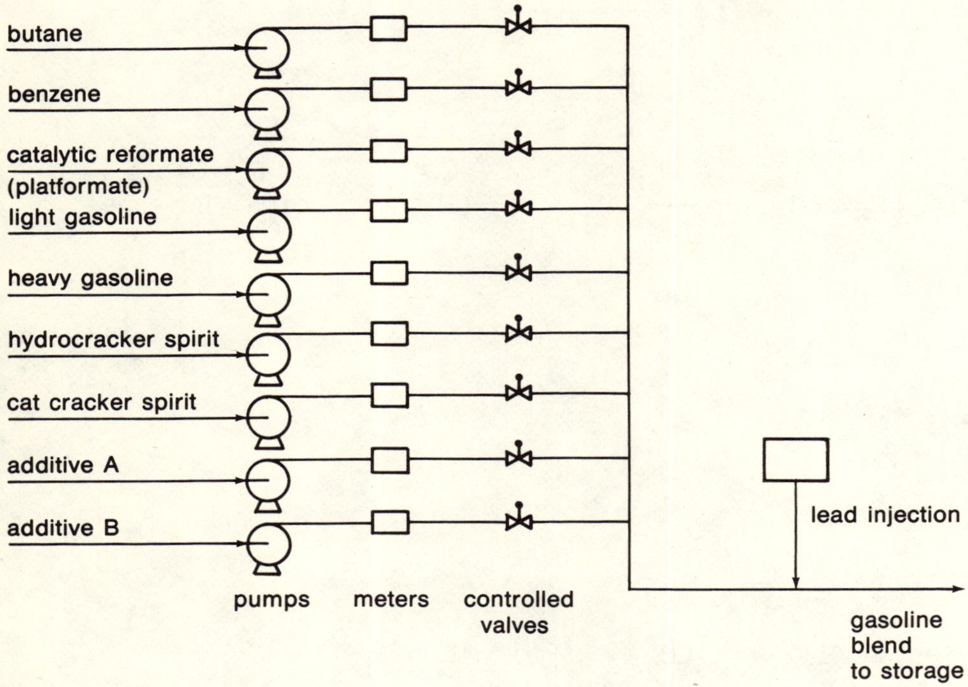

Figure 6-22. *The gasoline blending plant.*

Summary

In this chapter you have seen how the crude oil fractions are turned into a variety of fuels, lubricating oils, and miscellaneous feedstocks for the chemical industry. You also met a few more basic chemical engineering operations. You saw a lot more use of the operation called distillation, which can be operated in three basic modes:

1. Operation at atmospheric pressure for hydrocarbons that boil in the range from 80–660°F;
2. Operation at pressures above atmospheric pressure to separate hydrocarbons with boiling point ranges below 80°F and to give a "sharper" cut between products. What this means is that in, say, a binary distillation unit operating under pressure there will be very small amounts of B in product A and small amounts of A in product B.
3. Operation under a partial vacuum to distill products with such high boiling points that at atmospheric pressure the required operating temperatures would be so high that they would crack or decompose the feed into products chemically different from the ones desired.

You also learned how the bubble tower has another use as a stripper and how this was particularly useful in the operations of rectification or stabilization to remove low-boiling range hydrocarbons from a heavier product such as gasoline or kerosene. You learned, too, that the bubble tower can be used to carry out an operation called absorption, in which an undesirable component of a gaseous stream can be absorbed away from it in a solvent. I described mass transfer as it occurs in absorption towers, and you learned about liquid-liquid extraction, which is an operation where an undesirable component is extracted from one liquid stream by another: the extracting fluid (called the solvent) and the finished product (called the raffinate).

Finally, this chapter should have shown you a glimpse of the way that a process flowsheet is built up to do a certain job. The research and development chemists who support the chemical engineering operation in a company can tell the designer the temperature, pressure, and catalyst required to make specific reactions occur. It is then up to the chemical engineer to design a reactor on a large industrial scale which will reproduce those conditions and bring about the required reaction. After that the rest of the equipment on that site can be regarded as merely auxiliary to the main reaction. There will be pumps to move the liquids around the site just as compressors will be moving gases and vapors. Heat exchangers take heat from streams that no longer need it and give it as sensible heat to those that shortly will. Furnaces raise the temperature of reactants to the designated reaction condition. Fractionators separate reaction products into boiling point ranges, and strippers rectify or stabilize products by removing the lower-boiling point components from them. Distillation columns separate one product from another on the basis of their difference in normal boiling points, and absorbers or solvent extractors can remove undesirable by-products from a sales product by extracting them with a solvent.

The process designer's art is to choose enough of these ancilliaries so that the product is made and purified to the specifications required by the market demand and to do this at minimum cost for both constructed hardware and day-to-day operations. The professional chemical engineer has to be certain that the process configuration he develops to do this job has no detrimental effects on the environment and the quality of living in the vicinity of the plant or anywhere else.

Further Reading

British Petroleum. *Our Industry Petroleum*. London: The British Petroleum Company, Ltd., 1970.

Cook, G.A. *Survey of Modern Industrial Chemistry*. Ann Arbor: Ann Arbor Science, 1975.

McCabe, W.L. and J.C. Smith. *Unit Operations of Chemical Engineering* New York: McGraw-Hill, 1976.

Royal Dutch Shell. *The Petroleum Handbook*. London: Shell International Petroleum Company, Ltd., 1959.

Chapter 7

The Petrochemicals Industry

In this chapter you will learn about the manufacture of just a few of the many thousands of chemicals which are made for specific uses in business, industry, and in the home. The petrochemicals industry could best be thought of as "molecular architecture." The molecular structure of the required product is examined carefully, and the presence of particular molecular building blocks, some of which are shown in Figure 7-1, is noted. At this point the developers of a process to manufacture the desired compound, armed with a list of basic types of reactions that add or remove these basic building blocks, attempt to find a "route" backwards from the required product through one or more basic reactions to a low-cost basic material.

The petrochemicals industry uses some of the products from a crude-oil refinery to manufacture basic chemicals. Figure 7-2 shows some of the more common basic chemicals produced from petroleum and just a few of the many end-products made from them. In this chapter it is impractical and unnecessary to attempt a description of the chemistry and chemical engineering of even the few chemicals shown in Figure 7-2. Instead I will describe the most common basic reactions, give one or two examples of their use, and I will devote a little more detail to the production and use of what is perhaps the most important basic chemical produced today: *ethylene*.

methyl	—CH_3	hydroxyl	—OH
ethyl	—C_2H_5	amine	—NO_2
propyl	—C_3H_7	amide	—NH_2
hydrogen	—H	phenyl	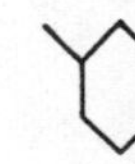
chlorine	—Cl	sulfonate	—SO_3H
carboxyl	—COOH	sulfate	—OSO_3H

Figure 7-1. *Some molecular building blocks for "constructing" specialized chemical compounds.*

Process Development

The petrochemicals industry is like molecular architecture, since a finished product is literally "built up" of basic building blocks in exactly the way that a house or hospital is a structure containing basic building blocks such as windows, doors, concrete slabs, and so on. After the research chemists and chemical engineers in the process development team have established a particular route of reactions, they will set about testing their suggested plan by building a small plant called a *prototype*. If all goes well with this plant, the company may take the plunge and build a full-scale production plant if it appears that the product can be sold at a price that will give a reasonable return on the investment.

Process plants for the petrochemicals industry are in many ways similar to those we met in the fuels and lubricating oils industry. Frequently, companies will have crude refineries, fuels and lubricating oils plants, and petrochemical plants next to each other and integrated into one huge complex.

After the chosen reactions have occurred in vessels called *reactors*, the products must be separated; you will see why in just a moment.

Separations usually involve the types of equipment described in Chapters 5 and 6, i.e., the distillation column, the fractionator, the liquid-vapor separator, the absorber and liquid-liquid equipment. The main difference is that in the petrochemicals industry these pieces of equipment will not have to process the same volume of material that a distillation column would, say, in a gasoline plant. However, the fuels industry has less problems than the petrochemicals industry with respect to the purity of its products. Although the final blend of a gasoline mix, for instance, is carefully controlled to meet exact specifications, the chemical species which go into it may vary in composition and sometimes can be completely replaced by

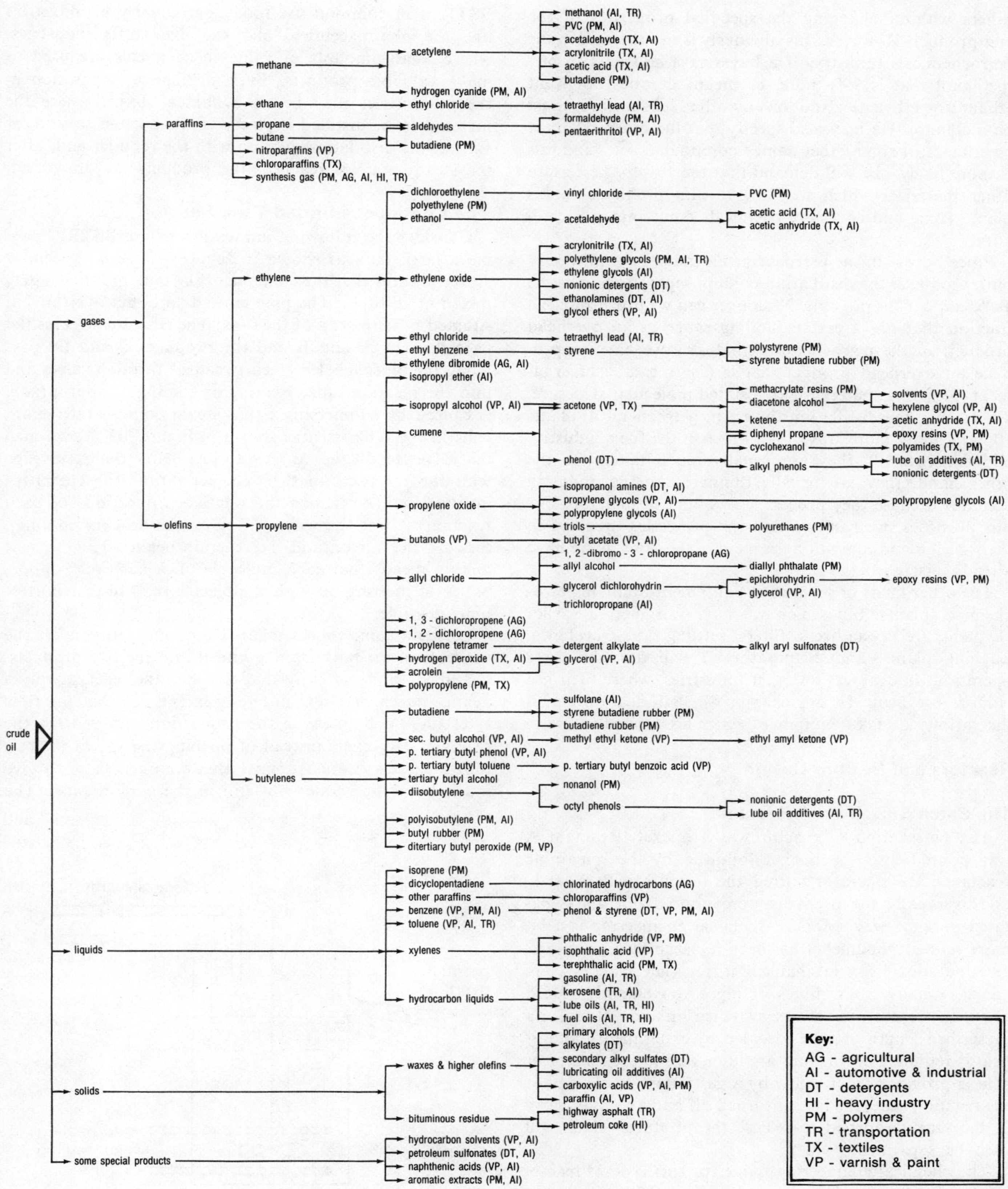

Figure 7-2. *Some products made from crude oil, and their applications.*

others without changing the specified properties of the end-product. However, this obviously is not the case in the petrochemicals industry. If a buyer says he wants to buy methanol that is 99% pure, he means he does not want ethanol or ethylene glycol or any other substance; it must be methanol. He may also specify that the other 1% must be ethanol or some other named compounds; or—and this is more likely—he will demand that the 1% *not* be certain other materials which would give him problems in his plant. Toxic and corrosive chemicals come into this category.

Since buyers in the petrochemicals industry require very pure products, the distillation column separations must be as "clean-cut" as possible. No longer can we just produce a fraction that has a certain boiling range as an overhead product; in the petrochemicals industry we have to produce an overhead product that is (for the sake of argument) 99% of a material with a stated molecular structure. To do this, the distillation towers in petrochemical plants contain many more trays than those in the fuels industry, and consequently they are usually much taller. On the other hand, they are usually thinner than fuel industry columns because they process less volume of material. For our purposes, we can say that the principles involved in their design and operation are exactly the same as we have already discussed.

The other kinds of separation equipment found in petrochemical plants, but which we have not met in earlier chapters, are crystallizers, filters, settling tanks, cyclones. leaching plants, and evaporators. I will describe these operations in chapters on other industries, where their use is more prevalent. In this chapter we shall deal only with the various kinds of equipment which are termed reactors.

Reactors and Reactor Design

The Batch Stirred Tank Reactor

The oldest kind of reactor was a bucket. Reactant A was poured into the bucket followed by the remaining reactants, the operator stirred the mixture with a stick, and eventually the products were poured out and separated in some way. Modern chemical engineering is a bit more refined: the bucket has been replaced by a large tank and the stick by a mechanical stirrer. The reactants are added, usually from chutes or pipes conveniently placed above the reactor, the stirrer switched on and left on for a prescribed length of time (decided by experiment), and, if heat is required, steam coils are immersed in the reactor or heat is provided from below by a gas flame. At the end of the reaction time, the heat is turned off and an outlet valve in the tank is opened to release the products which will then be separated.

The chemical engineering name for this type of reactor is the *batch stirred tank reactor* or *BSTR*. The word "batch" tells us that the operation is done in batches. The BSTR is in common use today particularly in industries like the pharmaceutical and the dye stuffs industries, where small amounts of valuable chemicals are used to make valuable products. Even distillation separation in these industries may be on a "batch" basis, where the material to be distilled is placed in the bottom section, or still, of the distillation column with the reboiler and, after the steady state is reached, the products are drawn off.

The Continuous Stirred Tank Reactor

Consider the following innovation to the BSTR: Imagine a large tank as shown in Figure 7-3. A large central stirrer ensures that the contents of the tank are thoroughly mixed at all times. The pipe labeled "product overflow" is situated near the top of the tank. The reactant streams (let us call them A and B and the products C and D) pass along the pipes labeled "feed streams," through valves and into the reaction tank. Heating or cooling, if appropriate, is caused by submersing either steam coils or refrigerant coils in the tank itself as shown in Figure 7-3. Now, open the valves to fill the tank up to just below the outlet pipe with the reactants in their correct ratios (the chemistry involved will determine the relative proportions of each reactant). Close the valves. Turn on the heat (or cooling), turn on the stirrer, and the reaction begins. The efficient mixing means that each molecule of A has every opportunity of meeting up with a molecule of B in the shortest time possible.

Let us assume that it takes five minutes to convert the contents of the tank from A and B into the two products, C and D. Now, suppose that we open the reactant valves again, but to precisely different extents, so that the flow rates of A and B are in the proportion required for the reaction. However, instead of opening the valves to their fullest extents, let us suppose that we open them to give much lower flow rates but still in the correct ratio. The

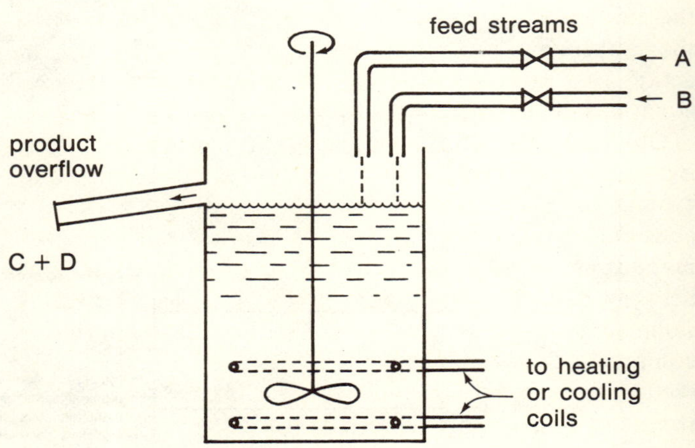

Figure 7-3. *The continuous stirred tank reactor.*

new molecules of A and B enter the highly stirred environment of the tank, which is, remember, almost pure C + D. The new molecules of A and B, because of the stirring, meet up with each other and react. If we leave the valves open a little, the level in the tank begins to rise. Almost immediately, the product overflow pipe begins to flow and we can take off the mixture of C and D for further processing. Since the reactants continue to come in, albeit quite slowly, the product can be removed on a continuous basis. Hence, this kind of reactor is called a *continuous stirred tank reactor* or *CSTR*.

If we opened the valves on the reactants A and B to their fullest extents without regard to the ratio of their flow rates, the CSTR would not work properly. So, the first thing we must do is gain sufficient control of the two feed streams such that the ratio of their flow rates is exactly what we require for this reaction. If there were to be slightly too much A, or too little B, the product stream would contain the excess amount of A. In practice, though, this more often than not is the desired state of affairs. If we cause the cheaper of the two reactants to be present to excess, we can be sure that all of the more expensive reactant will be used up. Of course, we must have a reliable and inexpensive means of separating the excess reactant from the products.

The other thing we have to watch is the flow rates of the two reactants. Not only should they be in the correct ratio (either exact or with the cheaper one in excess), but their rates of entry into the CSTR must be (a) slow enough that all of the more expensive reactant has ample opportunity to react yet (b) fast enough to produce the end-product at an economical rate. The best approach to this second constraint is to decide first upon the desired rate of production. By experiment or calculation, we can decide upon the amount of time the reactants will need to complete the reaction. It is then a simple calculation to determine the size tank which will allow the reactants that length of time between entering and leaving the tank. This period of time is called the *residence time* since it is the amount of time spent by the materials in residence in the tank. With the production rate determined, the reactant feed rates are simply calculated from the chemical equations which determine the course of the reaction.

Obviously, the CSTR is an ideal type of reactor for any industry which deals with continuous flows of materials around the plant. Not only is the end-product made at a constant rate, but there is also no need for any expensive manpower to turn on valves and clean out the tank between batches. The product stream can be continuously sampled by automatic means and the results fed back to a computer. If the computer has been programmed correctly, any change in the composition of the product stream will cause the computer to make the necessary adjustments. For instance, it may be that the heat needs to be turned up a little because it is a cold day, or perhaps the stirrer should be moved a little more quickly because a feed stream viscosity has temporarily increased. Similarly, the computer can be made to control the valves on the reactant feed streams.

Human intervention, then, is unnecessary except for nominal supervision that can be carried out by the plant operational staff. Except in the case of products or reactants which cause deposits or corrosion in the tank, cleaning is unnecessary except at infrequent intervals as part of general maintenance. There are thousands of CSTRs in action around the world today and you will learn the mathematical description of these useful devices during your formal chemical engineering education.

Tubular Reactors

It must be obvious to you that the CSTR cannot always be used. It was fine for two or more liquids and I am sure that you can see that it could still be useful if we had to stir in a solid in particulate form or to bubble in a gas. However, it could not be used for a reaction between two or more gases, nor would it be suitable if great amounts of heat had to be supplied or taken away from the tank, simply because the area available for heat transfer is limited to the walls of the tank or a simple coil, as we have already seen. The heat transfer is quite slow under these circumstances.

To accommodate these kinds of reactions, another kind of reactor is used: the *tubular reactor*, sometimes called the *plug-flow reactor*. Various geometries are possible for tubular reactors. The simplest would be one heated tube (see Figure 7-4) down which the reactants, again let us say A and B, are made to pass after efficient mixing. If the reaction requires heat (an endothermic reaction), then the tube can be surrounded either by steam or by the hot gases of a flame in a furnace. If we know the time it takes for A and B to react completely, and we know the production rate required, it becomes a simple calculation to determine the length of the reactor.

If the addition or removal of heat is important to the reaction and must be carried out at a high heat transfer rate, then a slow flow rate in a wide-diameter tube would be unsuitable because the heat transfer to or from the bulk of the gas would be slow, causing a long residence time and requiring a long reactor. It would be most impractical to have to build, for example, a reactor tube a mile long surrounded by a one-mile-long furnace. It would be far better in these circumstances to have the thoroughly mixed feed stream split and passed along a number of narrow-diameter tubes in which their velocity is high. This arrangement has good heat transfer characteristics and the required length of the reactor would be brought down to a reasonable value. Such a design is called a *multi-tubular reactor*.

If a catalyst is required, the tubes can be replaced by a wide-bore tower arranged vertically with the gas flow

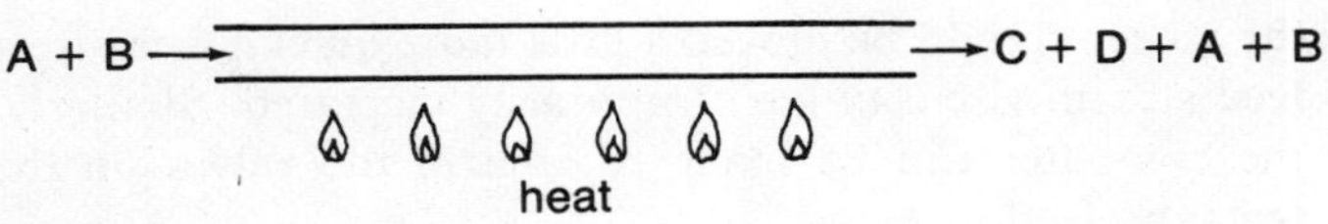

Figure 7-4. *The tubular reactor.*

passing through a porous support tray upon which rests a bed of the catalyst, as shown in Figure 7-5. Although this is essentially a tubular reactor, it is referred to as a *packed-bed reactor*. Problems arise with packed-bed reactors in cases where the reaction itself causes the catalyst to be spoiled somehow during the reaction. We saw in the example of gasoline manufacture, in the riser cracker, how the catalyst becomes coated in carbon, rendering it ineffective or "spent."

There are other examples of catalysts being spoiled, particularly when the reactant stream contains a trace of some "poisoning agent" such as sulfur. In such cases the spent or poisoned catalyst has to be taken out and either replaced with new catalyst or regenerated back to its full effective form. This can cause extra expense in plant construction since, in order to maintain continuous operation, the vertical packed-bed reactor has to be duplicated: while one is in operation the other undergoes cleaning, maintenance, catalyst replacement or regeneration. Also, unless mechanical devices are made to remove, replace or regenerate the catalyst, manpower is required to carry out these operations, and this is costly.

You have already seen a desirable alternative to the fixed packed-bed of catalyst which may involve problems such as those just described. This was in Chapter 6, where gasoline hydrocarbons were manufactured from heavier oils by cracking the molecules into smaller ones in the riser cracker reactor—which is a fluidized bed.

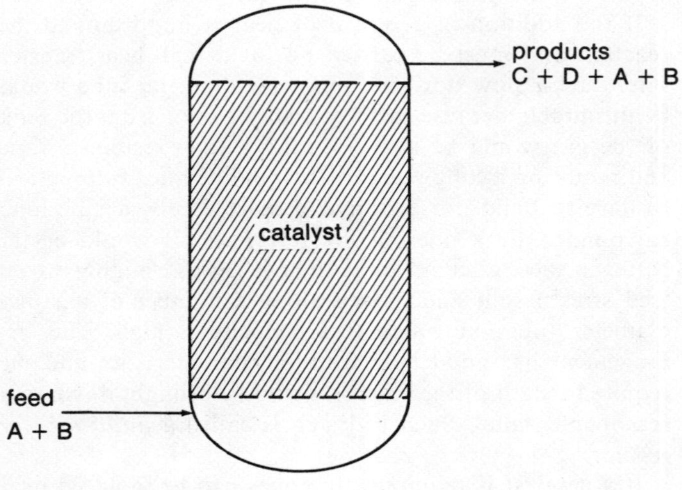

Figure 7-5. *The packed-bed reactor.*

The Fluidized-Bed Reactor

You now have seen a means of reacting liquids with liquids in the CSTR and gases with gases (catalyzed or not) in the tubular, multi-tubular or packed-bed reactors. Catalyzed gaseous reactions can be carried out in the fluidized bed, particularly where a uniform temperature is required in the bed of catalyst. But what of the reaction between gas and solid, i.e., where the solid is used up and needs to be replaced by fresh solids? Again the fluidized bed is ideal. Perhaps the best example of this type of reaction is the fluidized-bed combustor for coal and other solid fuels. This means of burning coal was not mentioned in Chapter 4 because, at this writing, the fluidized-bed combustor is not yet fully developed, although there are a number of research prototypes around the world. By the time you are an engineer the fluidized-bed combustor probably will have been accepted as a feasible or practical means of burning coal or it will have been rejected if problems inherent in its operation have proved insurmountable.

The fluidized-bed combustor is still in the "research and development" phase, and our discussion of it will be a good opportunity for you to have a glimpse of the problems that professional chemical engineers grapple with before the commercialization of a new idea. Before we talk about its problems, though, let us look at its principles of operation.

Figure 7-6 shows the hardware. The perforated plate at the bottom of the bed allows the upward flow of the combustion air which, as with other kinds of solid-fuel combustion, has been preheated in air heaters set in the path of the cooling combustion gases. The walls of the fluidized bed are made of water tubes in order to keep the wall cool and at the same time heat the boiler water. There is an ash take-off point in a position corresponding to the top of the fluidized bed when in operation. Because the superheater tubes must be in the hottest part of the flame (for reasons set out in Chapter 4), they are contained within the bed of burning coal itself. Now imagine starting up one of these plants. First, fill the enclosure with coal crushed down to 100% minus a quarter inch. Then, turn on the air flow and introduce a fuel gas into the air stream to burn in the bed long enough to get the coal started. Soon the entire bed will be on fire, and due to the thorough mixing in the fluidized bed, every particle of coal will be burning away as it moves rapidly around the bed under the effects of fluidization.

Eventually the bed will contain nothing but red hot ash. If you continue to keep the air flowing through the bed, the hot ash will be fluidized. Furthermore, if the coal feed continues in small amounts, combustion will be maintained, producing heat for the superheater tubes and the hot gases which leave the fluidized bed. Soon the level of the fluidized bed of ash will reach the overspill where the ash will flow out. From now on the ash will flow out at

Basis: 100 lbs of coal with 1% sulfur.

Dolomite is $MgCO_3$ + $CaCo_3$ (assume 50:50 mix).

Upon heating, both carbonates decompose to MgO + CaO.

Sulfur present = 1 lb = $\frac{1}{32}$ moles/100 lbs of coal

$$CaO + SO_2 \rightarrow CaSO_3$$
$$\tfrac{1}{64} \quad \tfrac{1}{64} \quad \tfrac{1}{64}$$

$$MgO + SO_2 \rightarrow MgSO_3$$
$$\tfrac{1}{64} \quad \tfrac{1}{64} \quad \tfrac{1}{64}$$

$\therefore$ $\frac{1}{64}$ moles each of $CaCO_3$ and $MgCO_3$ are required

= $\frac{100}{64}$ lbs of $CaCo_3$

plus $\frac{83}{64}$ lbs of $MgCO_3$

= 2.86 lbs of dolomite/100 lbs of coal

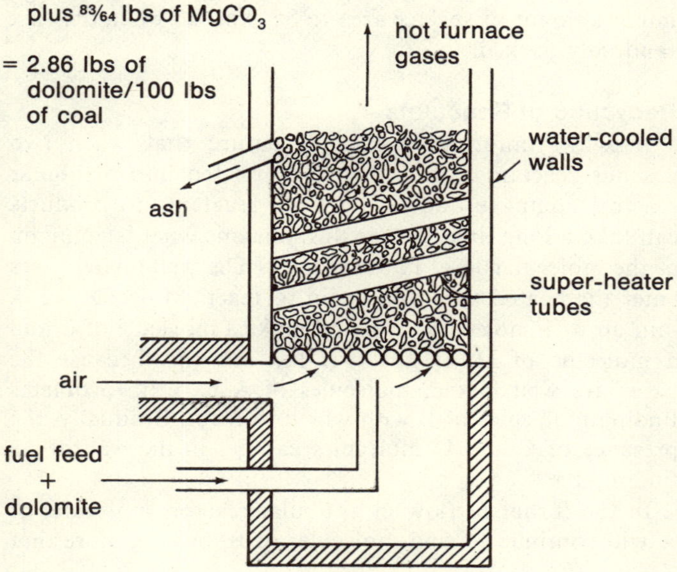

Figure 7-6. *The fluidized-bed combustor.*

the same rate that the coal is admitted. Notice that under normal operation the fluidized-bed combustor for coal is not a fluidized bed of coal at all! In fact, it is a fluidized bed of hot ash in which we put coal in order to burn it.

Now let us look at the problems with the fluidized-bed combustor. First of all, the violent motion of the ash and coal particles in the bed efficiently grinds down the surface of the superheater tubes. In other words, our first problem is the metallurgical one of finding a steel which will resist this kind of violent attack. We could, of course, put the superheater tubes just above the bed, but this would be less efficient overall because the steam would not be so highly superheated in that position.

The second problem is the carryover of particles of ash in the gas stream. This is no worse than the fly-ash carryover that occurs in conventional pulverized-fuel-fired boilers. However, the human condition is such that we want to make each piece of new technology solve *all* the problems of previous ones.

A major reason for developing the fluidized-bed reactor is the hope that it can be designed to burn coal cleanly—that is, without emitting certain pollutants whose subsequent removal from the combustion gases is costly. In conventional coal-burning systems the sulfur from the coal enters the waste gas stream and must be removed before the waste gas is vented to the atmosphere, where the sulfur would contribute to environmental pollution. Innovative engineers began to consider an alternative to post-combustion gas clean-up. Why not absorb the sulfur oxides as fast as they are formed?

It was reasoned that by putting in the bed certain sulfur-attracting solids, the sulfur dioxide could be removed from the combustion gases before they even left the combustion zone. It is known that limestone, when heated to these temperatures, breaks down to "quick-lime" (calcium oxide), which reacts with sulfur oxides to give either calcium sulfate or calcium sulfite. Figure 7-6 shows a fluidized bed with dolomite being delivered to the fuel feed. Dolomite is a mixture of the carbonates of magnesium and calcium, and its greatest advantage here is that it is inexpensive. Figure 7-6 also shows how to calculate the amount of dolomite required to react with all the sulfur in the coal. (This is the same kind of calculation used to determine the ratio of reactant flow rates to a CSTR.)

The fluidized bed may also solve another problem with coal combustion in conventional equipment. Coal contains small amounts of nitrogen. It is bound up into the molecular structure of the coal and during combustion the free nitrogen atoms, which do not join the combustion reaction, find themselves leaving the coal and becoming available for reaction with other atoms around them. Some of these nitrogen atoms will react with other nitrogen atoms to produce molecular nitrogen, N_2. Unfortunately, some will react with oxygen atoms to give varying amounts of the different oxides of nitrogen. There are many of these and combustion engineers refer to their presence simply as NO_x, a term signifying that we do not know how much of each oxide to expect; but they will, perhaps, all be there.

The trouble with NO_x is that it is believed to be a cause of a health hazard called "smog." The theory goes something like this: Small amounts of NO_x enter the atmosphere from vehicle exhausts, combustion gases from boiler plants and industry in general and slowly rise into the upper atmosphere, where they become mixed with water vapor. Sunlight falling on the mixture catalyzes complex reactions between the water and the NO_x. In areas like the Los Angeles basin, where high-altitude clouds often can roll back down the hills into the city, the result of these complex reactions between water and NO_x is a dense, yellowish fog—and because its origin is a mixture of smoke and fog, it has been called "smog." In the worst days of smog in London, and in my hometown, Sheffield, very bad smogs were usually referred to as "pea-soupers": a reference to their thickness and color.

So far, no reliable, inexpensive means of removing the NO_x from combustion gases has been found. One possibility is that if the fluidized bed is operated under pressure, the NO_x emission may be reduced. There is mounting evidence which indicates that a high combustion pressure tends to inhibit NO_x formation. If this turns out to be the case, though, it may be very convenient in using the fluidized-bed combustor to make the coal suitable for use as a gas turbine fuel.

Obviously, we cannot put burning coal particles through a gas turbine—at the speeds involved in turbines, the particles would "erode" or tear away the finely engineered turbine blades. But if we were to put the fluidized-bed system within a closed vessel such that the air can be compressed prior to admission to the combustion zone and the hot gases would leave under pressure, those gases could be passed through a gas turbine to make power. The hot gases exhausted from the gas turbine could be put through a waste-heat boiler (see Chapter 6) and then through an economizer and an air preheater. Figure 7-7 shows a flowsheet of such a process. The sulfur would be absorbed by dolomite in the fluidized bed. NO_x formation would be inhibited by the high pressure (if this turns out to be the case). The hot gases would drive the gas turbine and use up their pressure energy and some of their sensible heat. The remaining heat would be collected in the waste-heat boiler, the economizer, and the combustion air preheater. These, then, are some thoughts connected with the operation of the still experimental fluidized-bed combustor.

There are other kinds of reactors. Some are based upon the principle of the absorber that we saw in Chapter 6. The gas to be reacted is passed down a column and the liquid to be reacted is sprayed in from the top. To increase the surface area available for reaction to occur between the two phases the tower is packed with some inert, or non-reacting, material which causes the liquid to spread out in films over it and the gas to find tortuous and varying channels through it. The packing varies from industry to industry and from application to application. A common packing material is porcelain made into small regular shapes such as rings or a variety of ingenious shapes which cause a maximum amount of surface area to be exposed when they are randomly packed.

Recycling of Reactants

It is an unfortunate fact of nature that when two gaseous reactants are well mixed and fed into a tubular reactor, complete conversion from reactants to products can take a long time. To see this, imagine what is going on at the molecular level in the reactor. The well mixed gases enter the heated tube and begin to react. Molecules of A join up with molecules of B and make a molecule of C and a molecule of D. In the first few seconds, because the gases are well mixed, molecules of A have no problems finding molecules of B with which to react. Gradually, the presence of C and D molecules can get in the way of As finding Bs.

In the turbulent flow in a tubular reactor, molecules of A will continue to find molecules of B, but the more that

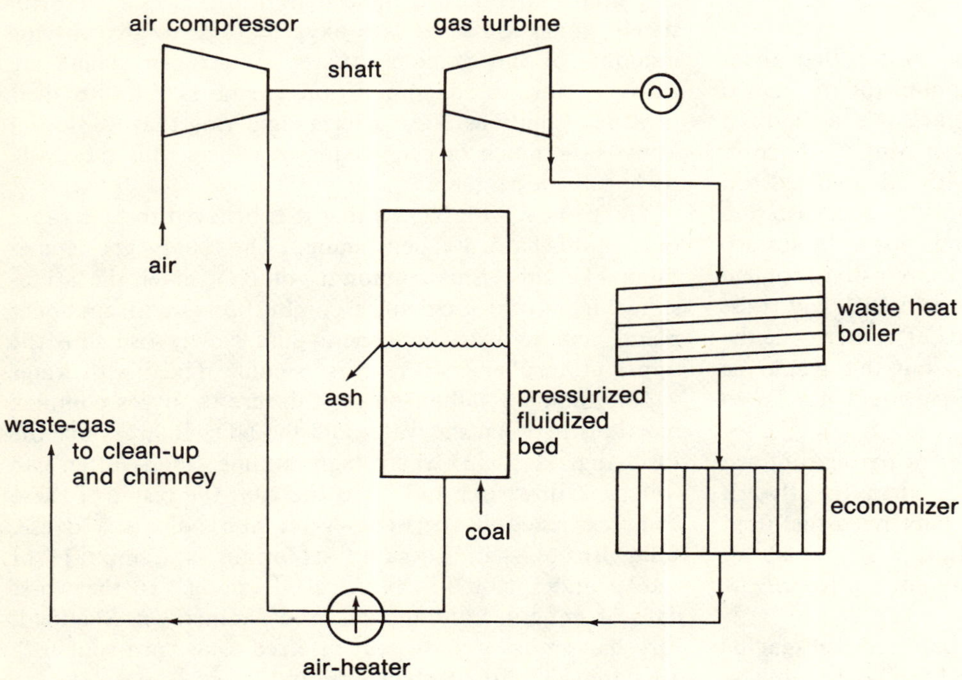

Figure 7-7. *The combined cycle method of generating electricity.*

do so, the more product molecules are formed, and hence the harder it becomes for the reaction to continue. After 90% of the reaction is complete you can see that the molecules of reactants will be few and far between and will take considerably longer to find each other for reaction than, say, in the first 10% of reaction. In the extreme, imagine just one molecule of A and one molecule of B remaining, it is conceivable that they would never find each other. This is a common problem in chemical engineering and it is frequently found that the last 10% of the reaction can take 10 times as long as the first 90%. In some reactions the dampening effect of the product molecules can begin to slow down further reaction beyond a percent conversion as low as 15-20%. Clearly, there has to be some way of dealing with this problem. Figure 7-8 shows two ways.

Let us suppose that the customer requires a product stream that contains no more than 10% of the reactants. The length of reactor required to bring this about is impossibly long and cannot be considered. It is found that the longest heated tubular reactor that there is room for on-site gives a 70% conversion. This means that the product stream contains 30% unconverted reactants. One solution is to situate equipment after the reactor to remove the unconverted reactants and return them to the feed stream to be fed through the reactor again. If an inexpensive and simple means of removing the reactants from the product stream is available, this is the preferred way of dealing with the low conversion rates.

The second method, also shown in Figure 7-8, is a little more difficult to grasp. In fact, it used to astonish me that it worked at all. The stream leaving the reactor is split into two—the ratio of the split being carefully calculated by the method described below. One stream is the product stream; the other is a recycle stream which is taken straight back and put through the reactor again. How can this possibly reduce the concentration of reactants in the product stream? We have to do a calculation called a *mass balance* before this becomes clear (see Figures 7-9 and 7-10).

First, let us hypothesize a reaction which is A being catalyzed to products B and C as shown in Figure 7-9. The reactor length is shortened as much as possible to save valuable space on-site, and the length that has been chosen gives a 70% conversion of A into B and C. In the first example the stream leaving the reactor is divided equally (50-50) into a product stream and a recycle stream which

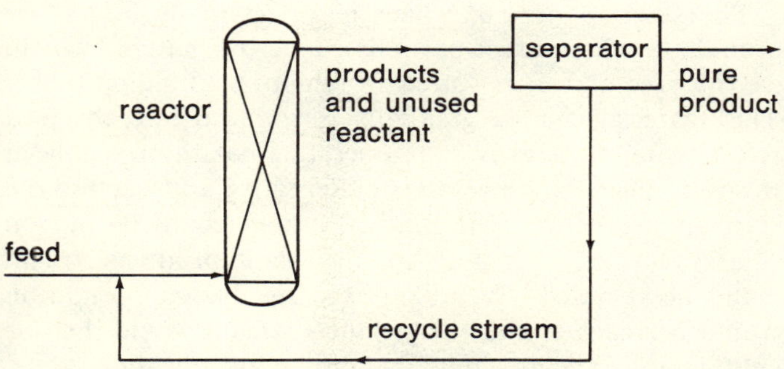

Method 1: separation of unused reactants for recycling.

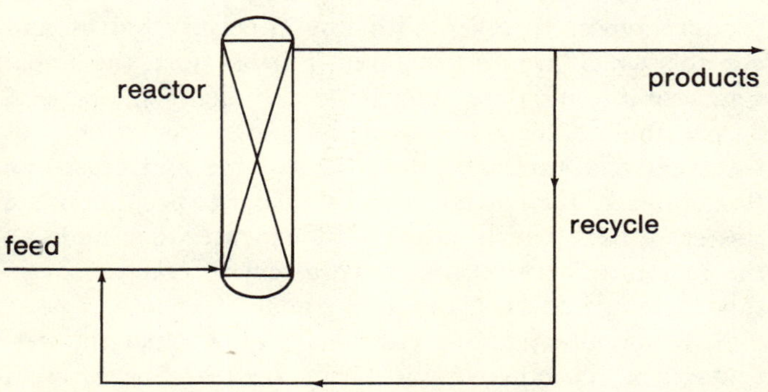

Method 2: splitting products stream and recycling a fraction of it.

Figure 7-8. *Two ways of improving systems with low reactor yields.*

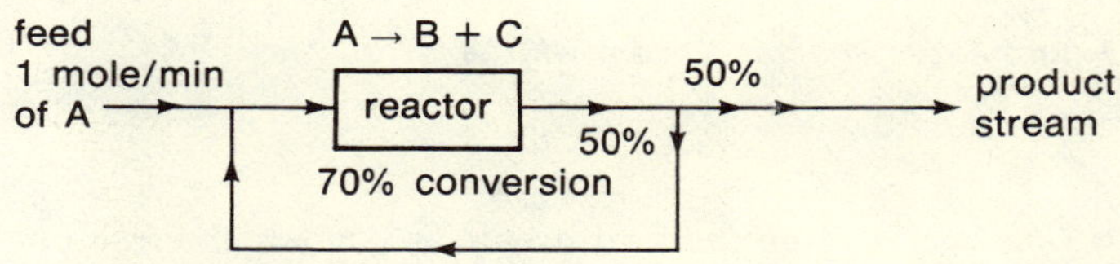

Time (minutes)	Moles entering reactor			Moles leaving reactor		
	A	B	C	A	B	C
1	1.000	0.000	0.000	0.300	0.70	0.70
2	1.15	0.35	0.350	0.345	1.16	1.16
3	1.173	0.580	0.580	0.352	1.40	1.40
4	1.176	0.700	0.700	0.353	1.52	1.52
5	1.177	0.766	0.766	0.353	1.58	1.58
6	1.177	0.790	0.790	0.353	1.61	1.61
—	—	—	—	—	—	—
—	—	—	—	—	—	—
Steady state	1.177	0.81	0.81	0.353	1.63	1.63

percentage of A in product stream with no recycle = 30%

percentage of A in product stream with 50% recycle

$$= \frac{0.353 \times 100}{(0.353 + 1.63 + 1.63)} = 9.7\%$$

Figure 7-9. *Reactor system with 50% recycle.*

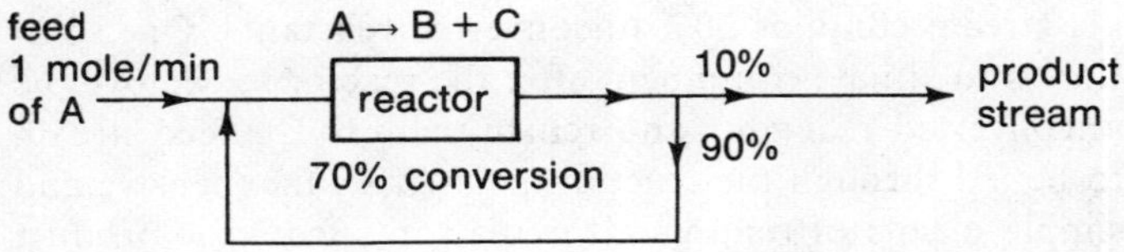

Time (minutes)	Moles entering reactor			Moles leaving reactor		
	A	B	C	A	B	C
1	1.000	0.000	0.000	0.300	0.700	0.700
2	1.270	0.630	0.630	0.381	1.519	1.519
3	1.343	1.367	1.367	0.403	2.307	2.307
4	1.363	2.076	2.076	0.409	3.303	3.303
5	1.368	2.727	2.727	0.410	3.685	3.685
6	1.369	3.317	3.317	0.411	4.275	4.275
7	1.370	3.848	3.848	0.411	4.810	4.810
8	—	—	—	—	—	—
—	—	—	—	—	—	—
—	—	—	—	—	—	—
Steady State	1.371	8.685	8.685	0.411	9.65	9.65

Percentage of A in product stream with no recycle = 30%
Percentage of A in product stream with 90% recycle

$$= \frac{0.411 \times 100}{(0.411 + 9.65 + 9.65)} = 2.1\%$$

Figure 7-10. Reactor system with 90% recycle.

is sent back to join the feed stream as it enters the reactor. Look at the table in Figure 7-9, remembering that a mole of A is the molecular weight of A expressed in pounds. The first column shows the time after start-up in minutes. The second, third and fourth columns show the number of moles of A, B and C, respectively, which enter the reactor during that minute in the feed stream. The fifth column shows the number of moles of A leaving the reactor, and the last two columns show the number of moles of B and C leaving the reactor.

In the first minute let us say that there is one mole of reactant A entering the reactor in the feed stream. It is converted to 0.7 mole of B and 0.7 mole of C because, as stated earlier, there is only 70% conversion. This causes 30%, or 0.3 mole, of A to leave the reactor. We now split that stream and return half of it back to mix with the feed stream. In other words, 0.15 mole of A and 0.35 mole each of B and C are returned to feed through the reactor again. Incidentally, the product stream (which also contains 0.15 mole of A and 0.35 mole each of B and C) is taken to a temporary storage tank where it will be kept for a while. After the reactor has been started up, the stored material will be used as feedstock as you will see shortly.

Since A is fed into this system at the rate of one mole per minute, in the second minute the new mole of A meets up with the recycle stream and 1.15 moles of A enter the reactor (together with 0.35 mole each of B and C). The 70% conversion of the 1.15 moles of A gives 1.16 (0.81 + 0.35) moles each of B and C and there remains 0.345 mole of A in stream G. Again this stream is split into two equal parts and one is returned; and so 0.173 mole of A is added to the new mole of A feed, and together with 0.58 mole each of B and C, this stream goes through the reactor again.

In the table you can see how this continues until a steady state is reached. At that point you can see that the product stream contains 0.353 mole of A and 1.63 moles each of B and C. This is a total of 3.621 moles, and since we have 0.353 mole of A, this represents a concentration of A of 9.7%—considerably better than the 30% we would have had without a recycle. The temporary storage of the product stream can be stopped and it can be sold directly if the 9.7% of A is acceptable in the product. By the simple addition of a recycle to bring back a portion of the products, a reactor capable of a 70% conversion of its feed has been improved to a reactor system capable of a 90% conversion.

The start-up period, which may last many hours—or even days, will produce partially converted material which, as you saw above, is stored in temporary holding tanks. This material can be gradually introduced into the feed stream after the steady-state operating condition has been reached. There is, however, an alternative approach to the start-up procedure. The reactor system could be put on "100% recycle", that is to say the whole products stream could be recycled. Eventually, the steady-state condition would be reached and the products stream would then be split and a stream would be drawn off for sale.

Of course, there are some snags with a recycle-reactor system. First, an extra pipe must be provided, probably with expensive insulation on it, to recycle the products. Also, there must be a T-junction on both the feed and product pipes, together with any necessary valves and controls which are also required. Furthermore, the actual flow rate through the reactor, instead of being one mole of A per minute, becomes over 3.5 moles per minute of reactants and products; thus the volume and cross-sectional area of the reactor must be greater to accommodate this extra flow, which means more expense when building the reactor. There must also be a recycle compressor in this system, and that too is an added expense.

Now suppose that the customer requires that the percentage of reactants in the product stream be brought down to less than 2.5%. In this case it will be necessary to increase the size of the recycle in order to meet specifications. Figure 7-10 shows the effect of changing the recycle from 50% to 90%; i.e., taking off only 10% of the products for sale. Look at the table in Figure 7-10. It takes a lot more arithmetic to reach the steady state, and you will see towards the bottom of the table that I have omitted great chunks of the calculation in order to save space. Eventually, however, the calculation reaches a steady state and this is found when one mole of A joins 0.371 mole of A in the recycle stream (together with 8.69 moles each of B and C) and passes through the reactor to give a product stream containing 0.411 mole of A and 9.65 moles each of B and C. The total number of moles here is 19.7, representing a percentage A of 2.1%. This is below our specified requirement. But look at the cost. The reactor has to be large enough to allow the passage of 19.7 moles per minute instead of one mole per minute. The recycle pipe must be about 20 times the cross-sectional area of the feed pipe. In some situations, though, this is perfectly acceptable and this method of improving reactor performance is used.

Common Reactions Used in the Petrochemicals Industry

Having learned about the various types of hardware called reactors, now let me turn your attention to the kinds of reactions for which they can be employed in the petrochemicals industry. I have described below the 12 most common reaction types. Go through them carefully and examine the associated figures until you get a feel for the versatility of molecule-building in our industry.

There is just one small piece of organic chemistry to know before going through these reactions. Each carbon atom has four *valency bonds* with which it can form links to other atoms. An oxygen atom has two bonds: hydrogen and chlorine have one each. The point to grasp is that in all compounds processed and made in the petrochemicals industry, molecular structures are governed by the constraints imposed by the number of valency bonds of each atom in the molecule. If you look through the chemical structures in the figures in this chapter and in Chapters 5 and 6, you will find that however simple or complex the molecule, carbon, hydrogen, oxygen and chlorine will have four, one, two and one bond, respectively. Occasionally a carbon atom will be linked to another carbon atom with two bonds; this is called a *double bond.*

Study all the reactions thoroughly and try to see how the atoms have been moved around. If you hit trouble in understanding them, ask your professor to explain the problem to you. As you read the short description of each reaction, look at the accompanying figure until you can see the mechanism involved.

Hydrogenation (Figure 7-11)

A molecule which contains a double bond is called an *unsaturated structure.* The molecule is strained, although the reason for that may not be clear at first. It may help to consider the carbon atom as a steel ball with four stiff steel springs projecting from it and positioned such that they are equidistant from each other. A methane molecule is formed if a hydrogen atom is attached to the end of each of the steel springs. Now consider an ethylene molecule where two carbon atoms link up to each other. The two springs have to bend around to each other to connect up. This causes considerable strain in the middle of the molecule, and although the bonds are quite rigid under normal circumstances, it is simple to flick open the bonds and attach other atoms to them. Thus, by adding hydrogen atoms, the reaction known as hydrogenation fills up the free bonding in unsaturated molecules.

Look at the example in Figure 7-11. Ethylene hydrogenates to ethane; benzene to cyclohexane. Hydrogenation, then, is a means of increasing the hydrogen-to-carbon ratio in an organic molecule.

Dehydrogenation (Figure 7-12)

This is a more common reaction than hydrogenation, the latter being used only in very special circumstances. It is far more likely that there will be a need to produce the unsaturated molecule with its double bonds because it is easy to do further reactions with a compound "eager" to be rid of those double bonds. By choosing molecules into which double bonds can be placed (by dehydrogenation) at strategic points in the molecule, it is possible to control the ultimate product of a series of reactions. Figure 7-12 shows a dehydrogenation reaction. Ethane is reduced to ethylene by pyrolysis in which the dehydrogenation reaction is made to occur. Isopropyl alcohol, itself a basic petrochemical, is reduced to acetone, a well known solvent.

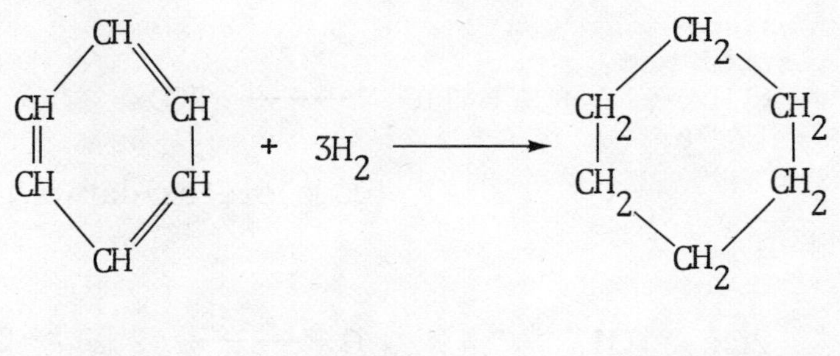

***Figure 7-11.** Hydrogenation reactions.*

$$(CH_3)_2CHOH \longrightarrow (CH_3)_2CO + H_2$$

isopropyl alcohol — acetone — hydrogen

$$CH_3{-}CH_3 \longrightarrow CH_2{=}CH_2 + H_2$$

ethane — ethylene — hydrogen

Figure 7-12. Dehydrogenation reactions.

Oxidation (Figure 7-13)

The oxygen atom has two bonds, or valencies, available for bonding to carbon atoms. This means that if a carbon atom is linked to an oxygen atom, the carbon atom has only two more bonds available for use. Oxidation is a reaction in which the ratio of oxygen to hydrogen is increased in a molecule. Careful oxidation control makes it possible to add just one oxygen atom to a molecule. More atoms may be added by prolonged oxidation, and in the extreme, the carbon and hydrogen atoms in the molecule could become completely oxidized to form carbon dioxide and water.

Figure 7-13 shows several oxidation reactions, the last two of which fall into the complete oxidation category. These are the chemical equations representing the complete combustion of methane and heptane. The most important reactions here are the oxidation of ethylene to produce ethylene oxide, itself an important basic petrochemical that can be used to make acrilonitrile, acrylic plastics, ammonia, ethanolamines (used as basics for weed killers), detergents and cosmetics, ethylene glycol (antifreeze) and other glycols which are used in the cosmetics, pharmaceuticals and other industries.

Another interesting reaction in Figure 7-13 is the partial oxidation of ethyl alcohol (ethanol) to acetaldehyde. This is the fate of the ethyl alcohol in liquor, wine and beer when it enters the stomach of a human being. The partial oxidation occurs and acetaldehyde is formed. If the imbibing occured at a party which was a long and well stocked one, sufficient acetaldehyde will be present to behave as an emetic, and vomiting occurs. I hope that you do not need to experiment and you are happy to take my word for this!

$$2CH_3CH_2OH + O_2 \longrightarrow 2CH_3CH{=}O + 2H_2O$$

ethyl alcohol — oxygen — acetaldehyde — water

(partial oxidation)

$$CH_4 + 2O_2 \longrightarrow CO_2 + 2H_2O$$

methane — oxygen — carbon dioxide — water

(complete oxidation)

$$C_7H_{16} + 11O_2 \longrightarrow 7CO_2 + 8H_2O$$

(complete oxidation)

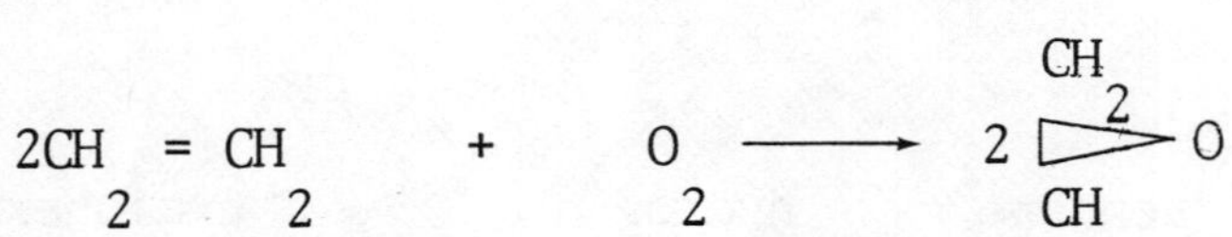

(complete oxidation)

Figure 7-13. Oxidation reactions.

Reduction (Figure 7-14)

This is, of course, the reverse process to oxidation, and its objective is to reduce the ratio of oxygen atoms to hydrogen atoms in a molecule. Thus we could turn acetaldehyde back into ethyl alcohol by adding hydrogen in a suitable reactor. Figure 7-14 shows the reaction.

$CH_3CHO + H_2$	→	CH_3CH_2OH
acetaldehyde + hydrogen		ethanol

Figure 7-14. *A reduction reaction.*

Chlorination (Figure 7-15)

This is an important reaction because there are many useful chemical products that have chlorine atoms in them. If the hydrocarbon to be chlorinated is saturated, i.e., it has no double bonds, then the chlorine atom replaces a hydrogen atom. This is a selective process, and by repeating the reaction under slightly different conditions, more and more hydrogen atoms can be replaced by chlorine as shown in Figure 7-15. The first example is the removal of one hydrogen atom from methane, replacing it with an atom of chlorine. The product is methyl chloride. Notice how the reaction can be repeated to give, successively, dichloromethane, trichloromethane (known commercially as chloroform) and carbon tetrachloride (known commercially as CTC).

If, on the other hand, the hydrocarbon to be chlorinated is an unsaturated hydrocarbon, then the chlorine atom attaches to either side of the double bond. The example shows how ethylene can be chlorinated to produce dichloroethane. This is particularly important since dichloroethane can be made into vinyl chloride, a vitally important basic chemical because it can be polymerized (see Figure 7-22) into polyvinyl chloride (PVC). An alternative way of making vinyl chloride is from acetylene which, having a triple bond, is easily chlorinated with hydrochloric acid to give vinyl chloride.

Alkylation (Figure 7-16)

We have already encountered one or two examples of this reaction in Chapter 6. The word "alkyl" signifies a building block which is made up of carbon and hydrogen atoms in the ratio given by the formula:

$$C_nH_{2n+1}$$

Examples of alkyl groups are the methyl group (CH_3), the ethyl (C_2H_5), the acetyl (CH_3CO), and so on. Alkylation is a general term meaning any reaction in which an alkyl group is added to a molecule. Methylation is a specific term to indicate the addition of a methyl group CH_3 building block to a hydrocarbon which of course must lose hydrogen to make way for the new alkyl group unless a double bond is broken open.

In the examples in Figure 7-16 you can see how double bonds are broken to add the new group. The isobutane and propylene are reacted to give isoheptane. This reaction is a useful one in producing gasoline-range hydrocarbons from light gaseous hydrocarbons. A more useful application of alkylation is to take an aromatic hydrocarbon to form an aromatic with a side chain. Figure 7-16 shows how an olefin (an unsaturated hydrocarbon in the same homologous series as ethylene) called octylene is reacted with benzene to give octylbenzene. Notice that the hydrogen which is taken off the benzene ring actually adds into the octylene side group and saturates it.

The first example shown in Figure 7-16 is a common one. It is the first reaction in a sequence used to make detergents. Alkylation of the benzene is followed by sul-

$CH_4 + Cl_2$ methane + chlorine	→	$CH_3Cl + HCl$ methyl chloride + hydrochloric acid
$CH_3Cl + Cl_2$ methyl chloride + chlorine	→	$CH_2Cl_2 + HCl$ dichloromethane + hydrochloric acid
$CH_2Cl_2 + Cl_2$ dichloromethane + chlorine	→	$CHCl_3 + HCl$ trichloromethane + hydrochloric acid
$CHCl_3 + Cl_2$ trichloromethane + chlorine	→	$CCl_4 + HCl$ carbon tetrachloride (CTC) + hydrochloric acid
$CH_2 = CH_2 + Cl_2$ ethylene + chlorine	→	$CH_2Cl - CH_2Cl$ dichloroethane

Figure 7-15. *Chlorination reactions.*

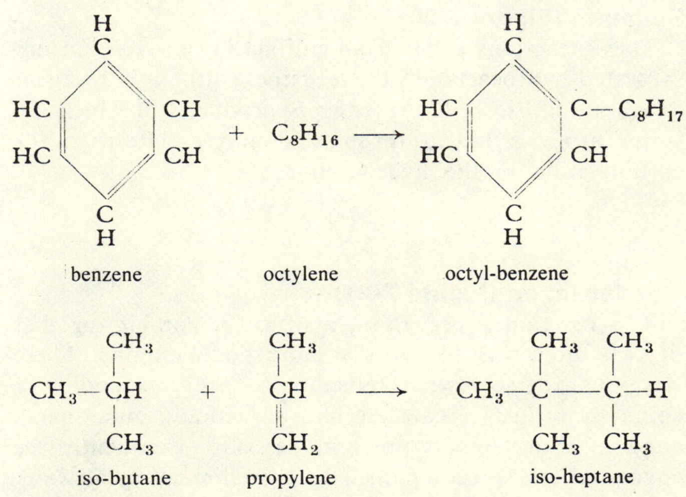

Figure 7-16. *Alkylation reactions.*

fonation and neutralization. Research into the bacterial digestion of alkyl detergents led to the discovery that if straight-chain olefins are alkylated, sulfonated and neutralized, then the resulting detergent is easily broken down by bacteria and contributes less to the pollution of rivers and oceans, the ultimate destination of surplus detergents in our sewage. These detergents are called "biodegradable."

Esterification (Figure 7-17)

This is a common reaction in which an alcohol is reacted with an acid to form an ester and water. A particularly important series of esters are called acetates and are made by adding acetic acid to various alcohols. One of these reactions is shown in Figure 7-17. Acetates are used in the production of several synthetic fibers and perhaps the most well known of these is artificial silk, called rayon.

Hydrolysis (Figure 7-18)

In this reaction water is made to react with an ester so that the reverse of esterification occurs. The products are an alcohol and an acid. Figure 7-18 shows an example.

Sulfonation (Figure 7-19)

Sulfonation is the result of mixing sulfuric acid with an aromatic hydrocarbon to form sulfonic acid. The sulfur atom attaches directly to a carbon atom in the hydrocarbon, as shown in the example in Figure 7-19.

Sulfation (Figure 7-20)

These reactions differ from sulfonation because an unsaturated hydrocarbon is the feedstock, although the agent is again sulfuric acid. An ester is produced in which the sulfur atom is linked through an oxygen atom to the carbon atom in the hydrocarbon.

Condensation (Figure 7-21)

This reaction is best thought of as the coming together of two molecules to form a more complex one. These reactions are somewhat reluctant to progress and it is common to use a catalyst and sometimes quite harsh conditions of temperature and pressure. Frequently the condensation is accompanied by the elimination of water from the compounds. Figure 7-21 shows an example of

$$CH_3COOH + CH_3CH_2OH \rightarrow CH_3COOCH_2CH_3 + H_2O$$

acetic acid + ethanol → ethyl acetate + water

Figure 7-17. *An esterification reaction.*

$$CH_3COOCH_2CH_3 + H_2O \rightarrow CH_3COOH + CH_3CH_2OH$$

ethyl acetate + water → acetic acid + ethanol

Figure 7-18. *A hydrolysis reaction.*

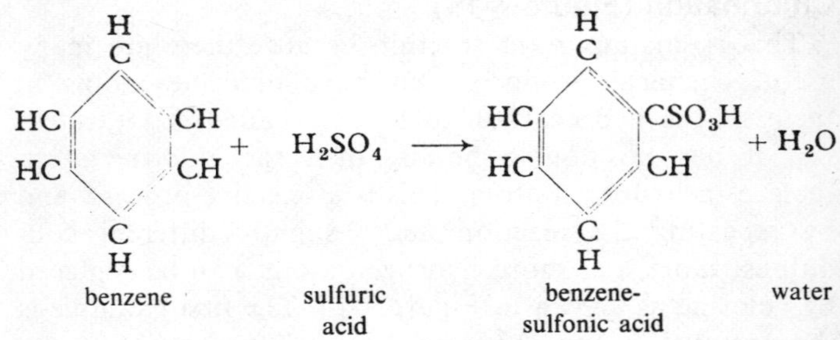

Figure 7-19. *A sulfonation reaction.*

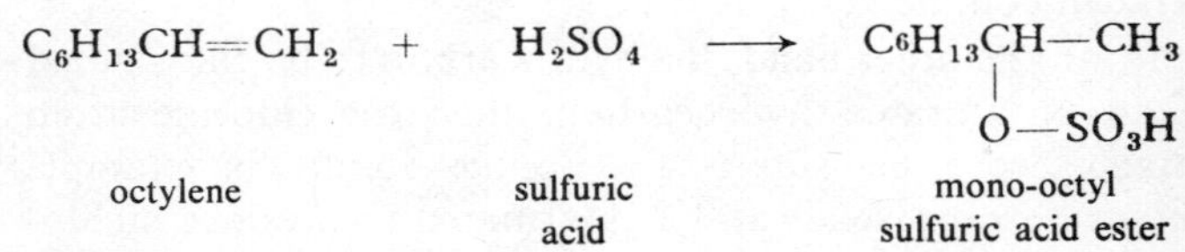

Figure 7-20. *A sulfation reaction.*

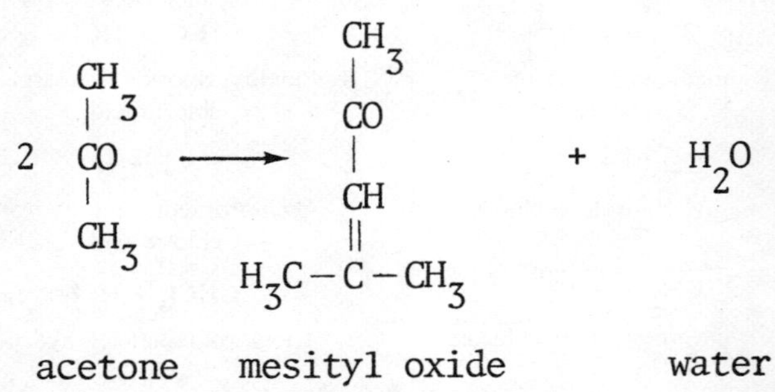

$$CH_2{=}CH_2 + H_2O \longrightarrow CH_3{-}CH_2{-}OH$$

ethylene water ethyl alcohol

Figure 7-21. *A condensation reaction.*

two acetone molecules condensing to form a molecule called mesityl oxide (used as a feedstock in making industrial solvents) with the elimination of water. Sometimes the second molecule is water itself, and in Figure 7-21 you will see the example of ethylene reacting with water to condense to ethyl alcohol.

Polymerization (Figure 7-22)

This must be the single most important type of reaction in petrochemical engineering. Polyvinyl chloride, polystyrene, polyethylene, polypropylene, nylon, Dacron (a tradename of the Du Pont Company) and polyurethane are well known examples of the extraordinary products obtained by the polymerization of simple organic compounds. There are many, many more, but the subject is far too complex to cover here except at a level of detail that will do no more than whet your appetite. In the department where I teach there are two complete courses on the subject of polymers and their processing. Indeed, at some universities it is now possible to take a degree in polymer science.

We have already seen how the ethylene molecule consists of two carbon atoms joined together by a double bond. Attached to each carbon atom are two hydrogen atoms. Thus, each carbon atom has all four of its valency bonds occupied. Because of the tension of the strained double bonds, they open up easily, allowing other bonding to occur at the two carbon atoms. This is the major factor which makes polymerization possible.

Before we examine the polymerization process and its results, note that the structure of the ethylene molecule is common to a whole range of organic chemicals. Figure 7-22 shows the chemical structures of ethylene; vinyl chloride, in which one of the hydrogen atoms has been replaced by a chlorine atom; styrene, in which one of the hydrogen atoms has been replaced by a benzene ring; propylene, in which a hydrogen atom has been replaced by a methyl (CH_3) group; and vinyl acetate, in which one hydrogen atom has been replaced by an acetate (CH_3OCO) group. Each of these molecules (called *monomers* in Figure 7-22) can be polymerized to produce a corresponding polymer. Here the Greek prefixes *mono*, meaning one, and *poly*, meaning many, are used to differentiate between the structure of the material before and after polymerization.

Another phrase I shall use is *high polymer*, which means that the monomer building block has been strung together to form a very long chain, only part of which is depicted in Figure 7-22. The number of monomers which go together to form a high polymer varies with both the kind of polymer and the purpose to which it will be put. The familiar clear plastic called polystyrene contains about 1,000 styrene monomer units per molecule. This means that the new molecule contains about 16,000 atoms instead of 16 and, if stretched in some way, would have a length of just under 10 millionths of an inch. This compares with the length of a molecule of styrene monomer of just over $^1/_{50}$-millionth of an inch. In other words, the polymer is about 500 times as long as the monomer.

Though these molecules are still very tiny (a human hair is 5,000 millionths of an inch in diameter), they are unwieldy and are difficult to move around or arrange into regular patterns to form crystals. This is because they easily coil up and become entangled with each other. It is worth stressing this point. If you consider substances composed of small molecules like salt (sodium chloride) or ice, you can imagine that when a few billion billion of them are placed together in the solid form they stack in just the same way that a few billion spherical beads would if poured into a football stadium. If a solid is melted, individual molecules are suddenly freed from the crystalline structure at a given temperature—the melting point of the solid. Furthermore, if we heat up the liquid, the molecules, as you saw in Chapter 5, are free to move around, and after the liquid boils the molecules have even more freedom in the vapor phase.

Now let us consider the polymer molecules. First of all, they cannot be thought of as small spheres or beads. If we wish to consider water molecules in an ice crystal as being, say, as large as a pea, then a polyethylene molecule (Figure 7-22) on the same scale would look like a climber's rope: half an inch thick and over 40 feet long. Imagine throwing a billion or so such lengths of rope into a football stadium. The way they stack will be most nonuniform and there will not be even the slightest similarity to a crystal. Even when heat is provided to the mass of molecules, there could never be one exact temperature which could be called a melting point, since the molecules are far too tangled up with each other to be able to flow into a liquid at any single temperature. What actually happens is that melting occurs over a range of temperatures, starting off with the polymer becoming softer and eventually finishing up with it in the liquid phase as the tangled chains become free from their neighbors. The liquid, or "melt", produced is extremely viscous because the long bulky molecules must be dragged over one another.

It is seldom possible to turn the polymer into a gas. Because of their length, the amount of energy required to give one of these super molecules sufficient velocity to escape from its fellows is enormous. Even if the north end of a super molecule, or macromolecule, of a high polymer happens to be traveling towards the surface of the melt at a high enough speed to break through, the south end could be standing stationary or wrapped around a dozen other molecules in a seemingly hopeless tangle. Assuming sufficient heat could be provided to the melt, the temperature would rise so high that the molecule would decompose rather than pass into the gaseous state.

The physical behavior of polymers is most familiar. In fact, the average person is far more familiar with polymers than any other chemical substances because many natu-

Monomer	Part of Typical Chain Structure of Polymer

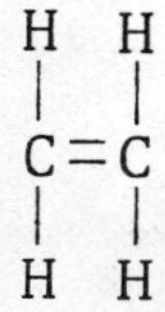

ethylene

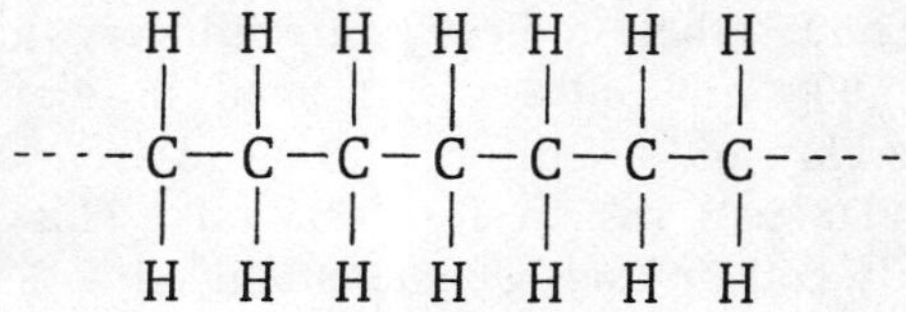

polyethylene

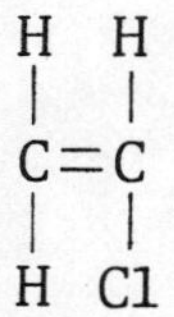

vinyl chloride

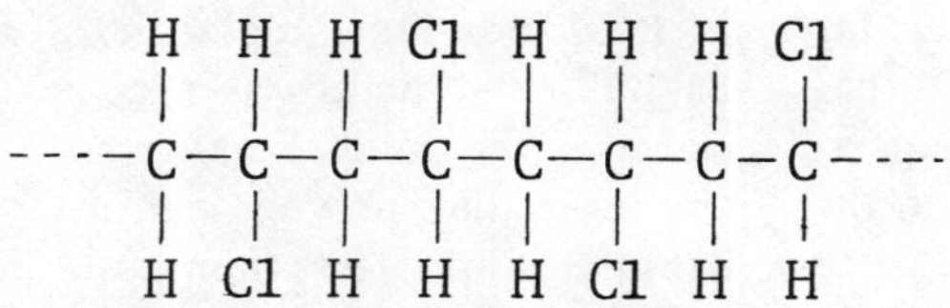

polyvinyl chloride (PVC)

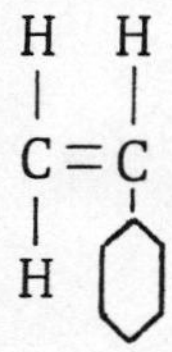

styrene

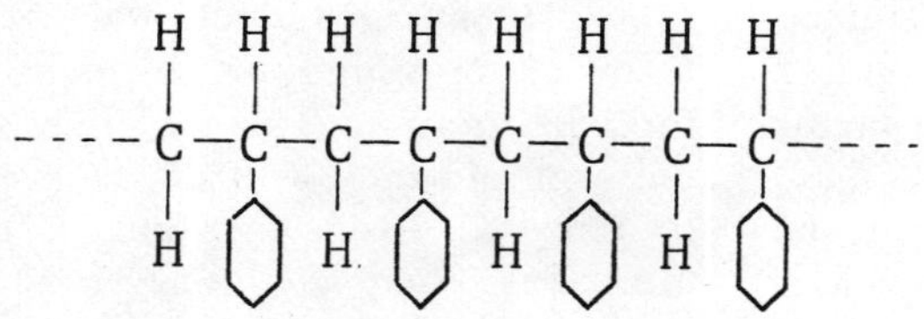

polystyrene

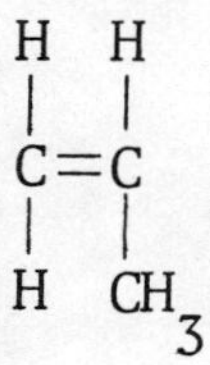

propylene

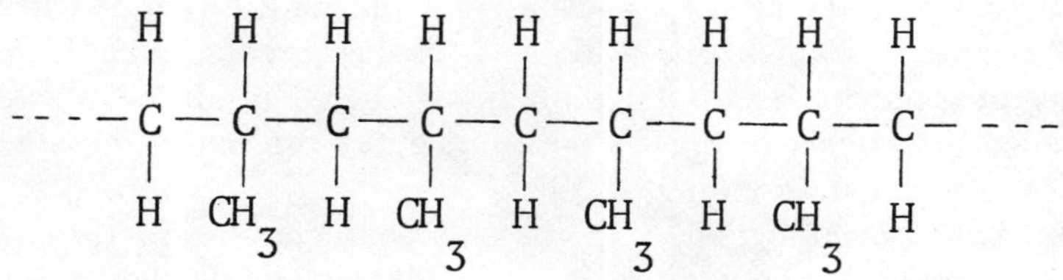

polypropylene

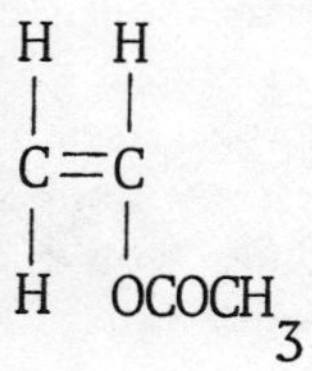

vinyl acetate

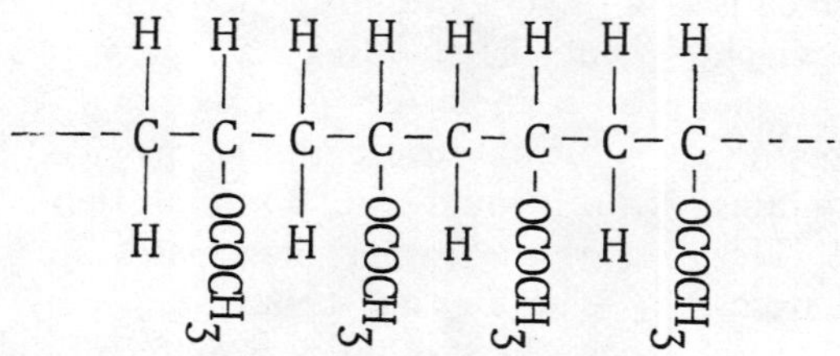

polyvinyl acetate

Figure 7-22. *Polymerization.*

rally occurring materials are polymers: hair, leather, rubber, cotton, wool, bitumen and wood, to name a few. You already know that leather cannot be melted. You already know that you cannot vaporize silk. You can melt rubber and bitumen, but the melt is very viscous, even more so than honey. And who ever heard of rubber being turned into gas?

In the last 50 years it has been possible to manufacture polymers from petroleum, and new "miracle" materials are now available for use in every walk of life. A substance called isoprene can be polymerized to give synthetic rubber, which has properties almost identical to those of naturally occurring rubber. As a matter of fact, if you can think of a way natural rubber can be improved, then it is perhaps possible to modify the polymerization process to produce a synthetic rubber which has all the properties of natural rubber plus the improved property that you have specified.

Obviously, the chemical nature of the single molecule or monomer unit incorporated in a polymer determines the physical properties of the high polymer. Although with modern knowledge polymers can be made quite easily and cheaply, a high polymer is no use unless it (a) can be melted in order to form it into the particular shape required and (b) will finish up with a specific combination of desired physical and chemical properties such as strength or electrical resistance or immunity to solvents. Engineering developments in custom-making polymers have been followed by technological developments in the various methods of molding and shaping. Thus, a completely new industry has evolved.

Thermoplastic Polymers

There are two main kinds of polymers: thermoplastic and thermosetting. The first of these—the subject of all the preceding discussion of polymers—includes polymers that soften when heated and eventually become liquid if heated long enough. The same liquid, if cooled, forms the original solid with no change in any of its properties. This process can be repeated over and over again. The name *thermoplastic* tells us that the action of heat upon the polymer makes it plastic and pliable. Polyethylene, polyvinyl chloride, polypropylene and polystyrene are such polymers. They are manufactured from substances which are gases (ethylene, vinyl chloride, propylene) or liquids (styrene, methacrylate). They are usually polymerized in liquid form, so the monomers which are normally gases must first be liquefied.

The polymerization process itself—in which hundreds or thousands of monomer units are linked together happens when the monomer is heated in the presence of a catalyst. The catalyst is important; it helps initiate the polymerization reaction by opening up the double bonds of the first two ethylene monomer molecules. After that, the reaction becomes self-sustaining, like a "chain reaction", as each ethylene molecule finds itself in the vicinity of a double bond that could be thought of as reaching out for valency bonds to link with. The length of the chains depends upon the reaction conditions and, in particular, the temperature. As the double bond breaks down, its energy is released—the reaction is exothermic, heat is given out—and this must be dispersed by stirring the mixture thoroughly and at the same time cooling it.

A detailed description of the processes of polymerization, and there are many, is far beyond the scope of this book. However, with your newfound knowledge of chemical processes you may be interested in just the two following methods. The first is used for manufacturing polystyrenes; the second, for polyethylenes.

Manufacture of Two Thermoplastic Polymers—Polystyrene and Polyethylene

Liquid styrene is pumped into a stainless steel vessel which is then sealed. Heating and stirring follow until about a third of the monomer has polymerized. The polystyrene is actually soluble in the styrene liquid and so the liquid becomes very viscous and stirring becomes difficult. If it were not stirred, the chemical energy released as heat of polymerization would raise the temperature too high and cause decomposition. To overcome this, the viscous liquid is pumped into the top of a vertical tower in which the temperature is controlled by internal cooling coils and by lining the outside of the wall with cooling jackets. The liquid dribbles down the column and the temperature is allowed to increase toward the bottom of the tower, where the polymerization is just about complete. A screw feeder device (similar in every way to the one we saw in Figure 4-7) carries the molten polymer out of the tower and forces it through small holes drilled in a disc called a spinaret. The polymer squirts out to form cylindrical strands. These are passed into a water bath to cool and solidify. And finally they are chopped into small pellets. If colored polystyrene is required, pigments are added to the liquid polymer before the *extrusion* into the cooling water. The pellets are packaged for sale or taken to the molding process, which I shall describe in just a moment.

Polyethylene is not soluble in liquid ethylene. If a similar method of polystyrene manufacture were used, the polyethylene would form a *slurry* (a solid-liquid mixture) which, as polymerization progressed, would become increasingly difficult to stir, and temperature control would be out of the question. This difficulty is overcome by dissolving the monomer in a solvent that has two essential properties: first, both the monomer *and* the polymer are soluble in it; second, it is not affected in any way by the temperature and pressure conditions in the polymerization process.

In one method of manufacturing polyethylene, the catalyst is solid chromium oxide whose effect is completely stopped by the presence of even the slightest trace of water. To avoid waste, the ethylene and the solvent (cyclohexane is commonly used) are scrupulously dried to remove any last trace of water. The ethylene is dissolved in the cyclohexane and then pumped at high pressure into a stainless steel tank where a small amount of the catalyst is added. The solution is stirred and polymerization begins. The polymer product enters into solution in the cyclohexane. The liquid solution, still under pressure, undergoes centrifugation and filtration to remove the solid catalyst. Then the pressure is released. At this point any monomer ethylene remaining bubbles out of solution as a gas and is collected to be recycled to the beginning of the process.

The solution of the polymer in cyclohexane is poured into a vat of boiling water. Immediately, the cyclohexane evaporates, is collected, purified and recycled. The polyethylene, however, remains behind as a solid which is collected and dried under a hot air blast and then made into polyethylene molding granules or pellets by passing it through an extrusion process similar to the one described for polystyrene. Pigments can be added if colored polyethylene is required. The pellets or granules are packaged for sale or they are taken to the next process—molding.

The Screw Extruder

Again, the details of the molding process require their own book. To satisfy your curiosity, I shall just briefly describe the equipment used. The reason I am reluctant to go into greater detail is that the liquid polymers, once melted in the equipment shown in Figure 7-23, behave in a manner quite different to the fluids we discussed in Chapter 5. The molding equipment is called a "screw extruder"; Figure 7-23 shows a cross-section through one. A finely engineered steel barrel is heated externally by electric heaters. A feed hopper contains the polymer pellets (also called molding powder or molding granules) that are caught in the "flights" of the central solid steel screw, which turns slowly, forcing the pellets through the heated section of the barrel. Notice that the "pitch" of the screw decreases, and so as the pellets move from right to left in Figure 7-23 they are compressed between the shaft, the flights and the barrel. The increasing pressure and the heat coming through the walls cause the polymer to melt.

At the end of its travel through the heated barrel, the liquid is forced through the "breaker plate", which gives a uniform flow pattern to the polymer melt, and then finally through a "die" which has a cross-sectional shape according to the requirements of the process. If the die is a circular hole, it will produce a polymer rod. The die can be replaced by a hollow mold. For instance, to make a polyethylene handle (the kind you find on the cupboard door) a steel mold is made in the shape of the handle and the extruded polymer is forced into the mold.

Often, to increase the speed of the molding of individual articles such as the door handle, the mechanical engineering of the process is arranged such that the mold is clamped into position, the screw extruder shaft is given a ramming motion, and a small charge of the polyethylene is forced into the mold. The ram is then returned to its original position, a new empty mold is clamped into position on the extruder and again the ram is given a thrust. This can be highly automated so that the manufactured items are turned out very quickly.

Of course, there is a complex heat transfer problem. It

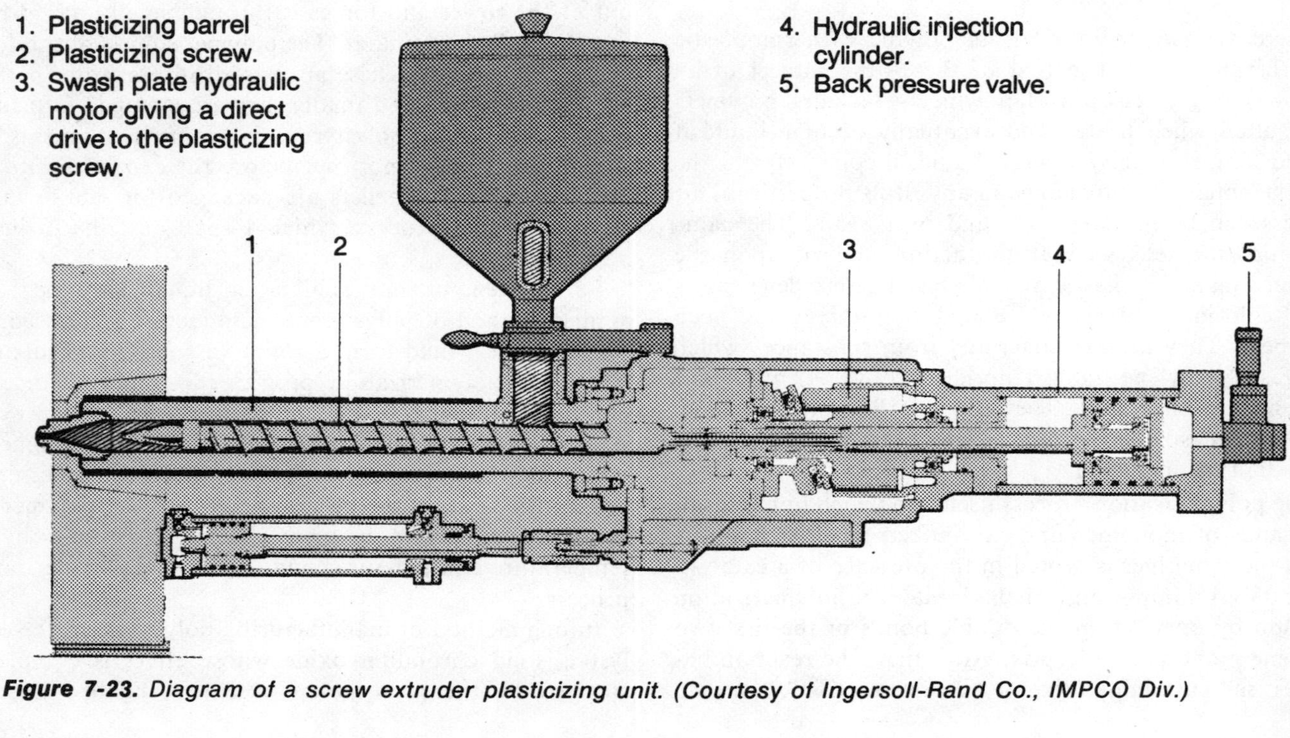

Figure 7-23. Diagram of a screw extruder plasticizing unit. (Courtesy of Ingersoll-Rand Co., IMPCO Div.)

is essential that the polymer be molten enough to flow into the mold without forming bubbles or cracks or any other surface defects. But the temperature of the polymer must not be allowed to go so high that any chemical decomposition occurs. Nor must the polymer get hot enough to become so fluid that it flows into the mold and out of the air escape holes. Similarly, the mold must be cooled so that by the time it is opened and the finished object released, the product is not still plastic enough to be damaged as it falls from the mold. On the other hand, if we cool too much, then the entering polymer freezes too quickly and forms surface defects in the product.

The designer of a screw-extrusion system has to overcome this difficult heat transfer problem. Chemical engineers who have specialized in heat transfer will be found in the design offices of these screw-extruder manufacturers. Their job is made even more difficult by the complex geometry of the shapes involved. It is no longer heat transfer through a wall such as we saw in Figure 3-4. Look at the shape of the extruder screw, for instance, or the shape of the polymer pockets between the screw, the flights and the barrel wall. The equations become almost too complex to solve. However, they *can* be solved by computer programs, which are themselves so difficult, long and complex that to my knowledge there are only two in the world. Their names tell you their jobs: one is called POLLY and the other is called EXTRUD.

Thermoplastic polymers are also used to make the familiar plastic bottle, plastic bowls and dishes, film, and sheet plastic. Where would the packaging industry be without polyethylene bottles and polyethylene sheet? To avoid lengthening this book unnecessarily, I have condensed the manufacture of plastic bottles into Figure 7-24. Similarly, Figure 7-25 shows the manufacture of poly-

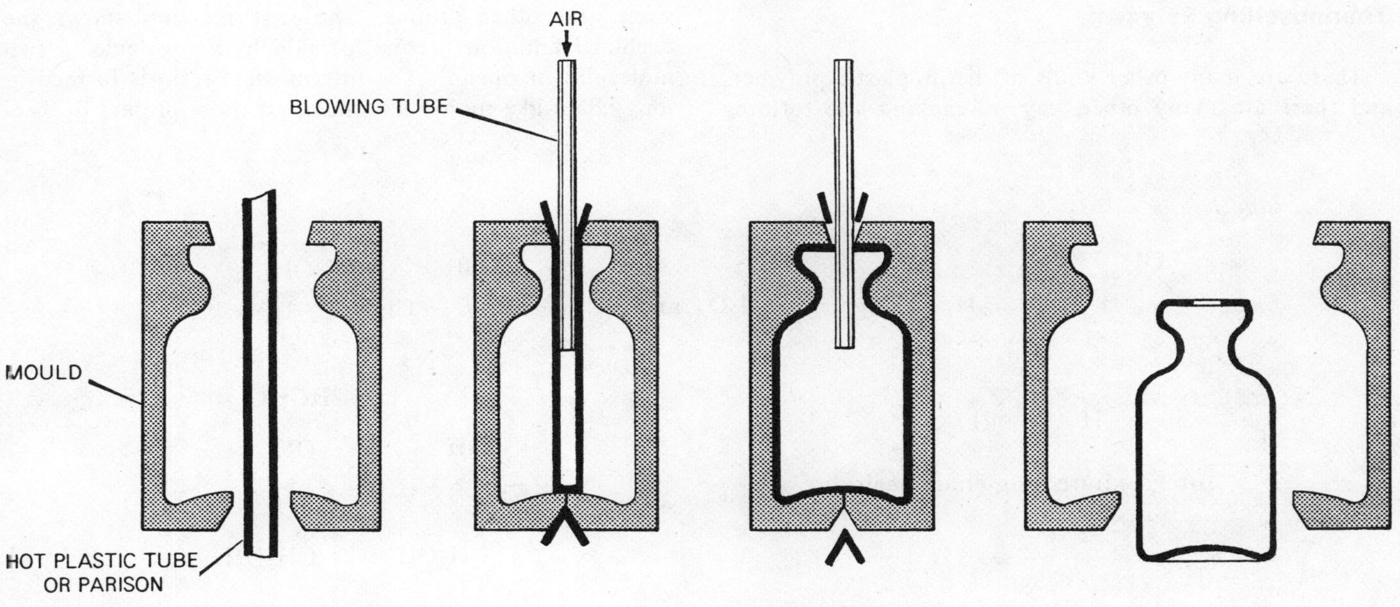

Figure 7-24. *The principle of blow molding.*

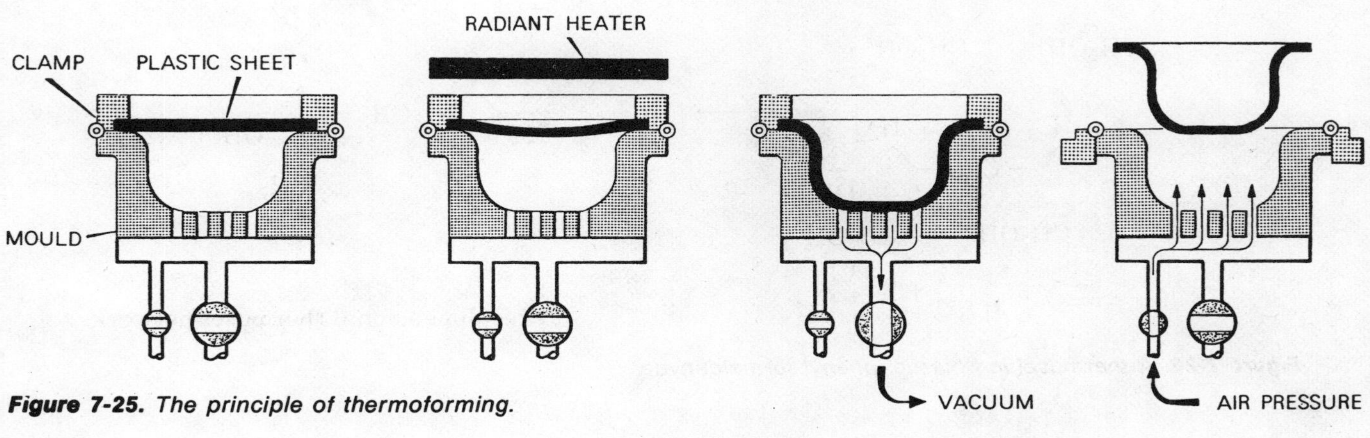

Figure 7-25. *The principle of thermoforming.*

ethylene bowls. Incidentally, as you look at these two figures try to imagine the heat conduction problems in the complex shapes. Again, the calculations are so complex that they must be solved by computer programs which are similar, although considerably more complex, to the ones you may be asked to write to solve the problems in Appendix II at the end of the book.

Thin film or plastic sheet is made by several methods, one of which is called *casting*. The polymer is dissolved in a suitable solvent and spread onto a moving steel band which is highly polished. The polymer solution moves along with the band and passes through drying cabinets where the solvent evaporates off and is collected for recycling. The dried sheet is continuously peeled off the band and passes through cutters which trim the edges. It is finally reeled onto a spool. Polyethylene is made into packaging sheet by this method and cellulose acetate polymers are made into photographic film this way.

Thermosetting Polymers

There are many other kinds of thermoplastic polymers and there are many other ways of making and forming them. But now we must turn to the other kind of polymers, called *thermosets*. As the name implies, the action of heat on these polymers causes them to set. What this means in practice is that once the polymer has been heated up, formed into the required shape and cooled down, no amount of heating will cause the polymer to melt a second time. If the temperature were to be taken high enough, the polymer would just decompose or burn.

The thermosetting polymers or resins were first developed in the 1920s and for a long time were the major synthetic solid materials used to make consumer goods. Early telephones and electrical fittings were made of Bakelite, a thermosetting resin. Today they are still being made and used, but not in anything like the quantities of thermoplastics being manufactured.

Figure 7-26 shows the production of phenylformaldehyde. You can see how the monomer is made into an intermediate long-chain resin which can be further polymerized to form a three-dimensional network of benzene rings and other groups. The first reaction shows the chemical addition of one formaldehyde molecule to two molecules of phenol. The intermediate resin is formed in long chain-like molecules similar to those present in ther-

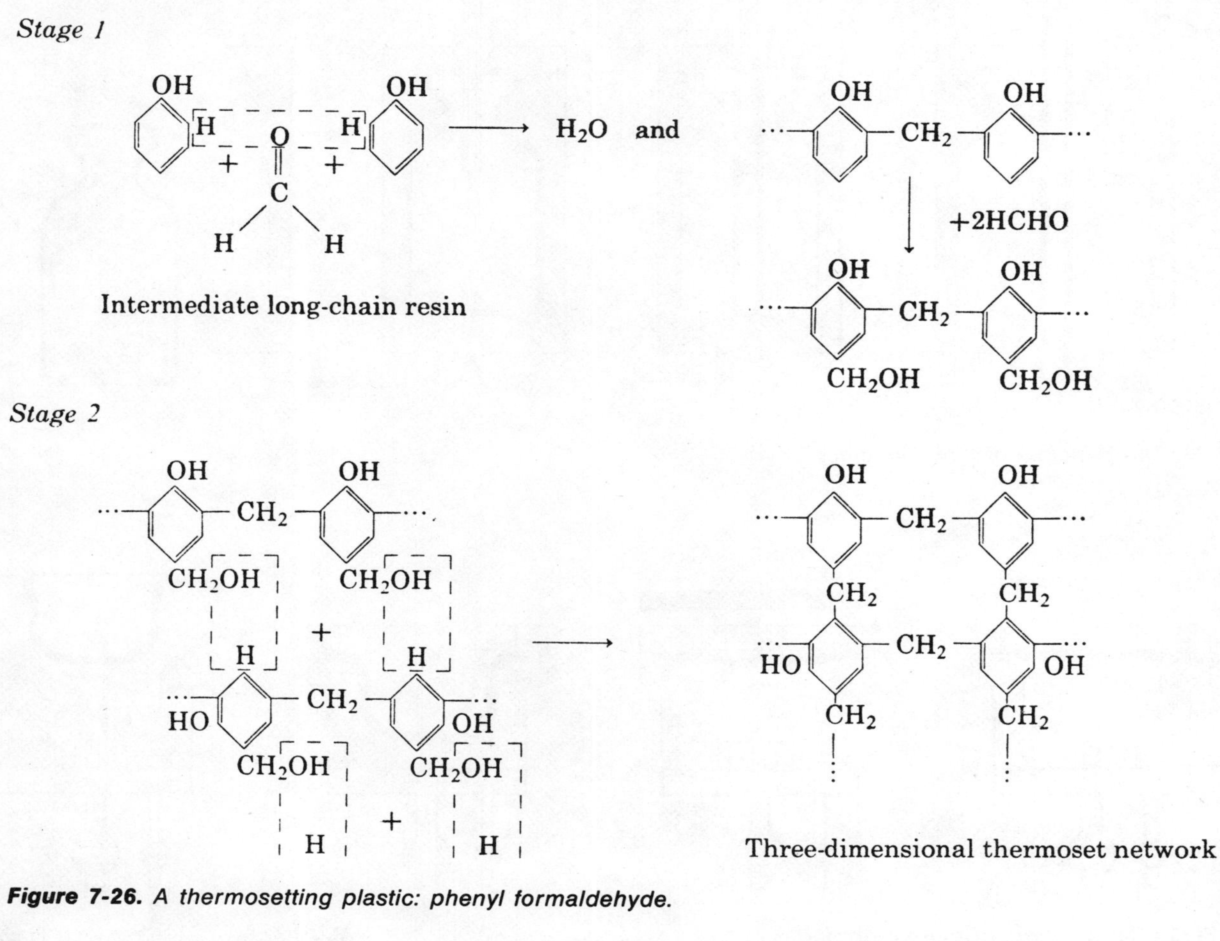

Figure 7-26. *A thermosetting plastic: phenyl formaldehyde.*

moplastics. It is liquid and can be handled easily. Most important, it can be mixed with color pigments or an inert powder which will act as a filler: a way of increasing the final volume at a cost lower than using more of the chemicals themselves.

The thermoplastic liquid resin is then placed in a mold and heated under high pressure. The mold is the same shape as the final product of this resin. The last reaction occurs and the rather complicated three-dimensional thermosetting network shown in Figure 7-26 is the result. After cooling, the resulting molding with its network structure cannot be softened by heat. Figure 7-26 shows just a schematic series of reactions for phenylformaldehyde; it is thought that the actual processes involved are somewhat more complicated.

Synthetic Rubbers and Fibers

There are other kinds of polymers. Synthetic rubbers, sometimes called elastomers, are made from the diolefins (or "dienes" for short). These are hydrocarbons containing two reactive double bonds, and the most common elastomer is natural rubber, which is called poly-cis-1:4-isoprene. Figure 7-27 shows a number of synthetic rubbers.

Finally, there are the synthetic fibers. The macromolecules of these polymers are long and rather inflexible. Strong, cohesive forces between the chains cause the molecules to be suitable in fitting together to form mechanical, almost crystalline, structures. Cotton, cellulose and other natural polymers show this property. It is as though the long molecules were a string of beads on a fairly stiff wire.

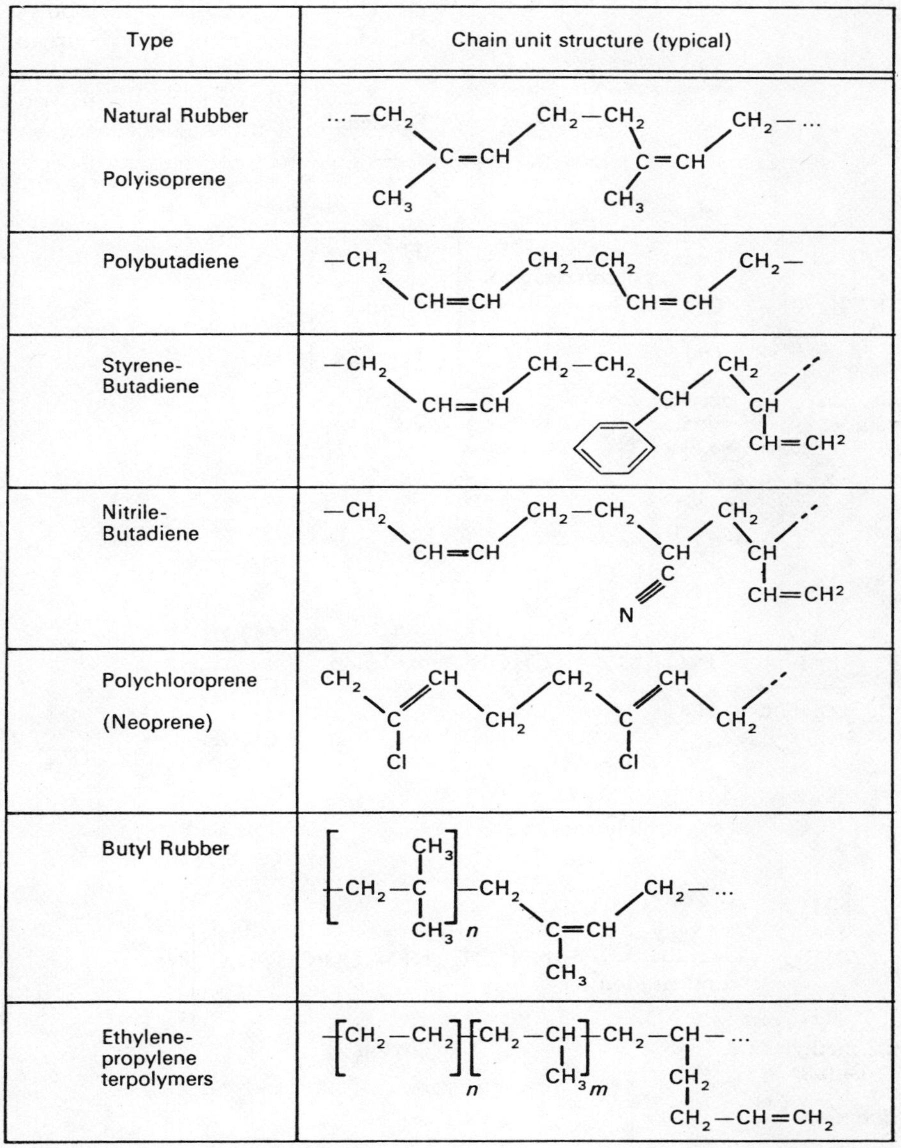

Figure 7-27. *Synthetic rubbers, or elastomers.*

The beads are shaped in such a way that when two strings are laid alongside each other, the individual beads tend to interlock with each other, forming links between strings. The result is a strong "bundle" of molecules.

An analogy to this would be the manufacture of rope. This is done by twining together tiny filaments of the rope material and twisting and plaiting the results to make a single rope. The filaments themselves are not very strong, but the frictional forces between them, and their collective strength, give the rope itself a far greater strength. Thus, the bundles of molecules in a synthetic or even a natural fiber have great strength, especially in the longitudinal direction due to the fact that they are "linked" to similar molecules lying in the same direction. Figure 7-28 shows the polymer structure of a few of the more familiar synthetic fibers.

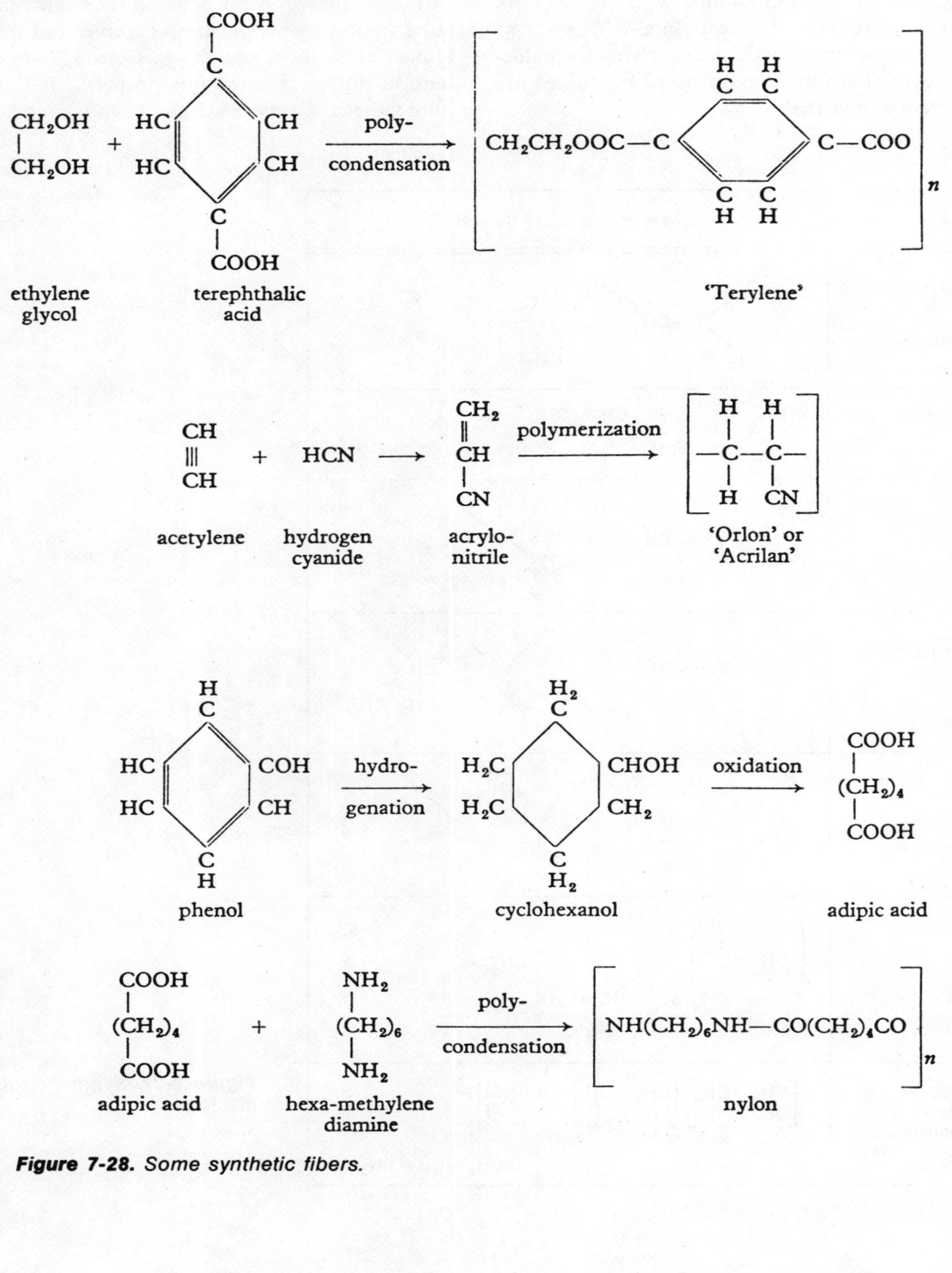

Figure 7-28. *Some synthetic fibers.*

Summary

In Chapter 7 you have learned of the chemical engineering unit operations involving the reactor. You learned about the most important kinds of reactors, and you also saw how reactor yields can be improved by recycling. You then had a glimpse of the petrochemicals industry: a world of molecular architecture which takes basic chemical building blocks and produces more complex end-products for consumer use. You saw, too, the most important way of stringing building blocks together: polymerization, a reaction that has been the basis of a multibillion-dollar industry around the world.

The petrochemicals industry is far more immense and wide-ranging than suggested here, and there is a world just as big, called the chemicals industry, which deals with the production of chlorine, ammonia, fertilizers, soaps, inks, dyestuffs, explosives, paints, cement, glass, phosphoric acid, inorganic chemicals generally, foodstuffs and pharmaceuticals. In fact, the chemicals industry comprises almost half of the total industrial effort of the United States!

Further Reading

British Petroleum. *Our Industry Petroleum*. London: The British Petroleum Company, Ltd., 1970.

Cook, G.A. *Survey of Modern Industrial Chemistry*. Ann Arbor: Ann Arbor Science, 1975.

McCabe, W.L. and J.C. Smith. *Unit Operations of Chemical Engineering*. New York: McGraw-Hill, 1976.

Royal Dutch Shell. *The Petroleum Handbook*. London: Shell International Petroleum Company, Ltd., 1959.

Chapter 8

The Sugar Industry

The sugar industry, though much less complex than the oils, fuels and petrochemicals industries (see Figure 8-1), is nonetheless an important one that depends upon some interesting chemical engineering principles.

"Sugar" is a general term referring to any member of a family of organic substances whose names end with "ose." Most sugars occur naturally in fruits and vegetables. Fructose, as its name implies, is commonly found in fruits. It is the principal constituent of nectar and shows up on the dining table as honey. Another sugar, lactose, is a naturally occurring sweet constituent of milk. You have most likely heard of three other well known sugars: glucose, dextrose, and sucrose. Sucrose has a major advantage over glucose and dextrose: it can be refined into a dry, crystalline form which has long shelf life, i.e., it does not deteriorate in storage. This makes it a most useful foodstuff since it can be packaged and distributed in low-cost paper bags. A second advantage is that when it is used as a food additive, it gives only the quality called "sweetness" to the food and adds no other flavor.

Sucrose, whose formula is shown in Figure 8-2, occurs naturally in many fruits and vegetables. But the two vegetables that produce the most sucrose are sugar cane and sugar beets. It is from these two plants that most sucrose (common sugar) is obtained.

Since the maximum sugar content of sugar cane is around 15%, and 17% for beets, the first step in a manufacturing process must be to extract the sugar and leave behind all the vegetable matter—stalks, roots and leaves. Since these are insoluble in water, and water is so inexpensive a solvent, the industry exploits the fact that sucrose is highly soluble in water by dissolving the sugar out of the vegetable matter, which then becomes a low-value by-product.

The raw material (cane or beets) enters the sugar mill straight from the trucks that bring it from the fields. It is unloaded onto chutes that deposit the material onto a conveyor belt. It passes beneath high-pressure water jets which hose off the soil, insecticides, and fertilizers. From this point in the sugar mill the processing route to obtain a clear sugar solution depends upon whether the raw material is cane or beet.

Sugar Cane Processing

After cleaning, the cane is crushed between three rollers, one above the cane and two below. A substantial fraction of the sugar is squeezed as juice which is collected beneath the rollers. The rollers themselves are very carefully designed. The top roller has ribs cut in it to improve its grip on the cane. The lower rollers have grooves which allow the juices to drain away from the cane very quickly. This is essential since the crushed cane is like a sponge, and as it expands after leaving the rollers it soaks up any liquids that are present.

There are usually three sets of crushing rollers, each one removing a little more of the sugar juice. However, all the sugar cannot be removed in this way and the crushed cane is taken to a machine which tears it into fine shreds. Next, water is sprayed into the shredded cane and this has the effect of diluting and washing out the sugar juice which has been exposed by the shredding process. A final pass through rollers removes the last few drops of liquid, which joins the other streams of sugar solution before being taken to the evaporators.

Sugar Beet Processing

Beets are processed differently. Their treatment involves a technique called a *counter-current mass transfer operation*. The sugar beet is cleaned and then cut into small V-shaped slivers in order to expose as much of the flesh of the beet as possible. These slivers are called "cossettes." The sugar is dissolved or "leached" out of the cossettes by hot water in a battery of about 12 soaking vessels. The cossettes are placed into the first vessel, allowed to soak,

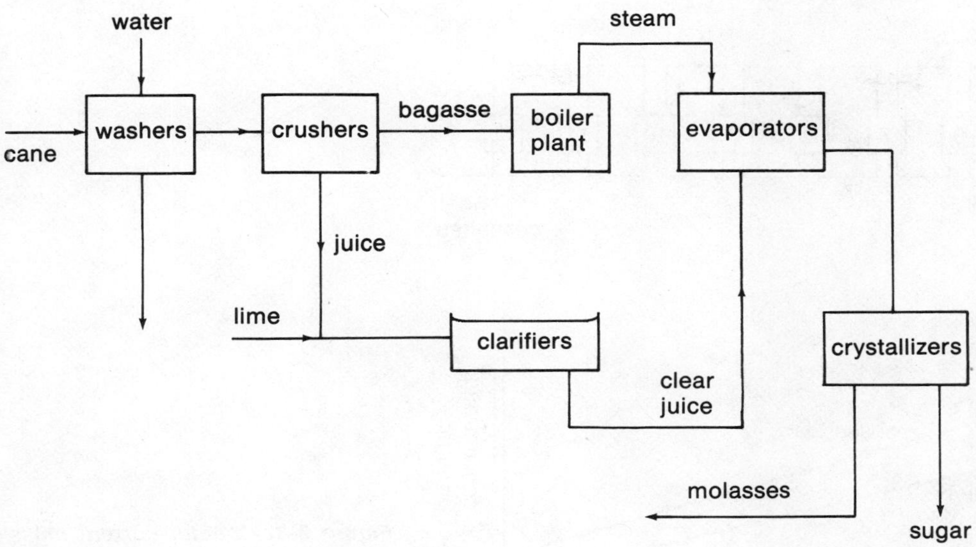

Figure 8-1. Flow of materials in a sugar plant.

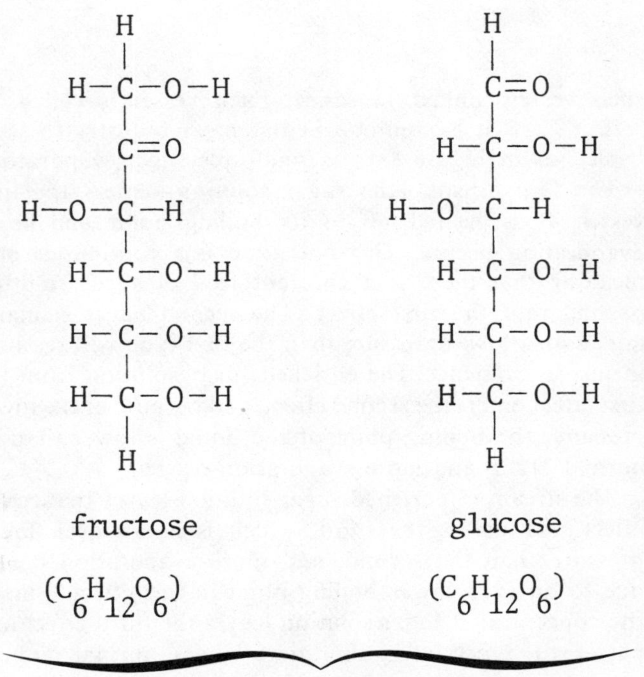

Sucrose and maltose are complex molecules made by combining glucose and fructose in different ways. They may be represented by $(C_6H_{12}O_6)n$.

***Figure 8-2.** Molecular structures of some sugars.*

and then transferred to the second. Again, after a period of soaking they are taken out and removed to the third, and so on until they reach the twelfth. Fresh, hot water is sprayed into the twelfth vessel onto the cossettes, which have little of their original sugar left. Yet there is enough sugar in the cossettes to cause a difference in concentration, or concentration gradient, between the cossettes and the body of fresh hot water surrounding them. This is sufficient for mass transfer to occur and the sugar diffuses from the slivers of beet into the fresh water, turning it into a dilute sugar solution. (We will come back to the destination of the spent cossettes later.)

The weak solution is passed into the eleventh vessel, where it meets cossettes that are somewhat richer in sugar than those in the twelfth vessel. Hence, there is still a concentration gradient between the beet flesh and the solution, and sugar diffuses across into the liquid. The still-weak sugar solution is taken into the tenth vessel, then the ninth, then the eighth and so on until it reaches the first vessel, where it is quite a strong solution. It is poured onto the fresh cossettes in the first vessel and, again, there is sufficient concentration gradient to cause diffusion of some of the sugar from the beet into the water. The water leaving the first vessel is a strong solution of sugar.

Notice the simplicity of this counter-current mass transfer. If all the cossettes had been poured or stacked into one large vessel and steeped in hot water, the concentration of sugar in both the beet and the water would have reached an equilibrium at the point where the concentration in the sugar was equal to that in the solution. Without a concentration gradient, no further diffusion would occur. Substantially less sugar would have been removed by such a method. The counter-current diffusional mass transfer method as described above is similar to the principle employed in the counter-current heat exchanger described in Chapter 5. Figure 8-3 illustrates this analogy.

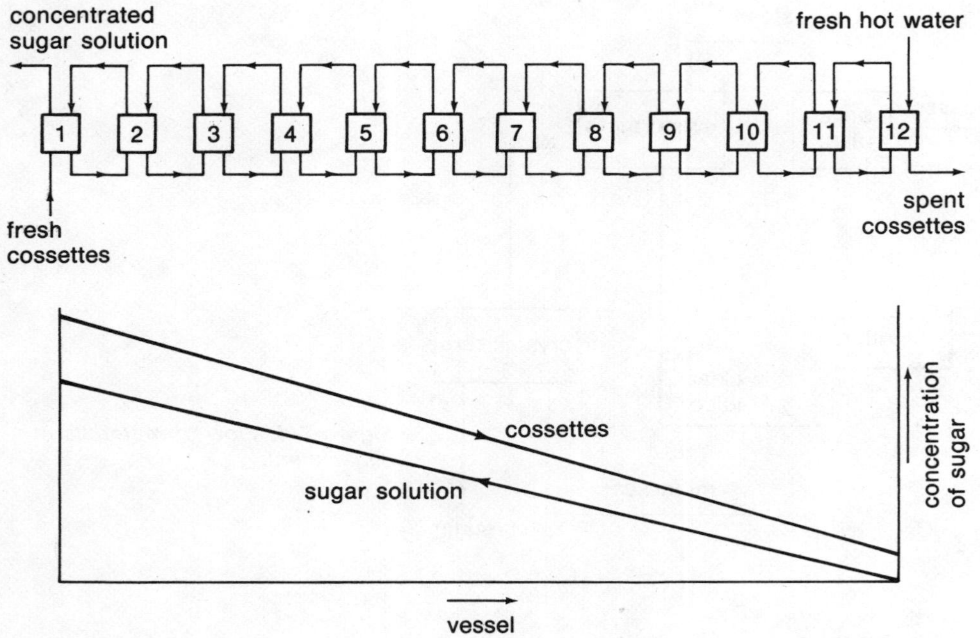

Figure 8-3. Counter-current mass transfer.

Clarification

Returning to our sugar mill, whether it be from cane or beet, we now have a sugar solution containing several other by-products that must be removed. The impurities are mainly organic acids. They are removed in a process called *clarification*, which consists of adding "milk of lime" (a suspension of calcium hydroxide in water) to neutralize the natural acidity of the extracts of the juice and form insoluble calcium compounds with the organic acids, causing them to precipitate. You can now see the significance of the name of this process—clarification—since it removes the heavy molecules from the solution whose cloudiness disappears, leaving it with a clear, colorless appearance.

During clarification, the solution is boiled in order to help with the reactions. The boiling has another, very important effect, and that is to kill any bacteria that may be present since, left alone, they would have a party eating all that sugar. As the sugar solution leaves the clarification vessel it is filtered through a fine-mesh to remove any final trace of the heavy organic precipitates.

Multiple-Effect Evaporation

The next stage is *multiple-effect evaporation*, and it is for the sake of describing this operation that a chapter on sugar was included in this book. The objective of multiple-effect evaporation is to thicken the sugar solution to the point where it can be taken to a crystallization unit. The sugar solution is concentrated by evaporating out some of the water in an evaporator vessel. Figure 8-4 shows three such vessels linked in series. Each vessel is called an "effect" and it is common in the sugar industry to have triple- (as in Figure 8-4) or quadruple-effect evaporators.

The feed stream—the sugar solution—enters the first vessel. It is heated up to its boiling point and some evaporation occurs. This operation is a continuous one, meaning that there is a constant feed of sugar solution passing into the first effect. The second effect is maintained at a lower pressure than the first (you will see how in just a moment). The enriched sugar solution from the first effect enters the second effect, and because of the lower pressure, the boiling point of the liquid is lower than its normal 212°F and more evaporation occurs.

The stream of enriched sugar solution leaves the second effect and enters the third, which is at an even lower pressure than the second, and more evaporation occurs due to an even lower boiling point in this effect. Finally the concentrated sugar solution leaves the third effect and passes to a crystallization unit. Under normal circumstances, if steam is used to boil water, then one pound of steam can evaporate just slightly less than one pound of water. Thus, in the first effect, each pound of steam entering the steam chest causes just less than one pound of water to be evaporated from the sugar solution and be driven off. This steam is not wasted, however.

As you can see in Figure 8-4, the steam that goes overhead from the first effect is passed through the steam chest of the second effect, and in this way we utilize its heat content or enthalpy. Similarly, the steam driven off in the second effect is used as the evaporation agent in the third. The steam driven off in the third is taken straight to a condenser which gives a partial vacuum as the steam

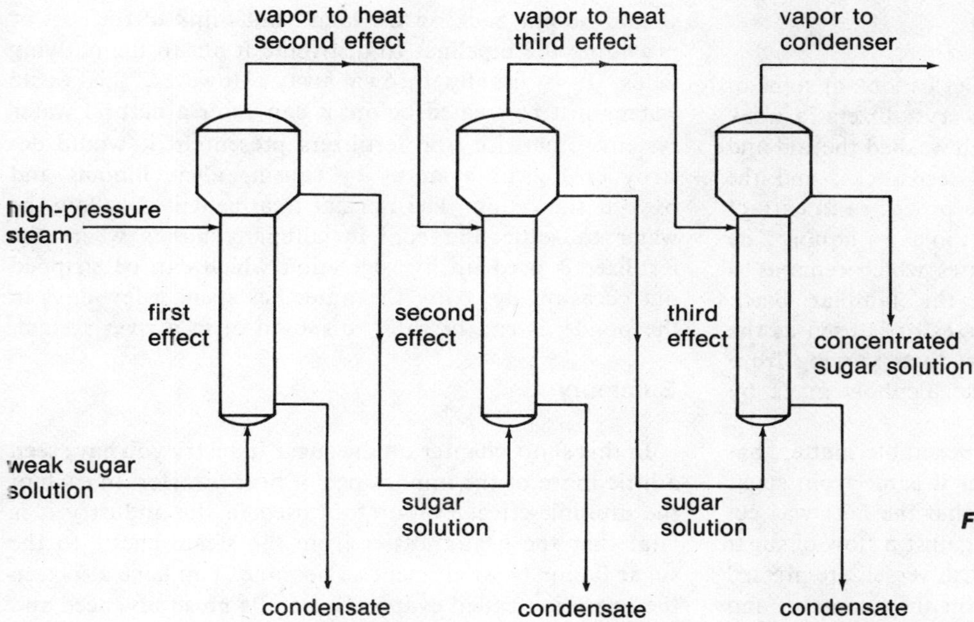

Figure 8-4. Multiple-effect evaporation.

condenses. The partial vacuum (around 2.5 psi) causes a lower boiling point in the third effect (around 135°F).

By carefully arranging the valves between the first and second, and the second and third effects, the pressures in the three effects can be maintained at atmospheric (14.7 psi) in the first, 6 psi in the second and 2.5 psi in the third. This means that the boiling points in the three vessels are 212°F, 170°F, and 135°F. It is then a simple matter to adjust the steam pressure in the three steam chests to cause the condensation point (which is the same as the boiling point) to be 227°F, 191°F and 157°F in the three effects, This ensures that in each effect there is sufficient driving force in the form of a temperature gradient to cause heat to flow from the steam across the steam chest walls simple tubes as in a heat exchanger) and into the bulk of the sugar solution, thus causing evaporation to occur. Notice also the economy we have achieved by a multiple-effect evaporation system. The initial pound of steam which we had to produce in a steam boiler on site has been responsible for the production of nearly three pounds of steam because of the "cascading" effect of the multiple-effect evaporators.

Crystallization

The product from the multiple-effect evaporators is a dark-brown, thick syrup. This is taken to boilers that are operated under a partial vacuum, which causes a lower boiling point, enabling boiling to occur without any danger of decomposing the sugar to caramel. The boiling has an effect of supersaturating the syrup, which means that the remaining water is carrying far more sugar than could normally be dissolved in it at that temperature. Operations in this part of the sugar mill exploit a chemical phenomenon called *super solubility*. The reason this supersaturation occurs is that after boiling in the low-pressure pans, the liquid is just about saturated with sugar. It is removed and poured into long stirred troughs to cool.

The solubility of sugar in water increases with temperature. Thus, when a saturated solution—one containing all the sugar it can—is cooled, there will be more sugar in the water than can normally be held. The reason the sugar does not come out of solution as crystals is a kind of reluctance caused by there being no nuclei around which crystals can form and grow. This *supersaturated* solution is cooled to a point where it contains almost twice as much sugar as it would normally at that temperature. Then tiny crystals of sugar called "seed" crystals are sprinkled into the solution and, suddenly, the crystals form, grow and are separated from the remaining liquor in a centrifuge (see Chapter 10). The liquor (a common word in chemical engineering for a solution which is left after crystallization) is made up of water which contains any remaining impurities and some sugar. It is, in fact, a weak solution of molasses.

The sugar crystals now have the familiar look of brown sugar and can be sold exactly as they are. However, the food companies require sugar for sweetening purposes only and do not wish to impart to their products the slightly bitter taste of brown sugar. They demand pure white sugar. Consumers, for some reason which entirely escapes me, also demand white sugar for use in the home. Some further refining of the sugar occurs, and the brownness is removed by successive dissolving and crystallization. The product is the familiar white sugar—pure sucrose.

By-Products of Sugar Production

Before leaving the sugar industry, let us look at some of its by-products: the liquor from the crystallizers (a weak solution of molasses), the water which washed the soil and fertilizers from the fresh vegetable feedstocks, and the crushed vegetable remains. There are processes to extract the remaining sugar from the weak molasses liquor. The pale-brown, weak solution of molasses which remains is concentrated by evaporation into the familiar black substance and either is sold as molasses or is used as the feedstock for other industries for products varying from rum liquor to cattle feed to organic alcohols made by fermentation using yeasts.

Another by-product, the crushed vegetable matter, has some value, depending upon whether it came from sugar beet or sugar cane. You remember that the beet was cut into cossettes which were cascaded against a flow of sugar solution. The cossettes leave the twelfth vessel, are pressed to remove most of the sugar solution they contain, and then are dried, packed and sold for cattle feed. It is more common to avoid the expense of pressing and drying the pulp material and instead provide a feedlot operation adjacent to the sugar beet mill, where cattle can be fattened on the beet tops and the wet cossette pulp. If this level of integration is employed, the final by-product of the sugar mill is beef.

If the raw material is sugar cane, the crushed fibrous waste material is called "bagasse." It has low nutritional value, being simply a fiber, bulk substance similar to bran, but it can be fortified with vitamins, amino acids, and other additives and used as an ingredient of cattle feed. However, most of the bagasse is burned in the power plant of the sugar company to produce hot water, steam and electric power for use in the sugar mill. Again, careful integration of the use of materials saves energy costs and increases profits. Incidentally, some sugar companies are able to burn their bagasse and produce more than enough power for their own needs. In these cases, the companies sell their surplus power to the local utility company.

Finally, the cleaning water now laden with fertilizers and soil must be regarded as a by-product. Ideally, it should be put back on the land—but think of the cost of providing the pipelines to distribute it out to the outlying fields. It is usually thrown away. However, this waste water must be treated before it can enter a natural water system; otherwise, the fertilizers present in it would destroy ecological balances by causing algae blooms and oxygen starvation. The normal treatment is to allow the water to settle and cool in company ponds where the fertilizer is used up by vegetation which can be stripped out occasionally. After the water has spent a few days in the ponds, it can be safely disposed of in a river system.

Summary

In this short chapter on the sugar industry you have seen a little more of the importance of heat transfer: in each of the multiple-effect evaporators used in this industry it is vital that the heat transfer from the steam chests to the sugar liquor be as efficient as possible. You have also seen the operation called evaporation in its most advanced and economical form: multiple-effect. Finally, you have seen that the operation of crystallization can be made to run on a continuous basis and a large scale.

Further Reading

Aykroyd, W.K. *Story of Sugar*. Chicago: Quadrangle Books, 1967.

Deerr, N. *History of Sugar*. London: Chapman and Hall, Ltd., 1949.

Honig, P. *Principles of Sugar Technology*. Amsterdam: Elsevier, 1953.

McCabe, W.L. and J.C. Smith. *Unit Operations of Chemical Engineering*. New York: McGraw-Hill, 1976.

Meade, G.P. and J.C.P. Chen. *Cane Sugar Handbook*. New York: John Wiley & Sons, Inc., 1977.

Sorenson, H.G. (in V.E. Baikov, ed.). *Manufacturing and Refining Raw Sugar*. Ch. 1. Amsterdam: Elsevier, 1967.

VanHook, A. *Sugar*. New York: Ronald Press, 1947.

Chapter 9

The Brewing Industry

This chapter reveals the importance of chemical engineering techniques in the international industry of brewing beer. You will encounter some new engineering principles in this chapter, principles used in other industries, too.

Brewing is perhaps the oldest chemical engineering practice of all. No one really knows when it first began as a craft, but it is known that in 4000 B.C. beer was brewed in Mesopotamia, where bread was mashed, malted, and fermented, and the resulting brew flavored with spices, dates, and honey. Mythology tells us that Isis, mother of the Egyptian gods, introduced beer to the Pharaohs. Egyptian brewers developed at least six types of beer, according to brewing manuals found in cuneiform and hieroglyphic inscriptions. The basis of this ancient Egyptian beer (around 3000 B.C.) was barley, and its name in ancient Egyptian was *boozah.*

The technology found its way to Britain, as did barley, and I surmise the name "booze" came with it. It is certain that the ancient Britons knew about brewing, and ale was served at all feasts and important celebrations. After the Romans left, the Saxons and Vikings invaded Britain and no doubt encouraged brewing, since the Norsemen were great lovers of ale, quaffing it from drinking horns before battle, after battle, in their inevitable feasts, and perhaps even during the battle.

While the brewing industry was growing in Britain, it was also developing in Europe, particularly Germany, which to this day has a reputation for making fine beers. My research in the literature of brewing turns up a most interesting story concerning quality control of the finished product—a vital part of chemical engineering to this day.

In the thirteenth century, there was no chemical or analytical way of assessing the worth of a beer. The government inspector had the authority to taste a brewer's ale and lower its price if it was not satisfactory. The official taster wore leather breeches, and the practical test employed was to pour some of the beer onto a barrel end and sit on it for a specified number of minutes. When the inspector stood up, if his breeches stuck momentarily to the barrel, the beer was up to standard!

In Shakespeare's time, brewing was improved and hops were introduced from France. The word "ale" continued to be used for a malt drink without hops flavoring, and the word "beer" began to be used for an ale flavored with hops. Early American settlers brought their brewing expertise with them, and hundreds of small breweries were established. Similarly, in Germany and Britain, the other great brewing countries, hundreds of breweries mushroomed and grew. However, this century has seen the growth of conglomerate brewing companies which have bought out and taken over the smaller ones.

It used to be that draft beer accounted for most beer sold commercially, and its condition was extremely fragile. The good taste of the beer depended on the skill of the manufacturers, distributors, and retailers in handling it. All too frequently, a batch of beer would be spoiled by careless handling or storage at the wrong temperature. Today's huge breweries have brought science and chemical engineering to their aid, and more and more beer is artifically conditioned, preserved and protected against any unskilled handling it receives after leaving the brewery.

Malting

Following our usual pattern, let us trace the path of raw materials into a brewery and through to the finished product (Figure 9-1). We will pause only to examine particular details that show some new aspect of chemical engineering.

The raw materials of a brewery are water, barley, and hops. Imitation beers are made from ingredients other than barley: rice, oats, other grains, or any source of starch (the floury material in grains) can be used. For the purposes of this description, however, I shall refer only to the use of barley.

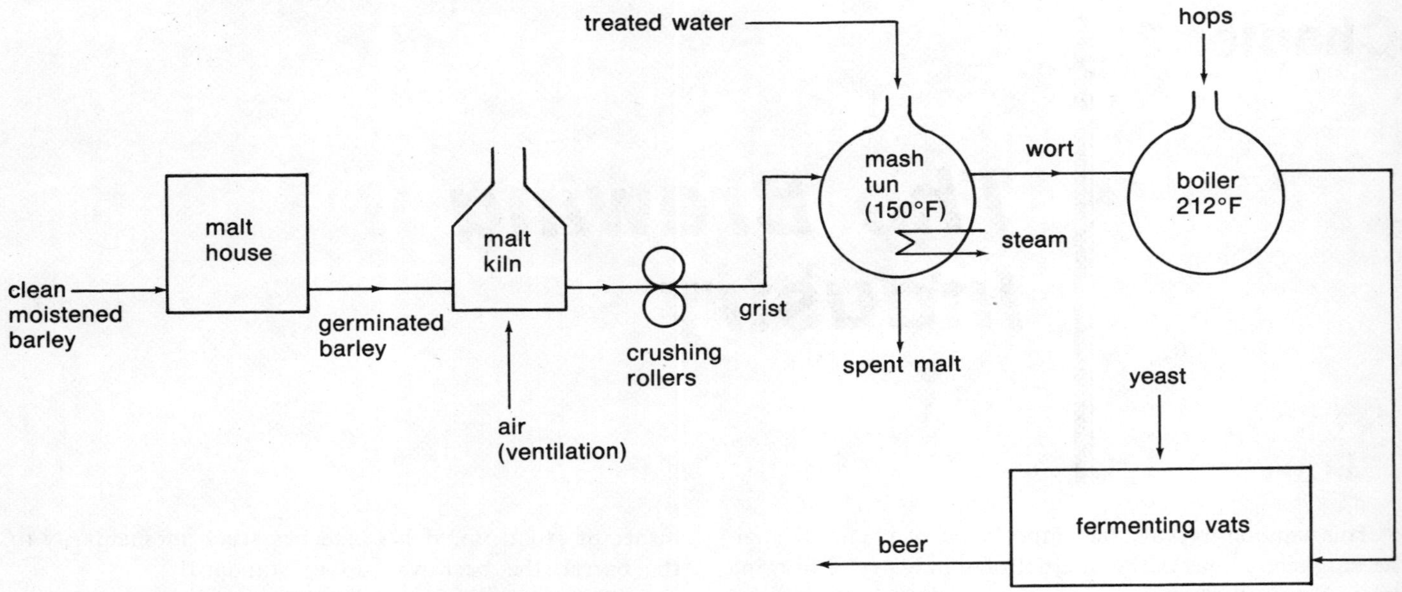

Figure 9-1. *Flow of materials in a brewery.*

The barley enters the brewery and is cleaned of dust and any other alien material, such as stalks, mud, pebbles, parts of tractors and other things that find their way into the grain. It then undergoes a malting process in the "malt house."

There are two ways of malting. The "floor method" is the older, by about 4,000 years! The grain is steeped in water for three days so that it absorbs moisture, swells up, and softens. The surplus water is drained off and the grain is heaped into the malt-house floor, which has been designed to allow careful control of the temperature and ventilation conditions. As you would expect, the grains of barley germinate. Operators at the malt house, called maltsters, turn the grain with shovels to ensure even malting, a word which simply signifies the germination phase of the barley grains. In the second method of malting (the drum system) the steeped grain is malted in revolving cylinders which allow for a more accurate control of temperature and ventilation.

As the barley germinates, a tiny shoot called the *acrospire* starts to grow from the base of the grain and carbon dioxide gas is given off. The malting process is allowed to continue for about 10 days at 60–70° F. The temperature is important, as is ventilation. Obviously, the design of the malt house or the malting drum needs heat transfer experts to determine the best means of providing the gentle heat required.

After about 10 days, the acrospire will be about three-fourths as long as the grain. At this point, growth must be terminated by a drying process known as *kilning*, and again, chemical engineering skills are required to design the necessary equipment. First of all, heat has to be provided to raise the temperature to about 140° F. Also, air must be circulated around and through the grains in order to carry away the excess moisture. This means that the dewpoint of the air entering the system must be as far below its actual temperature as possible in order to pick up the maximum amount of water from the malted barley. As the grain dries out, the temperature is raised slowly in order to wither the shoot. Since this must be done in a very exact manner, it is essential that, whatever the means of heating, the temperature be controllable to within 2° F.

The final temperature depends upon the kind of beer to be brewed. If the product is to be a pale beer, then the malting process will stop at about 180° F. After a few hours at this temperature, the malt is dry, the acrospire has withered, and the color of the malt is a light gray-brown. If the final product is to be a dark beer, then the temperature is raised to about 220° F, and in that case the final appearance of the dried malt is almost "burnt", the acrospire is black, and the body of the grain is a dark brown. Of course, there are all sorts of delicate variations of temperature, timing, and method in between to produce a wide spectrum of different flavors of beer. Again, because it is a tricky process, malting requires careful attention to the design of heat transfer and drying equipment.

The roasted malt is then lightly crushed between rollers. This comminution is nowhere near so severe as you saw with coal in Chapter 4: in this case, it is merely to expose the insides of the grain by cracking the cellulosic shell, which has become hard and brittle due to the malting process. The crushed malt is normally called *grist*. Because the craft of brewing is so old, this jargon has gradually found its way into everyday language, as in the phrase "grist to the mill."

Mashing

The grist is taken to the most important single operation in brewing—*mashing*. It is in this process that we meet an operation called the "inversion of starch to sugar" (represented by the equation in Figure 9-2). The catalyst required here is biological—not a simple metal like platinum or an acid as in Chapter 7.

Biological catalysts are called *enzymes*, which fortunately occur naturally within the grain and are unaffected by malting. There are several enzymes present and they all play a part in the mashing process. Enzymes are given names which end in "ase." The enzyme *cytase* catalyzes a reaction in which the protective cellulose coating of the barley breaks down into water and carbon dioxide, and thus exposes the starch. The enzyme *diastase* catalyzes the starch inversion reaction shown in Figure 9-2. The product of the reaction is a sugar called maltose (almost all the sugars are fermentable).

Mashing occurs in a vessel called a *mash tun*, usually an onion-shaped copper vessel, open to the atmosphere in order that certain gaseous by-products are vented away from the mash where they cannot affect its final flavor. The grist is mixed with other sources of starch such as flaked oats, maize, grits, rice, or even crushed corn. The grains are mixed with hot water in the mash tun, and a temperature of 150° F is maintained for about eight hours.

The temperature cannot drop below 145° F, or the enzyme activity may cease or cause slightly different reactions which will result in an "off taste" in the final beer. If the temperature rises above 155° F, some of the enzymes may be broken down or decomposed by the heat; in any case, side reactions may occur which again impart an off taste to the final beer. Thus, the temperature is crucial and it is up to the designer to find an effective means of providing controllable heat to the mash tun.

A common means of doing this is to heat the mash tun using steam coils immersed in the hot water. A number of thermometers also submersed in the mash relay the temperature in the mash tun to a controller which operates on the valve of the steam line. If the temperature goes slightly above 150° F, the steam flow is cut back; if the temperature drops slightly below 150° F, the steam valve is opened. The length, diameter, and thickness of the steam coil itself is a matter of careful design. After the size of the mash tun has been decided (from the required production rate), the volume and weight of liquid it contains can be calculated. Hence, using a knowledge of radiation and convection heat transfer, the heat loss from the mash tun can be calculated. It is then simple to calculate the amount of heat that needs to be added to the mash by the steam coils.

After an eight-hour period in which the starch-to-sugar inversion takes place (Figure 9-2), the resulting liquid is

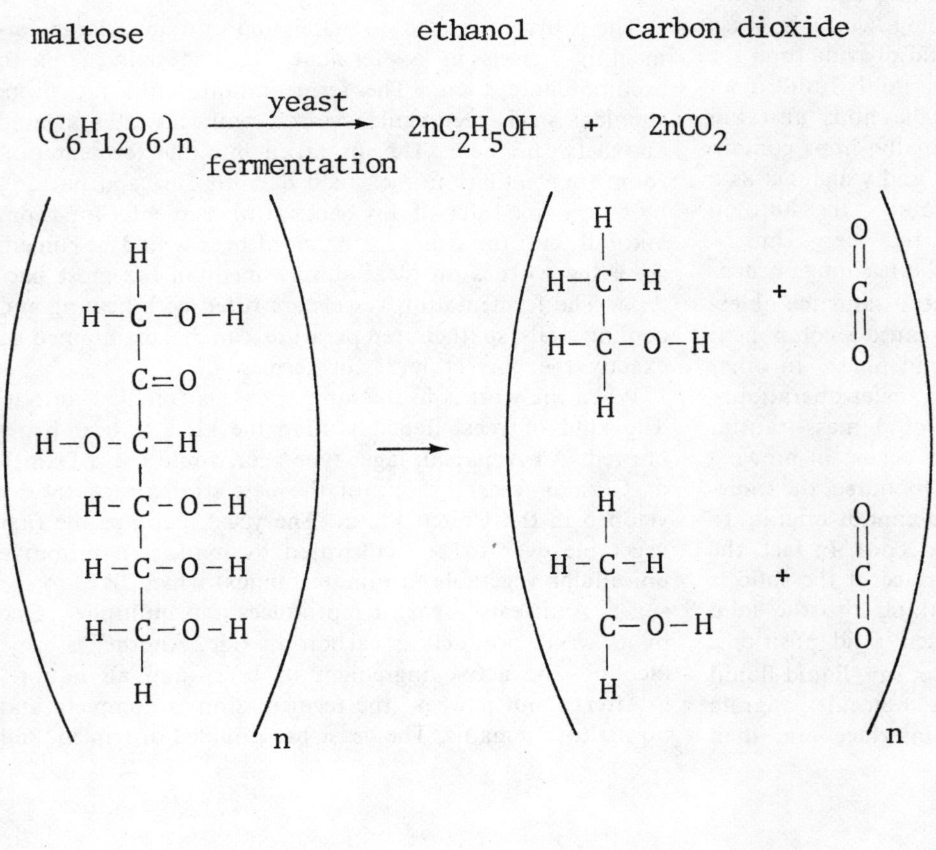

***Figure 9-2.** Fermentation of maltose to alcohol.*

called the *wort*. Since it contains maltose, the wort is a sweet, brown liquid. It also contains a large number of proteins, vitamins, and other soluble and insoluble materials from the grain. A valve at the base of the mash tun is opened and the sweet wort is run off and taken to another vessel very similar to the mash tun in appearance.

When the mash tun is drained of its liquid, all that remain are the insoluble shells or husks of the grain, and these are sprayed with boiling water to remove any last traces of maltose and other soluble components of the grain that improve the rich beer flavor of the final product. This operation is called *sparging*. In some brewing processes, a slightly different technique is employed to gain the last few soluble compounds from the grain. The mash tun bottom valve is opened and the sweet wort is led to a boiler where it is heated up to boiling point by a steam jacket. The boiling wort is then returned to the mash tun. This is kept up for about 30 minutes and eventually the entire contents of the mash tun are raised up to the boiling point.

Boiling and Hopping

Whichever method is used, the wort is brought up to its boiling point and hops are added at the rate of one to five pounds of hops per 50 gallons of beer, depending upon the required bitterness of the final beer.

The hops serve several purposes. First, they prevent any further enzyme activity. Second, the hops cause the heavy complex protein molecules to undergo a reaction which turns them from soluble proteins into insoluble proteins, and they precipitate out of the boiling wort. If these proteins were left in the beer, they could provide food for bacterial action, which would almost certainly result in an off-taste to the beer. As a bonus, the hops also kill bacteria and thus sterilize the beer. Finally, hops contain oils which are extracted by "leaching" and which act as a preservative in the beer. And, of course, the hop oil imparts the familiar bitter flavor to the beer. Thus, a principal operation during boiling is the leaching. Leaching also occurred in the mashing process—and the objective in both mashing and boiling is to cause a component from the solid phase to enter the liquid phase. In other words, we have another kind of mass-transfer operation.

Leaching is perhaps the least studied of all mass-transfer operations, though the leaching which occurs in brewing processes is well enough understood. Of course, the molecules in the solid are fixed rigidly and cannot migrate to the surface of the solid of their own accord. In fact, the resistance to their transport to the surface of the solid is really the resistance to the liquid soaking into the solid and dissolving the molecules from their rigid positions. Thus, leaching is not truly analogous to, say, liquid-liquid extraction (see Chapter 10), where the molecules migrate through one film resistance to the interface and then through another film resistance into the bulk of the raffinate.

Another reason for the boiling process is to remove a little of the water in the mix and so bring the wort to its specified concentration. The boiling process is carried out at atmospheric pressure, so the boiling point of the wort is 212°F. The boiler is usually open to the atmosphere and the fumes are allowed to escape.

In my home town of Sheffield, England, the steelworks were downwind of several breweries. In and around the steel and iron furnace shops, the temperature could be as high as 120°F, and as often as our work would allow, we would take turns to go outside for a breath of fresh air. When the wind was in the right direction and the breweries happened to be boiling their wort, the delightful smell of brewing beer drifted down to us. You have no idea how mouthwatering this can be, particularly on a hot day. Fortunately, back in those days, steel furnace workers were issued free beer when they picked up their salt tablets!

After about one hour in the boiler, the wort is pumped to another vessel called a *hop vat*. This special vessel has a false bottom; the wort is poured in and the spent hops settle out onto the false bottom to form a natural filter bed through which the hot wort drains on its way to coolers and fermentation. The spent hops and the remains of the grist and other starch sources are collected for sale to fertilizer companies.

Fermentation

The wort is cooled to 60°F and run into large fermenting vessels in a "ferment hall" specially built to accommodate them. The fermentation tanks are deep, stainless steel, rectangular vessels with catwalks around and between them. The air taken into the fermentation room for ventilation is cleaned of both dust and bacteria by a very fine filter. If any bacteria were to enter the room and fall onto the wort, the batch of beer would be ruined, since the wort is an ideal culture medium for most bacteria. The fermentation vessels are fitted with heating and cooling coils so their temperature can be maintained at exactly the correct level for fermentation.

When the wort is in the tanks, yeast is "pitched" into it. The kind of yeast depends upon the kind of beer being brewed. An American lager-type beer would use a Danish or German yeast or one of the new strains recently developed in the United States. The yeast, perhaps the first vegetable ever to be cultivated by man, is a primitive unicellular vegetable (a minute fungus) which likes to eat sugar. As it eats sugar, it reproduces and multiplies. One of its waste products is carbon dioxide. Another is ethyl alcohol—the active ingredient of beer—and all liquor.

After about a week, the fermentation is complete and no maltose remains. The yeast has doubled or tripled, and after it is skimmed off the beer, it is washed, dried, and stored. Some yeast is used in the next batch of beer and the remainder is sold to bakeries. The beer is further filtered and then taken to a dispatch room where it is bottled or put into barrels. Some final additives may be put into the beer prior to bottling or kegging. One of these is isinglass, an extract from fish bladders. It has the curious property of flocculating tiny fragments of solids (remains of the yeast, the grist, and the hops) and causing them to settle, leaving the beer sparkling clear in appearance. The process is called "fining." Somtimes dry, unboiled hops are added to the beer at this stage simply to impart the characteristic bitter flavor to the beer.

Just prior to bottling or kegging, the beer is chilled and saturated with carbon dioxide. It is run through coils in a refrigeration process and into a vessel which contains carbon dioxide under pressure. The carbon dioxide dissolves in the beer, and then the carbonated beer is bottled or kegged and taken away for distribution.

The process of brewing is a simple one as far as chemical engineering is concerned. The operations of leaching and drying are important, but the major skill required is that of heat transfer in order to manage the energy requirements of the brewery. You have seen in earlier chapters how chemical engineering techniques are used to bring about chemical reactions on a massive scale in order to manufacture high-purity products required by the consumer. In this chapter you have seen that the chemical reaction may even be one which involves biological catalysts (enzymes) or biological processes (fermentation). These operations could perhaps be classified as "biochemical engineering."

Although brewing and wine-making processes have been used for thousands of years, their principles have not been economically applied to other manufacturing industries. But, the economics are changing and it is almost certain that before the year 2000 we will see major applications of biochemical engineering in the process industries. Already, bacteria are being put to use as waste treatment agents. Other bacteria are being "trained" to manufacture rare and expensive drugs. Some members of the algae kingdom are being used to treat sewage and other wastes. More recently, researchers have begun to study how high-grade, low-cost protein can be made from a member of the blue-green algae family. Other researchers are studying an algal species which manufactures a hydrocarbon wax that could be cracked (see Chapter 6) into gasoline and other hydrocarbon products.

There are many more examples, from a wide range of industries. There is a common basis, though: Wherever these new techniques are used, chemical engineers will need to adapt conventional design methods to the requirements of living organisms—as the modern brewer has learned to do.

And so some of the skills of chemical engineering are used in the brewing industry. Perhaps that is why members of the chemical engineering profession consume somewhat more than their fair share of beer, at least in my experience. On the other hand, I suspect that this is *not* the true reason, which is likely to be more obscure.

Further Reading

Cook, A.H. *Barley & Malt*. The Brewing Industry Research Foundation. New York: Academic Press, 1962.

Cook, A.W. *The Chemistry and Biology of Yeasts*. The Brewing Industry Research Foundation. New York: Academic Press, 1958.

Hitchcock, T. "Hitchcock on Brewing," *The Brewer*, Volumes 6 and 7, 1976.

King, F.A. *Beer Has a History*. London: Hutchinsons Scientific & Technical Publications, 1947.

Kirk-Othmer, *Encyclopedia of Chemical Technology*. New York: John Wiley & Sons, Inc., 1978.

Preece, I.A. *The Biochemistry of Brewing*. Edinburgh: Oliver & Boyd, 1954.

Weeks, M. *Beer and Brewing in America*. New York: U.S. Brewers Foundation, 1957.

Chapter 10

The Waste Treatment Industry

There are by-products for just about every product made by the process industries. Until about 30 years ago, by-products were usually dumped or buried or left in liquid effluents which found their way to rivers, lakes, and the sea; or they were allowed to become airborne, in which case they drifted away in the atmosphere.

Of course, if the by-product was valuable, it was sold, perhaps in an impure form, to be purified and further processed by some other industry. If the by-products were combustible, they were burned. Rivers became industrial sewers, and wildlife was robbed of many habitats. Industrial dust formed oily silt which settled in bays and oceanic canyons to smother underwater vegetation and cause violent disturbances to ecosystems. By-product compounds containing nitrogen and/or phosphorus found their way into lakes and rivers, where they promoted algae and other vegetable matter to grow at extraordinary rates. The rapid growth caused the dissolved oxygen in the water to be used up, and not only were other forms of life extinguished, but the new growths of algae and water weeds themselves died for the same reason.

Oxides of sulfur literally poured into the atmosphere, causing serious health problems for bronchitic people and unpleasant throat and lung irritation for others. Iron, steel, and other unpainted metals in the vicinity of an industrial area rusted and corroded rapidly to produce eyesores and cause the expense of early replacement of the structures. Fly ash from utility and industrial boilers was shot up into the sky on a column of hot gases from the plant. Eventually, the fly ash would settle on drying laundry, clean window sills and windows, the mucous surfaces of the esophagus and lungs, and the sidewalks and streets, where it would be stirred up again by each car or passerby and continue to menace society.

Clearly, this kind of behavior by our industry was unacceptable. Today, the same behavior would be immoral, since we not only know the terrible effects of industrial pollution but we also have the technology and know-how to prevent pollution. In those days, not only was industry ignorant of the dire effects of its pollution, but so was the general public, with few exceptions. Just as industry considered it unnecessary to be concerned with the environment, and accordingly would never have raised its selling prices to find the money to keep it clean, the consumer would never have considered the environmental effects important enough to pay a premium on the price of goods.

That premium, by the way, is not trivial. Although in most cases the cost of pollution control at the factory or plant will add perhaps 40–50% to the price of the finished article, there are cases where the retail price of a product is twice or even three times what it would have been without pollution control. Of course, this extra *has* to be passed on to the consumer. A company could not stay in business if it spent more making a product than it could retrieve in the marketplace. Nor does the increased cost to the consumer end at increased prices. As a taxpayer, the consumer also must foot the bill for maintaining federal, state, and local agencies of administrators, engineers, scientists, and inspectors who act as watchdogs on our industry.

Today, the environmental impact of industry is very different from the past. The public is aware of the danger of environmental pollution and demands that federal agencies act as watchdogs to protect our environment. Industry has developed a professional pride in containing obnoxious by-products and doing so at minimum cost. Newer industries have sprung up to use by-products as raw materials which they process further into other useful, marketable products. An army of waste treatment experts is available in engineering consulting companies and similar organizations. Today's level of environmental pollution is incredibly low considering that manufacturing industries are many times more productive than they were 30 years ago.

Yet pollution still occurs. Sometimes it is caused by accident, as when an oil tanker runs aground in a storm. In such cases, industry and environmental agencies work together to correct the damage as quickly as possible with the many clean-up technologies now available. Sometimes pollution is caused by ignorance. For example, people have been venting nitrogen oxides to the atmosphere for decades completely unaware of the effect that these oxides have on the formation of smog in certain areas. However, research teams studying the complicated subject of our environment are gradually piecing together just how this complex ecological system works. We must hope that whenever a hazardous impact upon the environment is identified and its cause isolated, technology will be able to solve the problem and, if possible, undo some of the damage wrought in earlier years.

Finally, there is willful pollution. Such is the case when irresponsible sea captains discharge oily ballast water into the ocean, ruining beaches, maiming and killing wildlife, and interfering with the fishing industry. In these cases society can at least punish criminal offenders.

Because of the increase in awareness and knowledge of the environmental impact of industrial pollution, it is unlikely that it will ever be a threat in any way other than a minor or temporary one. But the pollution problem has been replaced by another, more subtle one. The cost of today's relatively pollution-free environment has been enormous for consumers. And the cost of removing the last few percent of known pollutants will be staggering: perhaps many times the cost of removing the first 90%! Our problem now is to ask ourselves, "Just how clean do we want our environment to be and just how much are we prepared to pay for the level of cleanliness we decide upon?" It is we who must pay: we must pay the salaries of civil servants who regulate and control pollution; we must pay twice as much for products because half of their manufacturing costs are spent to catch the last few percent of effluents leaving the factories. As the public realizes this, the more hysterical demands for pollution control will abate and an equilibrium will be established between the cost of consumer goods and the pollution control laws.

The Size of the Industry

With such a vast range of products, you can imagine the range of by-products and potential pollutants from the process industries. A thorough study of waste treatment in all the process industries would run to many tens of volumes. But as a future professional of the industry, you must recognize now the importance of waste treatment, and so I have included this necessarily brief chapter.

The waste treatment industry uses just about all the other major chemical engineering unit operations not yet mentioned in the industrial processes described in this book. And so, here is my compromise—a compendium of methods of removing pollutants from gases which must be returned to the atmosphere or from liquids which must be disposed of in rivers or oceans. Remember: the best and most important waste treatment method of all is selling the pollutant mixture to another industry for reprocessing and sale. This means of making pollutants disappear could be called "industrial integration."

To help you compare waste treatment methods, I have categorized them into solids from liquids, solids from gases, liquids from gases, liquids from liquids, and gases from gases.

Removing Solids from Liquids

The primary operations concerned with removing solids from liquids are filtration, sedimentation, and centrifugation. There are others, and you will learn them in good time; but they would not contribute anything to this chapter.

Filtration

Filtration, sometimes used as a means of removing solid wastes from a water stream before it is returned to the sea, is more often used as a method of recovering valuable solids from a waste stream effluent. As you probably know, it works on the principle of introducing the "slurry" or liquid–solid mixture onto a filtration medium which is often called the mesh or screen. The remaining solid residue is called the "cake," and the clear liquid is called the "filtrate." I only intend to discuss three types of filtration in this book. There are many more, and if you would like to see the ingenuity of our industry's designers, look through some of the suggested reading at the end of this chapter.

The *rotary drum filter* (Figure 10-1) is cheap to run, reliable, and can handle a large flow rate of slurry on a continuous night-and-day basis with little maintenance. If you examine Figure 10-1 carefully, it will be clear how it works. The feed slurry enters the trough in which the drum rotates. The liquid is sucked through the mesh and leaves the drum via a hollow shaft, which also provides the turning motion for the drum. The cake builds up on the outside of the mesh and is exposed to the air as the drum rotates. Here is an opportunity for some washing with water sprays if the cake needs to be free of any soluble salts which were in the slurry.

The cake is removed by the knife or scraper on the right, which is positioned carefully so that it scrapes most of the cake off the mesh but leaves a tiny amount behind to start the filtration again as the drum rotates and the mesh goes below the slurry surface. Why is this small

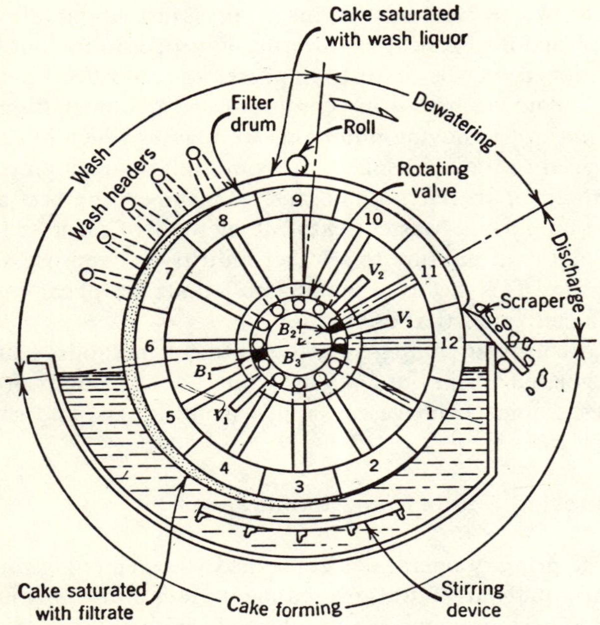

Figure 10-1. Cross-section of rotary-drum filter.

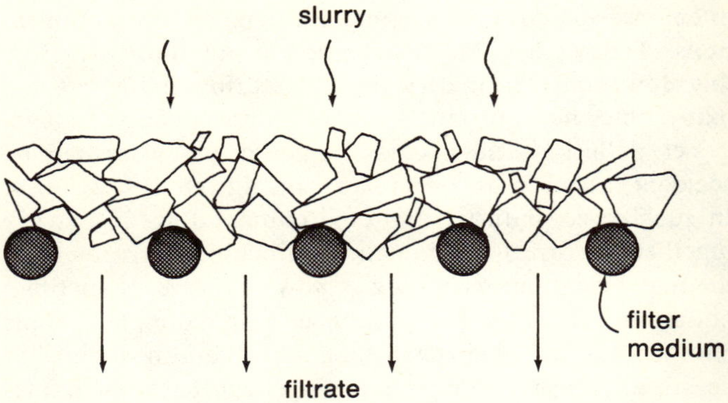

Figure 10-2. Principles of filtration.

amount of cake left? Filtration is not the result of catching the particles on a mesh with smaller holes. On the contrary, the holes in the mesh are usually several times as big as the particles and, as shown in Figure 10-2, the true filtration action is caused by the particles piling up, or "scaffolding," on the mesh.

There are advantages to having a mesh size larger than the particle size. First of all, it is cheaper, since the finer the mesh, the more expensive it is. Second, a fine mesh increases the pumping power required to push the liquid through it. And third, the mesh can be made stronger, since if the holes are bigger, the wire between the holes can also be thicker. The disadvantage of having holes larger than the particles is that there is a short delay before the scaffolding occurs. To minimize this delay, the scraper knife is adjusted to leave a thin layer of cake on the filtration mesh.

I have said already that the rotary drum filter is most reliable. The only difficulty is when the slurry settles out and forms a sediment in the bottom of the shell. Eventually, the sediment builds up so much that it interferes with the even deposition of cake on the filter mesh. This can easily be remedied by arranging to have a small stirrer in the bottom of the shell. The agitation keeps the particles in suspension until they are filtered.

Another kind of filter is the *filter press*, or *plate-and-frame* filter. Figure 10-3 shows an example. This kind of filter is only used for filtering solid wastes from liquid wastes when the solid is valuable. It is uneven in operation because it has to be stripped down for the cake's removal at frequent intervals.

The frame type "A" in Figure 10-3 has a channel cut from the large gallery in the top left-hand corner to the rectangular cross-section. The frame type "B" in Figure 10-3 has a channel cut from the central rectangular cross-section to the large gallery in the lower right-hand corner. The filter press is now assembled with filter plates in the order A, B, A, B, A, and so on. Between each plate a sheet of filtration material is trapped and held in place. Two blank end plates, shown as types "E" in Figure 10-3, are then put on either end of the press and long bolts are passed through the bolt holes and securely tightened into position. The upper gallery is connected through the end plate, E, to the source of the feed slurry. The slurry floods the upper gallery and passes through the channels into the interior of type "A" frames. The liquid (filtrate) passes through the filter mesh into the space made by type "B" plates and then through the lower channel into the large gallery below. The filtrate is taken off either as waste or for whatever purpose it will next serve. The solids are left behind as cake in the type "A" frames.

After several hours of operation (the actual time depends upon the amount of solids in the slurry), the filter press is full of solids and the effort to continue pumping the slurry through it becomes expensive. At this point the feed stream is stopped and disconnected and, if necessary, a clean water wash is pumped through the filter press through the upper gallery to exit from the lower gallery. After washing, the press is stripped down by removing the bolts and carefully peeling off the frames and scraping out the filter cake.

The nature of the equipment means that the process cannot be automated and, accordingly, labor costs are high and can only be justified when the product cake is valuable. The operation cannot be continuous, since there is a filtration time, a washing time and a disassembly time. But there is a puzzle to this: Is it best to continue pumping slurry through until the filter press is absolutely full of cake before washing and stripping down? Or is it better to stop when the filter cake is still uncompressed and, there-

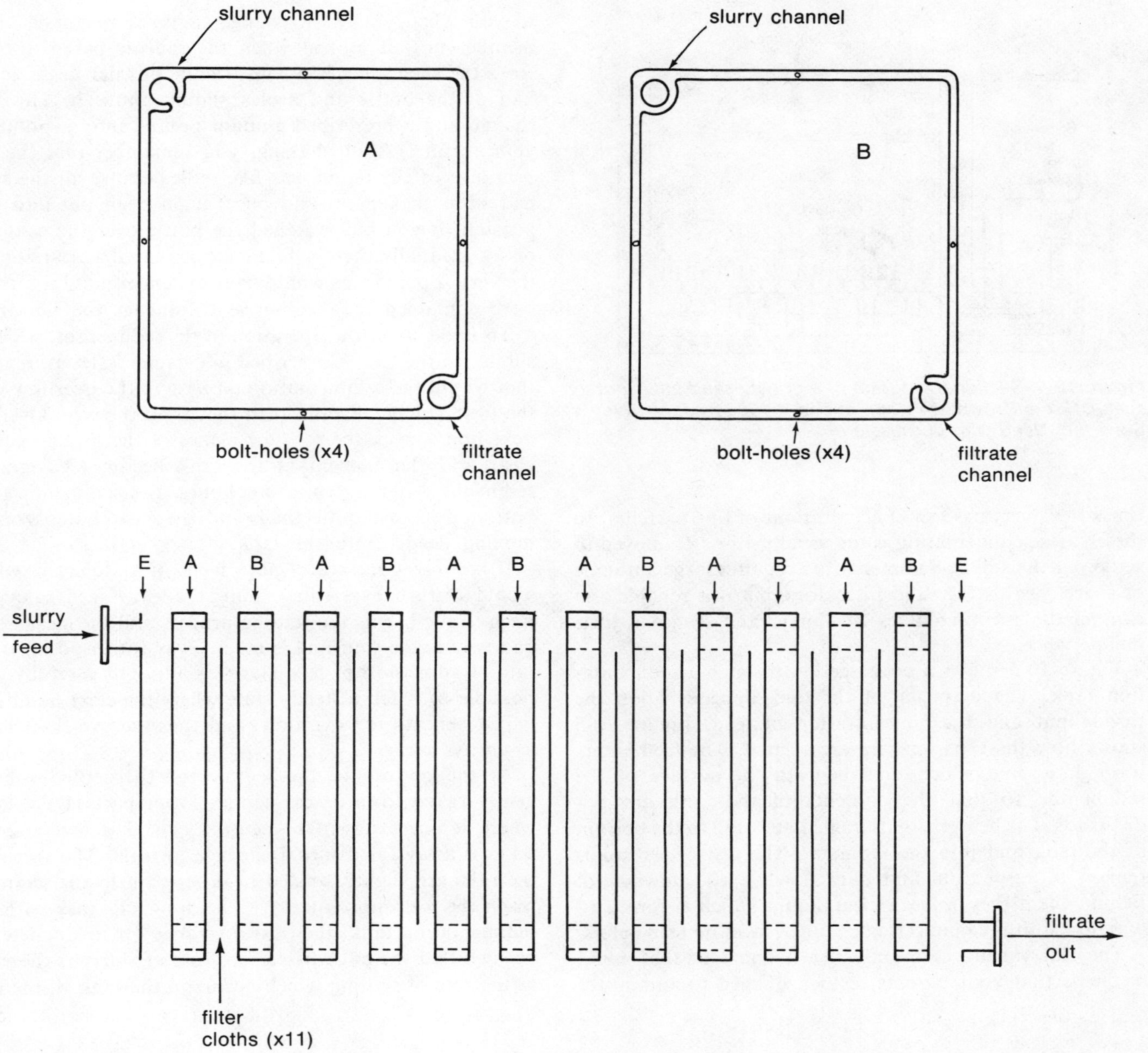

Figure 10-3. Plates and frame of an open-delivery filter press. (From Brown, Unit Operations. *© 1960 by John Wiley & Sons, Inc. Used with permission.)*

fore, pumping pressure for filtration and washing is lower? A mathematical method has been developed which enables the system designer to answer those questions, and no doubt you will learn the method later in your chemical engineering education.

The most common kind of filter used in the waste treatment business is the *gravity feed filter* which uses a bed of sand as the filtration material. These filters handle huge volumes of water, are very reliable, and are cheap to run. Figure 10-4 shows such a filter. The sprinkler arms bring the feed slurry and deposit it on top of the bed of sand, which is supported on a course filter screen. The liquid trickles through the sand, leaving behind a cake on top of the sand bed. When the filter becomes clogged, the slurry is taken to a standby filter and the clogged one is cleaned either by disposing of the used sand and cake together and replacing it with fresh sand, or by washing the sand in some auxiliary process.

Sedimentation

The filtration methods described above work better for more concentrated slurries of larger particles. Yet, some-

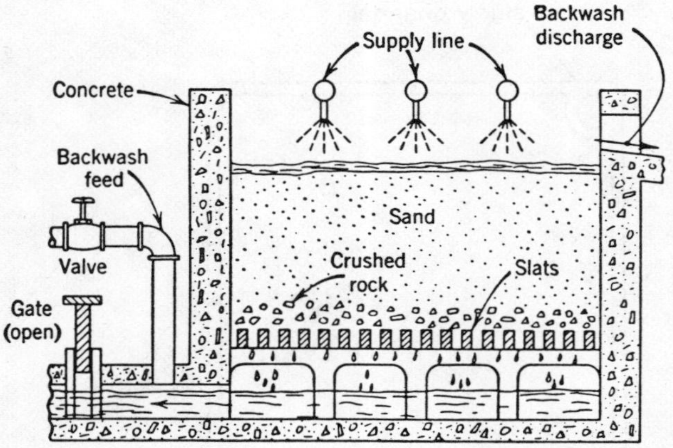

Figure 10-4. *Sectional diagram of a gravity sand filter (not to scale). (From Brown,* Unit Operations. *© 1960 by John Wiley & Sons Inc. Used with permission.)*

times the slurry has low concentrations of fine particles. In these cases sedimentation is the usual method employed to settle out the solids. Sedimentation requires large amounts of space, just as the sand filter does, but it is reliable and can handle great volumes of slurry, and it needs little maintenance.

Figure 10-5 shows a cross-section through a sedimentation tank. The diameter of the tank depends upon the throughput and the concentration of feed. Figure 10-5 shows how these parameters are related. The feed enters through a pipe which dips beneath the surface of the sedimentor so that the contents of the tank are not disturbed. The heavier solids immediately fall to the bottom of the tank and join the sediment. The rest of the solids remain in suspension and only slowly settle downward. Behind them they leave a clear liquid which is forced to overflow into the connecting channels as more feed enters.

The slurry behaves exactly like a bottle of that awful medicine that your parents seemed obliged to administer to you when you were young: "milk of magnesia." Remember how it looked when the sadistic parent took it from the shelf? It was a dense, white powder in the lower half of the bottle and a clear liquid above it. This was shaken and a prescribed amount poured into a spoon for your torture. After shaking, you remember that the appearance of the liquid was like milk because all the solid had been shaken up and once again been put into suspension. If you had watched the bottle over the next day or so, gradually there would have appeared a clear zone at the very top and this would have expanded until the solids were contained in a compressed zone at the bottom.

To understand the operation of the sedimentor, which is similar to the CSTR described in Chapter 7, let us imagine that we fill the sedimentation tank up to the overflow with the fresh slurry feed and then switch off the feed. Over the next hour or so, the top few inches of the liquid become clear and the bottom of the tank becomes covered in sediment. After a few more hours, the sediment at the bottom becomes quite thick and the clear liquid zone is moving deeper into the tank.

If we introduce some more feed—not violently, which would disturb everything, but at a slow feed rate—the fresh slurry begins the settling process and the clear liquid passes over the weir and is taken away for disposal. If the rate of admission of feed slurry is adjusted carefully, it is possible to reach a steady state where the clear liquid line is just beneath the overflow, and thus the overflow liquid is always clear.

At the bottom of the sedimentor the settled solid is gently raked towards the center, lower part of the tank, where it oozes down the central well. The screw feeder takes it away for disposal, storage, or sale. The vanes on the rake are slightly angled in order to help the sediment reach the well more quickly. The designer's task with the sedimentation tank, then, is to arrange for its diameter to be such that the anticipated feed rate of slurry is the same as the rate of settling which occurs in the tank. If the feed

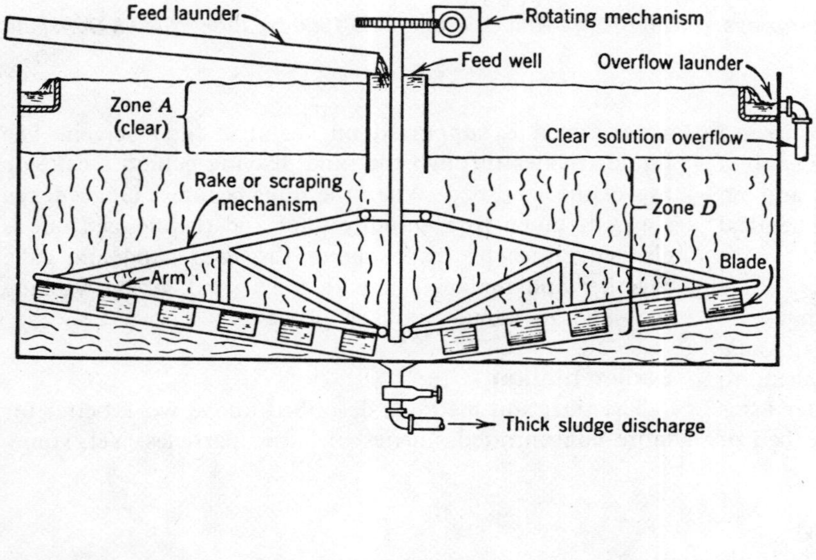

Figure 10-5. *Sectional diagram of a continuous thickener. (From Brown,* Unit Operations. *© 1960 by John Wiley & Sons, Inc. Used with permission.)*

rate were faster (or the diameter of the tank smaller), then the overflow liquid would not be clear. If it were slower (or the diameter of the tank greater), then the overflow would be clearer but we would have spent unnecessary money building a tank too big for its purpose, and the power required to drive the rake would be a little higher than it needs to be.

Centrifugation

Another method of separating solids from liquids is the centrifugation machine which, although highly unlikely as a pollution control device, is still one more way of separating solids from liquids. It does have a use in waste treatment, however, since solids which are extremely poisonous, but only present in small amounts in an effluent, would have to be removed by an extremely efficient device. The centrifugation method meets this demand.

In designing a centrifuge, we are exploiting the phenomenon commonly known as "centrifugal force." In fact, there is no such thing. If you stand at the circumference of a carousel, you are flung off unless you hang on tightly. The force that is in your arms holding you on acts towards the center of the carousel. It is called a *centripetal force*, and this word is taken from the Greek meaning "center-seeking." In other words, any object moving in a circle needs a centripetal force to keep it in a circular motion. The moon has the earth's gravity. A turbine blade has the tension in the steel. The boy on the carousel has the muscle tension in his arms. Now remove that force. The object flies off at a tangent from the circle. We say that the object is flung outward by a centrifugal force. In fact, it is flung outward because there no longer is a centripetal force.

Figure 10-6 shows a centrifuge. The slurry is fed into a basket which rotates at high speeds. A particle contained in the liquid in the basket also experiences a rapid rotation about the axis of the basket. There are no centripetal forces holding it at a particular radius and so it begins to migrate outward. Eventually, it reaches the wall of the basket, and there it remains until the basket is removed and the cake is scraped out. The clear liquid leaves the centrifuge in a hollow shaft above the basket. Careful design has led to the development of continuous centrifuges in which the solid is constantly scraped off the surface of the basket and collected. I tried several times to produce a drawing that would make this clear, and found that without a detailed engineering drawing and many sectional views the end result was more confusing than enlightening. However, if you now understand the general principle of the centrifuge, my objective has been served.

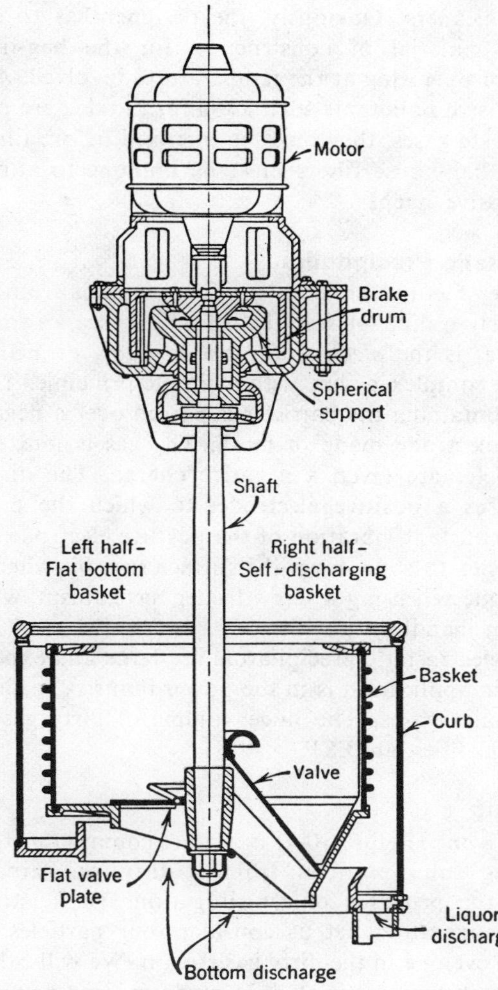

Figure 10-6. *A composite diagram of a suspended batch basket centrifuge, showing for the left half section a flat-bottom basket, and for the right half section a cone-bottom basket. (From Brown,* Unit Operations, *© 1960 by John Wiley & Sons, Inc. Used with permission.)*

Removing Solids from Gases

As in the three methods of removing solids from liquids, the equipment described in this section exploits the centrifugal effect and the filtration mechanism. In addition, you will see how electrostatic principles can be put to work in waste treatment.

Gas Filtration

Filtration of gases to remove dust is an old practice that lost popularity some years ago. But in the last decade or so, *gas-filters* have come back into favor, particularly in utility boilers which burn finely powdered coal, since these filters offer a means of removing the fine fly ash from the gases before they enter the atmosphere.

The filters are contained in a building called a "bag-house." When they are full enough that the pressure drop across them begins to build up and pumping costs increase, the gas stream is diverted to a stand-by bag-house while the first is cleaned. Bag-houses are simply gigantic vacuum cleaners. Obviously, the designer has to ensure that the material of construction for the bag-filter is capable of operating at the temperatures involved. Also, if any corrosive pollutants such as sulfur dioxide are present in the waste gases, they must be removed before filtration —or the bag-house filters must be immune to attack by the corrosive agent.

Electrostatic Precipitation

Another method of removing fine particles from gases, particularly useful for fine fly ash which escapes from the bag-house, is the *electrostatic precipitator*, or ESP. The ESP is a complex device with a simple principle. The gas stream containing the particles is passed over a negatively charged electrode made of mesh. The gas is ionized and the particles are given a negative charge. The dirty gas then passes a positive electrode, to which the particles stick. A constant vibration of the positive electrode causes the particles to slide down the surface to a pit where they can be collected using a screw feeder mechanism, which is useful for handling powders.

Since electrostatic precipitators are large and expensive, their main application is in the power industry to clean up utility boiler gases. The huge volume of dirty gas to be cleaned justifies an ESP.

Cycloning

The *cyclone* (Figure 10-7) is a more common method of separating solid particles from a dirty gas stream. It exploits the principle of centrifugation. To illustrate its mode of operation, let us consider four particles which enter the cyclone in the dirty gas stream. We will call them A (a large heavy particle), B (a medium-sized particle), C (a small particle), and D (an infinitesimally tiny particle). The gas enters the cyclone with a high velocity and on a tangential track as shown in Figure 10-7. It spirals downward, following the center shell of the cyclone. Particle A is flung to the wall almost as soon as the gas stream enters the cyclone, and it descends slowly under the effect of gravity. Particle B follows the flowlines of the dirty gas stream as it spirals downward in the separator. As the separator gets narrower toward the bottom and the velocity increases, particle B migrates outward with centrifugal effect until it, too, hits the wall and begins to descend under the effect of gravity.

The gas flow now enters the inner cycle and spirals tightly and quickly up the center of the cyclone to the exit throat. Particle C, which had been following the flowlines, now finds itself under the effects of even greater centrifugal effect because of the tight spiral. It migrates to the outside and passes from the central vortex to the outside vortex, and eventually will find its way to the wall where it, too, will descend. Particle D is so small that it follows the flowlines of any turbulence or eddy in the gas stream. Particle D actually leaves the cyclone in the "clean" gas.

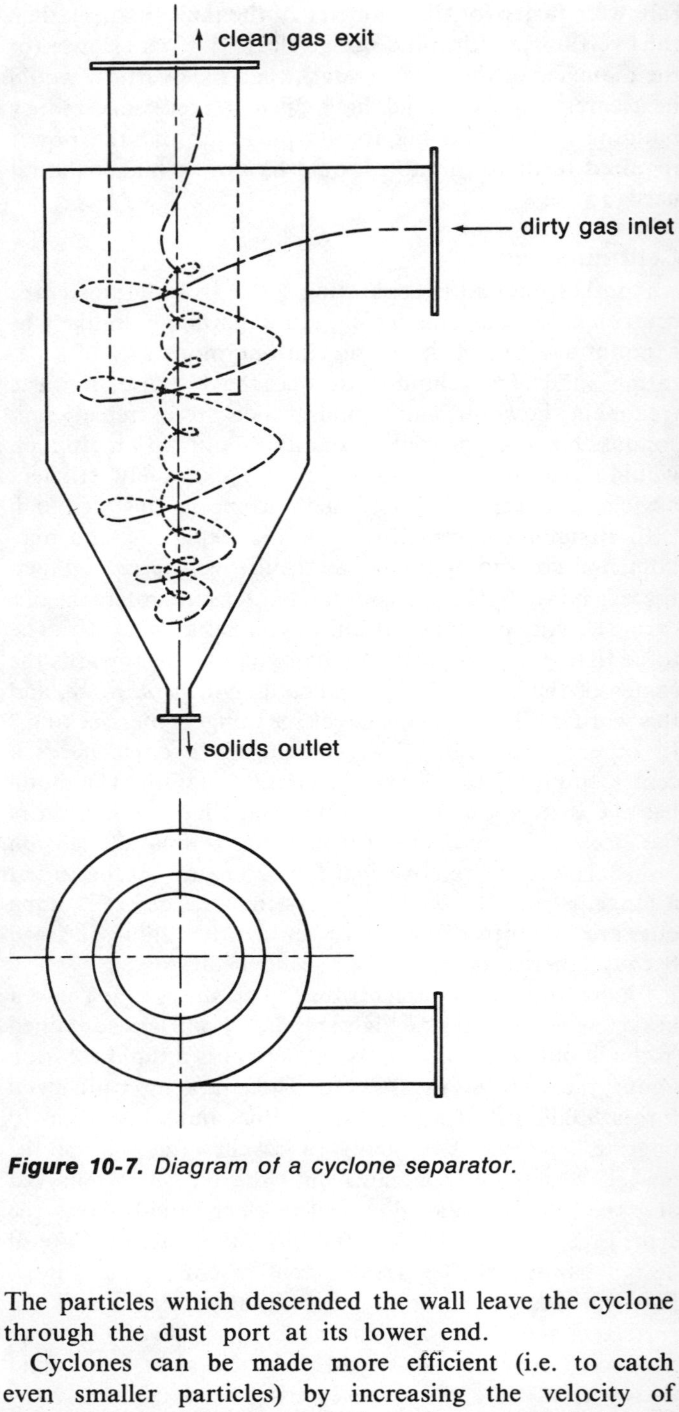

Figure 10-7. *Diagram of a cyclone separator.*

The particles which descended the wall leave the cyclone through the dust port at its lower end.

Cyclones can be made more efficient (i.e. to catch even smaller particles) by increasing the velocity of the entering gas. However, to do so increases their cost of operation because the larger compressors required use more electricity and, as always, there is a breakeven point beyond which further improvement is only gained at unjustifiable costs. Again, a balance must be sought which is economical and reasonable yet which allows only the most microscopic particles to escape and cause only trivial impact on the environment.

There are many other means of separating solids from

gases. The range of equipment to remove solids from dirty gases is as wide as human ingenuity and the technology of each is now well developed, so that a solution can be tailor-made for every situation.

Removing Liquids from Gases

The nature of some processes is such that the gases leaving them contain liquid droplets suspended in the form of a fine mist. The cyclone shown in Figure 10-7 can and is used to remove liquids as well as solids from gas streams. Since the liquid droplets cling to the wall once they have impacted against it, the cyclone is an efficient separator.

A less expensive alternative (because of its smaller size) is the *spiral vane separator*, shown schematically in Figure 10-8. In this equipment, the wet gas flows along the pipeline and enters the separator, which is an "in-line" device. The flow passes through the conically arranged vanes which impart a swirling motion to the gas. The droplets of liquids are flung to the outside of the vessel where they cling and run down to the bottom to be collected as shown. Dry gas leaves the separator and continues to flow along the pipeline. The liquid droplets coagulate into a stream which leaves the separator from the bottom nozzle.

The spiral vane separator is also used as a means of removing dust from a gas. First, the gas is passed through a vessel in which a water spray operates. The fast-moving gas picks up water droplets which form nuclei for the collection of the dust. When the gas passes through the separator, the water droplets are separated out, and with them, the dust.

Removing Liquids from Liquids

The most difficult clean-up operations are those entailing the separation of liquid from liquid. A most common case is the removal of oil and organic chemicals that have been spilled into water.

Surface Tension Separation

Let us first look at an oil spill on the ocean—a seemingly impossible clean-up task. The oil would kill wildlife if left on the water. However, the use of strong detergents or any other chemicals would be equally disastrous to some wildlife.

Early research teams concentrated on finding a physical property which had widely different values for oil and water. They finally settled on "surface tension." Oil, with its greater surface tension, tends to cling to solid surfaces; water, with a lower surface tension, tends to run off. The equipment devised to fight oil spills is a "boom" of steel discs mounted on a floating jig which is towed behind a ship. The discs, looking almost like huge circular plow blades, rotate slowly through the oil-contaminated water. The oil clings to the discs and is removed from the seawater as the discs rotate clear of the water. Scrapers, looking a little like windshield wipers, are mounted against the discs to scrape the oil off, collect it, and pump it into waiting tanks on the ship. It is a slow process and these oil spill clean-up devices have to be trailed through the affected area many times before any noticeable improvement is achieved. This is the only method presently available which does not involve the use of detergents.

Biochemical Waste Treatment

Oil and chemicals also come into contact with water in chemical plants. Natural and accidental leaks and spills find their way onto the ground. Rain or water spillage causes the chemicals and oil to accumulate in the plants' drainage ditches. At some point on site the drainage ditches come together and their contents would, if clean, be allowed to flow into a natural watercourse such as a river or the ocean. However, if they contain oily or noxious chemicals, it would be irresponsible to put them into a natural environment. This is the worst kind of problem to solve: the flow rates involved can be fairly large and yet the amount of pollutants contained in the wastewater is small. If a chemical means such as a detergent is used to react with the pollutant, there is still the problem of removing the substances formed by reaction.

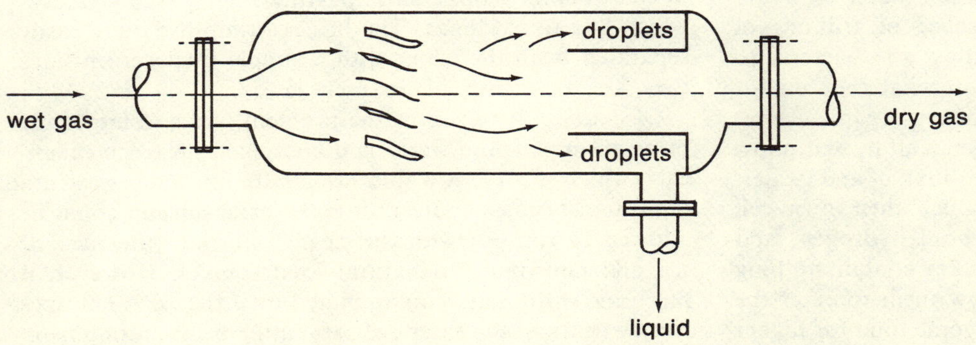

Figure 10-8. *Spiral vane separator.*

A major breakthrough in waste treatment occurred when it was discovered that certain bacteria liked to eat these chemicals. This technology was developed, and today many companies take advantage of these micro-organisms. As another branch of biochemical engineering, waste treatment using bacterial agents seems the most promising way of dealing with a wide variety of wastes. Almost every waste product becomes gourmet eating to some bacterium or other.

It is simple to find a suitable bacterium, or a number of them, for a particular pollutant. A mixture of water, common bacteriological nutrients, and the particular pollutants is placed in a dish in the open air. After a few days the dish will have been visited by hundreds of different air-borne bacterial species, and the ones which like to eat the particular pollutant will have set up home in the mixture. It is then a simple matter to isolate the culture and determine its species. Laboratory studies can be carried out to find out more about the bacteria which have stopped by for a meal. Their growth rates can be measured under the conditions which exist in the plant, and attempts made to speed up its growth rate by changing the conditions. For instance, the growth rate may be speeded up by bubbling air through the mixture. Or the addition of nitrates could also increase the rate of metabolism and hence the rate of disappearance of the pollutants.

Armed with this kind of knowledge, a chemical engineer, regardless of the fact that he is dealing with a life-form rather than a normal chemical reaction, can design a reactor in which this digestion process can occur. Of course, the reactor will need to be much larger than an ordinary chemical reactor, as you will see. If you remember the reactor descriptions of Chapter 7, you will realize that the CSTR is an ideal container for this biochemical reaction. However, the large volumes of liquids need large CSTRs and it is not unusual these days to find 40-acre ponds being used in this way. They are usually quite deep and are stirred by having air pumped in through underwater tubes. The air bubbles serve another purpose in addition to providing oxygen to the growing bacteria: as they rise, they cause a violent stirring or agitation.

The dirty effluent enters the ponds, and gradually the bacteria eat up the pollutants until the water is clean. Clean, that is, except for the presence of trillions of bacteria. Whether or not we can allow the bacteria to enter natural waterways depends upon what they do for their "living." If they eat heavy oils, heavy organic chemicals, and pollutants containing nitrogen, sulfur, and phosphorus, then they produce carbon dioxide and water, which are harmless. They also produce their own cell structure of sulfur, phosphorus, carbon, hydrogen, and oxygen in the form of amino acids. If they contain nothing more, then it is probably safe to allow them to enter the natural ecosystems where they will become food for higher life forms such as algae, plankton, or shrimp. (There is still some question as to the net result of an artificial abundance in any link in a food chain.)

Sometimes the waste water contains heavy metals or poisons which the bacteria may absorb to become another source of pollution to higher life forms. It is essential, then, that these bacteria be removed from the water before allowing it to flow out to sea; thus the water treatment problem is now one of removing a solid (the bacteria) from a liquid (the clear water), and filtration or sedimentation could be used. However, the bacteria are usually small, and efficient filtration or sedimentation is expensive.

There is a more natural way. As the water leaves the pond containing the bacteria, it can be fed into a pond containing a species of algae which likes to eat the bacteria. Algae are usually larger than the bacteria and easier to filter out of the water, leaving a clear, safe filtrate. Of course, the algae themselves must be dumped—they can be burned and the ash used as landfill, used as fertilizer, or fermented to make methane.

Even if the bacteria do not contain heavy metals or other noxious substances, it still may be undesirable to put them into natural water systems because they act as food for higher life forms, and thus disrupt the "natural" food chain. As a matter of fact, federal regulations issued by the Environmental Protection Agency (EPA) insist that the ecology must be unaltered in *any* way by industry. Clearly, if this legislation is to be enforced, the bacteria produced in the waste treatment ponds cannot be allowed to enter rivers. There are trade-offs all the way round, as you should see by now. Unfortunately, many regulators do not see the balances in these trade-offs.

In order to anticipate the future enforcement of this rather ill-considered legislation, one forward-thinking company in Texas has already pioneered and proven a completely biological method of cleaning up waste water and plant run-off. All the water is allowed to collect in a large pond system containing various species of bacteria which eat the organic and oily chemicals. The bacteria-infested water leaves this pond and enters another in which various species of algae are growing. The algae eat the bacteria. The algae-infested water then enters a third pond containing brine shrimp—tiny shrimp-like creatures which like to eat algae. The brine shrimp are quite easily separated from the water with a simple filter screen since they are an eighth of an inch in size.

Of course, there is now the problem of disposing of the brine shrimp. If the water had contained no toxic chemicals (which if present would eventually find their way into the brine shrimps' bodies), then the brine shrimp could be allowed to run out with the clean water into the river or sea and contribute to natural food sources. However, if the brine shrimp are contaminated or if the EPA enforces its new rules to the extent of preventing brine shrimp from entering the environment, then they should be filtered out of the effluent and sold either as fertilizer or as a feedstock for glue manufacturers.

Before leaving this topic I would like to point to the fantastic future biochemical engineering has. The bacteria, the algae, and the other tiny micro-organisms are like micro-miniature process plants in themselves. They work with extreme efficiency and carry out the most complex processes using little energy in doing so. Their use is not new. Bacteria and algae were the first inhabitants of our planet. They have been cleaning up after higher animals for the last few billion years! Man began to exploit this natural activity many years ago for sewage treatment. The sewage plant you will find downwind of your town exploits bacteria. Rotting vegetable matter in forest and garden alike is broken down and eaten by bacteria.

In the next few years the major breakthroughs in our industry will occur when new uses are found for different species of bacteria and algae. Already, certain algae can be used as enrichment agents for rare metals. Consider, for example, a plant that purifies uranium and has a waste stream of water containing, say, 12 parts per million of a uranium salt. By adding selected nutrients such as nitrates and phosphates to the stream and running it into a large pond, the uranium is removed from the water by the algae, which take up and store the uranium in their "bodies." They can then be filtered out of the effluent and roasted to produce concentrated uranium. The water from the pond, now free of uranium, can be run away to a natural water system.

Algae can also be regarded as a source of protein. One species, for instance, eats carbon dioxide (found at no cost in the atmosphere) and nitrate, and, using the photosynthetic process, produces a rich protein which is two thirds of its dry body weight.

With such a brilliant future, it is difficult to leave the subject of biochemical engineering. But remember this: the technology is in its infancy. Many new processes will be developed, and I expect that some of you will be responsible for these innovations.

Removing Gases from Gases

There are many means of separating gaseous pollutants from gaseous waste streams. However, the principles they employ have already been described in other chapters of this book. For instance, absorption can be used to remove a pollutant gas from a waste gas stream. A common use of absorption in this way is in the utility industry, where sulfur dioxide is removed from combustion gases before allowing them to enter the atmosphere. The operation is called *scrubbing*, but it is simply an absorption process. A slurry of calcium carbonate (crushed limestone) and water is passed down a tall tower which contains trays or packing, or both, to increase the wetted surface area. The warm combustion gases are blown up through the tower, countercurrent to the descending slurry. The sulfur dioxide reacts with the limestone to form calcium sulfite and calcium sulfate, which is filtered out of the water and used as landfill or sold as gypsum to make sheetrock. The cooled, clean gases are reheated by a steam heat exchanger to give them sufficient bouyancy to rise up the chimney.

Summary

In this chapter you learned a little of the history of environmental pollution and the way it is dealt with today. You learned how a balance or equilibrium must be established between the concerned public, the manufacturing industries, and governmental agencies, and how, if the equilibrium is disturbed, the cost is borne by all of us.

The chemical engineering techniques described in this chapter are: filtration, sedimentation, and centrifugation for removing solids from liquids; cycloning, filtration, and electrostatic precipitation for taking dust from gases; vane separation for "demisting" gases (removing fine liquid droplets); surface tension separation for cleaning up oil spills; and bacterial methods to digest pollutants from wastewater. Finally, you have had a second look at absorption as a means of removing a gaseous pollutant from a gas stream.

Further Reading

Abrahamson, D.E. "Environmental Cost of Electric Power." *Scientific American*, Vol. 223, No. 3, Sept. 1970.

Brown, G.G. *Unit Operations*. New York: John Wiley & Sons, Inc., 1960.

Busch, A.W. *Aerobic Biological Treatment of Waste Waters*. Houston: Gulf Publishing Co., 1971.

Commoner, B. *Science and Survival*. New York: Viking Press, 1966.

Kohl, A.L. and F.C. Rosenfeld. *Gas Purification*. Houston: Gulf Publishing Co., 1979.

Sax, J.L. *Defending the Environment*. New York: Alfred A. Knopf, 1971.

Singer, S.F. *Global Effects of Environmental Pollution*. New York: Springer-Verlag, 1970.

Thomas, W.L. *Man's Role in Changing the Face of the Earth*. Chicago: University of Chicago Press, 1956.

Chapter 11

The Process Plant Construction Industry

So far in this book I have talked about the basic principles and operations of chemical engineering. These have all been examined from the conceptual or process side of the industry. However, all operations must be contained in vessels. The definition of the functional behavior of the plant is contained in the process flowsheet, but the physical hardware—the vessels, connecting pipes, foundations, structural supports—must be designed to ensure that the plant will function according to the specifications of the chemical engineers who developed the flowsheet. Distillation columns, absorbers, heat exchangers, reactors, separators, filters, pumps, compressors, coolers, furnaces and other ancillary equipment must be "tailor-made" for their individual purposes.

The streams of chemicals which pass from vessel to vessel are carried through pipework which also must be designed. Similarly, when controlled conditions are specified in the flowsheet, instruments, signal cables, dials, gauges, electronically operated valves and hosts of other control equipment and instrumentation are necessary, and they, too, must be designed.

The vessels themselves will be made by the fabrication industry: companies specializing in particular kinds of vessels such as heat exchangers or distillation towers. Site construction and erection of the plant will be entrusted to a construction company. A process plant construction company, on receipt of a firm order, is also given the process flowsheet. That company is then responsible for the design and fabrication of the vessels by an appropriate subcontractor; the design and implementation of the necessary control and instrumentation systems; the selection and sizing of pumps, compressors and other ancillary equipment and their order from specialty subcontractors; the specification and ordering of pipes and pipe fittings; and the design and construction of site foundations and civil engineering structures which hold the vessels and pipes in place.

The same company is also responsible for the plant erection, leak testing, start-up and transfer of management to the customer company who will then operate the plant for the rest of its useful life. These activities are not normally classified as chemical engineering and are carried out by electrical, mechanical, and civil engineers. But it is essential that the designing chemical engineer be familiar with these tasks, since decisions made at the process flowsheet design stage will affect all later activities.

This chapter contains a glimpse of the development of a new plant from its initial conception through to the time when it is fully operational. It is a long chain of separate design activities carried out by a wide range of specialists, each design office taking the output of the last, adding their own expertise, and passing the information onto the next in the chain.

The Creative Process Called Design

Before explaining how a plant is designed and built, I would first like to draw your attention to the creative activity called "design." To any engineering design problem, there are an infinite number of solutions. The differrence between the competent professional and struggling amateur in a design environment is that the former can arrive quickly and confidently at a satisfactory solution; i.e., one that meets all the customer's specifications at the lowest cost possible.

Consider the design of a house. Imagine that the home buyer specifies to the architect that the new house must contain three bedrooms, two bathrooms, one dining room, two living areas, a study, a kitchen, a patio, a double garage, and a utility room. Suppose the owner also asks for a particular view from, say, a master bedroom window and requires a breakfast patio to the east of the house but close to the kitchen. How many designs meet those specifications? I would say an infinite number. The architect will add some more constraints or specifications to the design problem because he will know of any local, state or federal regulations which affect the design of the house. For example, any rules governing the siting of sewage arrangements from the house may affect the position of the kitchen or bathrooms in the plan.

Almost subconsciously, the designer will impose yet more constraints which are a rather complicated function of his own education and past career. For instance, the architect may not like designing Georgian homes, or may be going through a phase in which he prefers living rooms with high ceilings. As extra constraints like these are added, they reduce the number of possible solutions. Even so, after all the identifiable constraints have been determined and understood, there may still be several thousand designs which would do. Some would be impractical: a kitchen in the southwest corner and a dining room in the northeast corner would be a combination leading to inefficiency. If bathrooms are a long distance away from each other, then plumbing costs increase. Bedrooms should be as far away as possible from sources of noise such as the utility room or the kitchen.

These practical considerations further reduce the designer's freedom when he begins the creative process of arriving at a design which is efficient, meets all the constraints imposed upon it, comes within the budget of the client, and which has an attractive appearance. Before reaching a final decision, the architect will produce a number of sketches which will be evaluated for their economic suitability (i.e., cost to build), considered for their energy efficiency, and then discussed with the client before one is chosen. In other words, a number of solutions are suggested and analyzed using some procedure which is meaningful to both the customer and the designer. The sequence is to identify all constraints, create possible solutions, calculate their technical efficiency and likely cost, discuss them with the client, select a favored design, detail that design, and draft and document the final construction specifications.

These processes of design are similar in all fields of engineering, including ours. However, because chemical plant design is really a succession of design tasks linked together in a chain of total design activity, the design process occurs many times at different parts in the chain. Figure 11-1 shows the various design offices involved in a chemical plant design. In the left column, each office is named (titles will vary from company to company). The right column contains the name of their end-product (again, the name depends on the company). Written above the arrows connecting these boxes are some of the considerations that affect that particular design activity.

You can see the sequential nature of this design chain. The process designers develop the process flowsheet by studying the reports from the research chemists who have experimented with the reactions involved at two levels. Originally, bench-scale experimentation would have been carried out to learn of the conditions most favorable for the reaction. Next, a "pilot plant" would have been built to find what problems occur as the reaction process is scaled up. At both levels, the people in charge of the trials carefully note all observations and measurements. Information about the kinetics of the reactions, the physical properties of the reactants and products, and any other data, is all tabulated, documented and handed over to the process design office.

The designers, all chemical engineers, have available to them a number of tools: the design handbooks, compilations of physical properties, company design manuals and data sheets, the library (itself a gold mine of recorded experience) and the knowledge and craftsmanship of themselves and their colleagues. They are experts in the unit operations used to prepare the reactants and separate the products. They can work out the flow rate in every *stream* in the plant and calculate temperatures, pressures, and throughput rates for reactors, columns, absorbers and other equipment. The result of their work, the process flowsheet, is then used as the basis for what is sometimes called a "data explosion."

Vessel designers, usually mechanical engineers, take the process information and work out the details of heat exchangers, towers, columns, and other vessels. Also at this time reboilers, condensers, pumps, compressors, furnaces and boilers are designed or sized by comparing the required duty with an established product line from the specialized manufacturer.

Electrical engineers will be called in to work with chemical engineers to determine a system for the control, operation, and instrumentation of the plant. Chemical engineers will "size" the pipes by choosing diameters calculated from flow rates and tables of reasonable fluid velocities. Mechanical and civil engineers will take the hardware specifications and the pipe-sizing information and attempt to arrive at a suitable three-dimensional arrangement of the pipes and vessels in the space allotted to this plant. Their end-product will be a scale model of the plant and thousands of sketches (called *isometrics*) of the pipes and their positions. They will also have a *plot plan* as their output.

This information is then passed to the purchasing office, which employs civil engineers and clerks to carry out a *materials take-off* to work out the exact requirements for pipes, steel, gaskets, nuts and bolts, paint, insulation, and so on. These are then listed by the purchasing office, which also will place the final orders. When I take you through the design chain in more detail, you will see that these orders will have been made some time before on a speculative or estimated basis, but they are now confirmed, perhaps with some amendment.

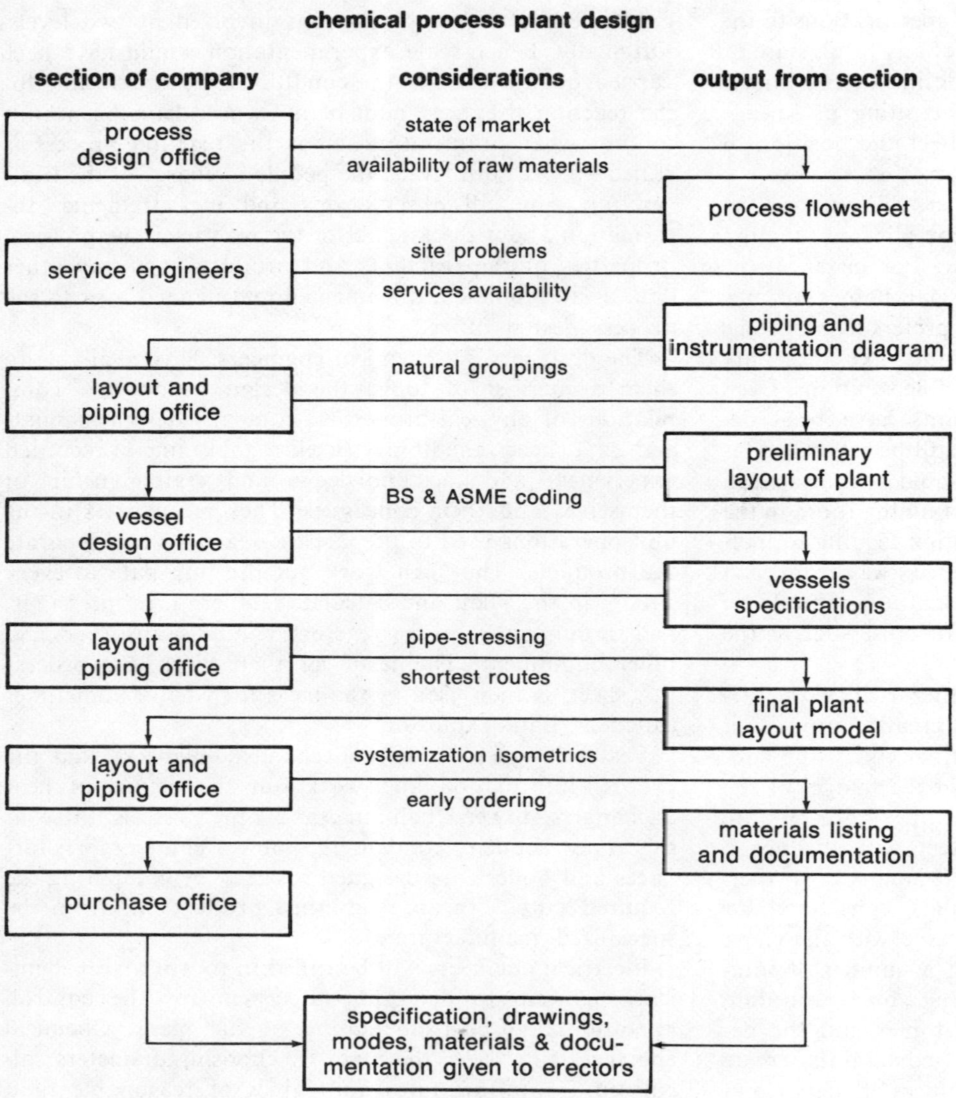

Figure 11-1. *Chemical process plant design.*

Design Documentation

Notice that all the engineers in the design chain must carry out a *documentation phase* in which they draw and specify their part in the design activity. This written and drafted documentation is the only means by which the future success of the plant's construction and operation can be assured. It must be clear, concise, and definitive. The reasons for every decision should be recorded. All calculations and sources of data should be readily available for future reference.

It is the responsibility of the expert who carries out a specialized engineering function to be the expert at setting down the outcome of that function in a clear and unambiguous manner. The so-called design process does not end until the documents describing the design have been planned, written, revised, and accepted by their future users. Clarity eliminates delays by those who must either manufacture, fabricate, erect, or operate the plant. Industrial employers claim that although the technical abilities of fresh graduates are unquestionably suited to modern times, it is generally true that their skills of written communication need considerable development before they can become useful members of design teams. This situation is to be regretted and is, in any case, unnecessary. Those engineers who have taken the trouble to develop a crisp, individual style of writing have found that it is just as easy to write clearly and concisely as it is to write vaguely and badly.

I ask you to remember this piece of logic: When you put your signature on a document that describes an engineering task you have successfully completed, you are signing your name to a total piece of work that includes not only your engineering but also your written description of it.

Just as your signature signifies your pride in assuming responsibility for the engineering contents of the document, it also signifies your responsibility for the manner in which it is described. The engineering content of your work will be checked by engineers with more experience than yourself. Perhaps they will be better paid, too. But do you really want the written component of your work checked and corrected by clerical or secretarial staff who probably stopped their education in high school and certainly are more lowly paid then you are? If by reading this book you have a resolve to become an expert writer as well as an expert chemical engineer, then my efforts in writing it will have been worthwhile.

Supply and Demand

Before we follow the progress of a project through the various plant design offices, it may be better to ask first why a new plant should be built. The answer is, of course, that it will be built if it has a high probability of making money.

A chemicals manufacturer exchanges goods for money, and the price of the goods is determined by the supply-and-demand balance at the time of the sale. In a strong demand market the price of a commodity will settle at a level which gives the manufacturer as much profit as can be added to the manufacturing costs without leaving him in a position where competitors can undercut his price—competition between companies helps stabilize prices. However, when the market is depressed or there is little demand for a commodity, its price will fall as each competing manufacturer tries to sell his inventory of that commodity.

In any case, if the manufacturing costs can be brought down, higher profits are possible while still remaining competitive. Certain items which go to make up that cost are fixed or are beyond the company's control. For instance, the insurance premium on the plant will probably be a fixed cost. Furthermore, the raw materials price depends upon market demand and cannot be controlled by the buyer. But the cost of labor depends upon the number of men needed to run the plant, and careful design can reduce this number. Similarly, the cost of energy to fuel the plant and the cost of maintenance are both costs which can be reduced by careful planning at the design stage.

The amount of money that must be borrowed to build the plant, i.e. its capital cost, has to be paid back out of the plant's earnings together with interest on the loan. Obviously, the capital cost of the plant will be minimized by careful design. I hope it is becoming clear that the two most important parameters in any design are its capital and running costs. The most important unit is the dollar. At the end of design and construction, and after successfully bringing the plant "on-stream", the objective is to pay off the loan, pay all the employees, meet all manufacturing costs and then make a sound profit which even after income taxes (currently 48%, almost half of a corporation's income) will still be worth putting in the bank and distributing between the shareholders.

When it is said that a company is meeting market demands, it is not meant to imply that the company operates the chemical plant in the sense of carrying out a duty. The company enters the marketplace by offering to supply a commodity. It is putting itself in a trading position with the consuming public. In other words, it is offering a deal. Any customer who accepts that deal knows that after the deal both he and the seller will be satisfied with the deal, otherwise one of them is a fool; either the company is a fool for selling inferior goods at an elevated price, a certain way to lose future customers, or the customer is a fool for paying more than the worth of the goods instead of buying at lower cost from a competitor.

The laws of supply and demand underlying this discussion assume that prices settle at a natural level. If prices are forced to a different level by artificial means such as government intervention, then all sorts of anomalies occur. Although government intervention is always claimed to be for the public good, it always seems to lead to a situation which is unfair to both buyers and sellers. Since these *are* the public, that justification for intervention is false. Fortunately, this has been realized, and price regulations by governments will be applied with considerable caution in the future, one hopes.

Cost management, then, is an important aspect of plant design, since it will affect all operating costs for the life of the plant, and an increase in the initial cost of the plant adds to the loan that must be paid off with interest.

With these thoughts in mind, let us consider the initiation of new plant construction project. The first decision is taken by a person who is equipped with shrewd foresight and the courage of his convictions. He has to predict the future, since it takes at least five years to begin making a new product in a major plant. Then the plant will run for another 15 years or so. He has to recognize a future demand in the market which can be met using the skills of his company. He will be aware that, even though his company is first in the market, other companies will soon begin their own design and construction and create competition. He has to bear in mind that the new commodity may be made from raw materials which are currently cheap but whose prices will rise as demand for them increases due to his own company's efforts. Often the raw feedstocks for his new plant will be by-products from his other plants, in which case he can control their prices.

Not only does this person have to be a prophet—he also must be a leader. He has to convince the company shareholders that this would be a good venture, i.e., it will be well worth going into serious debt because the return on investment will be high. Remember that shareholders are only interested in a return on the money that they have lent the company. They are extremely cautious when a company making a safe but mediocre profit wishes to borrow half a billion dollars to put up a new refinery. If his case is good enough, he will persuade them to take the plunge.

He then has to convince a banker. Bankers provide the lifeblood of our industry. They put in cash, and they, too, expect a return on their investment. They, too, have developed caution over the years. No one would ask a banker for half a billion dollars without an extremely good scheme to get the lenders' money back while making a profit himself. If the borrower goes bankrupt, then there will be one less potential client for the banker in the future. In the next section you will see how the profitability of a proposed plant can be assessed by prospective backers.

Economic Evaluation of a Design

To determine the profitability of a plant, some preliminary design is necessary, but it is customary to use methods which give a rough estimate of construction and running costs from the process flowsheet alone. These figures will be used in a calculation of profitability (called a *discounted cash flow* calculation), which will be the major factor considered when the decision is made either to drop the project or go ahead and carry out a complete detailed design prior to building and operating the plant.

The capital costs of the plant may be divided into two categories. The first is called *fixed capital investment*, which is the amount of money needed to design, build, and equip the entire plant. This sum of money includes the cost of all the pressure vessels, distillation and absorber columns, heat exchangers, pumps, compressors, pipes, valves, structural engineering, buildings, land, workshops, warehouses, utility provision, waste disposal, laboratories, office accommodations, and so on.

The second category is called *working capital investment*, and this amount of money is invested in four ways. First, the stock of raw materials and supplies has to be bought, and this is charged to working capital. Also in this category is the cost of building up an inventory of the finished product prior to its being sold. Similarly, the working capital includes an amount of money to cover accounts receivable (unpaid bills) for the first few months after start-up, when there will be a natural delay in receiving payment for goods.

A fourth kind of working capital investment is the amount of cash that must be kept on hand for monthly payment of salaries, wages, raw materials purchases and other miscellaneous items. The working capital investment will make up about 20% of the total initial investment in the plant. These costs can be estimated before detailed mechanical, electrical and civil engineering design if methods based on experience are used.

Running Costs

The costs of running the plant can be divided into six categories. The first includes all those expenses called *fixed costs*. These are expenses which remain practically constant year after year and do not vary (or only slightly at most) with an increase or decrease in the rate of production. Into this category falls repayment (including interest) of the loan which was obtained to build the plant. In addition, local land taxes, depreciation, research and development, patent royalties, and insurance are all fixed costs.

Direct production costs are costs which are incurred by production alone. These include raw materials costs, wages for the work force, managerial salaries, maintenance, repairs, supplies such as oil and grease and cleaning materials, any payroll overhead such as pensions, vacation allowances, social security, life insurance and so on, electricity and other utilities.

A third category is called the *general plant overhead*. This includes costs for general maintenance, safety services, security, a fire service, restaurant and recreation facilities, an engineering and drafting office, and the medical and first-aid services.

A fourth category is the expense of *administrative and office overhead*, which includes executive and clerical salaries, office supplies, and telephones.

Expenditures incurred in marketing the plant's products are called *distribution costs* and include salaries and supplies for sales offices, shipping and packaging costs, and advertising.

A final type of running cost is made up of what are called the *contingencies*. This is a safety factor on each individual cost that will take care of unforeseen emergencies or those economic developments that cannot be planned for ahead of time. A contingency on any particular cost will depend upon the stability of that cost or the amount of detailed knowledge available on it. For example, the labor costs for running the plant for its useful life will be calculated by looking at the inflation of wages of process plant operatives for the past few years and predicting how they will continue to rise in the next few years.

From these estimates, the expected wages bill can be computed for each year in the life of the plant. Then, working with the supposition that, due to a good labor relations record in the chemical industry, there will be no sudden shocks or rises in the future, the labor wages contingency may be set to, say, 5%. This reflects the estimator's belief that even if wages do rise faster than predicted, they will not be more than 5% more than he has estimated.

On the other hand, in the uncertain days in the early and middle 1970s, the contingency that was being used for the price of crude oil was as high as 75% in some companies. Even that turned out to be too low, since the price of crude oil inflated by over 400% during those few years.

With both the capital and running costs carefully estimated, it is possible to calculate the *return on investment* for the project. To show how this is done, I must explain the meaning of three more terms: *depreciation*, *cash flow*, and *discounted cash flow*, itemized in the cash flow table (Table 11-1).

The fixed capital cost of the plant is assumed to be its worth or value when first built because this is how much it would cost to replace it. At the end of its life the plant will have a scrap value, which is usually assumed to be zero, since the cost of demolition and site clearance will probably cancel out any revenues from scrap dealers. The reduction in value from the initial cost to the final zero is called depreciation and is a measure of the increasing obsolescence of the plant: after all, besides the general wearing out of the plant, it will only be useful as long as its product is marketable.

There are many reasons why a time will come when the product is unmarketable. Competitors will build and operate newer plants. If they use newer technology, their price may be better and force off the market the more expensive products of older plants. Also, fashions and demands change. A new polymer is only "a wonderful new fiber" until an even more wonderful new fiber is invented. Federal regulations may cause a plant to be obsolete. For example, the product known as tetraethyl lead, a gasoline additive, may no longer be salable in just a few years due to federal policy. Similarly, if the Food and Drug Administration is successful in its fight to ban saccharin, all plants manufacturing this particular sugar substitute will become obsolete.

Depreciation

Depreciation may be calculated by several methods, all of which give different answers. The company chooses the method of calculation which will reduce the income tax payable on profit made by the project. Following are two widely used methods of calculating depreciation. The first of these is called the *straight-line method* in which the value of the plant is assumed to drop by the same amount each year. Thus, if the fixed capital investment is $100 million and the plant cost is to be depreciated over the first 10 years of its life (a common practice), then the plant is said to depreciate on a straight-line basis at $10 million per year, as shown in Figure 11-2.

The second method of depreciation is called the *sum-of-years-digits method.* This method assumes that the most rapid wear and tear will occur early in the plant's life.

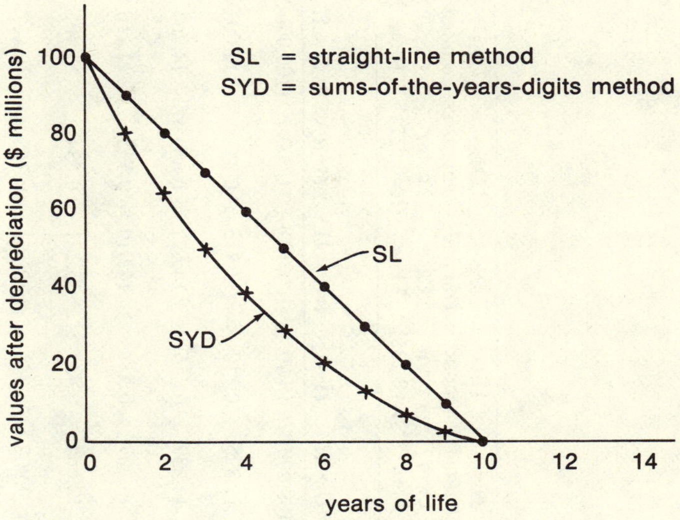

Figure 11-2. *Two methods of depreciation.*

Therefore, its value will fall more quickly in the first few years. Using the same figures as in the example above, a $100 million plant with a life of 10 years will deteriorate, not $^1/_{10}$ of its value each year, but $^{10}/_{55}$ in the first year, $^9/_{55}$ in the second year, $^8/_{55}$ in the third year, and so on, down to $^1/_{55}$ in the tenth and final year of depreciation. The value 55 is obtained by adding together the digits in all the years of the depreciation period:

$$1 + 2 + 3 + 4 + 5 + 6 + 7 + 8 + 9 + 10 = 55$$

Figure 11-2 shows the depreciation graphically. You must remember that these two depreciation methods, and the others available, are simply accounting devices. They have no physical significance. The sum-of-years-digits method is just a means of accelerating the depreciation of the plant. It does not really mean that $^{10}/_{55}$ of the corrosion of a heat-exchanger shell would occur in the first year. On the contrary, wear and tear will probably become more rapid toward the end of the plant's life. The reason for these different accounting methods is that they offer varying options on income tax savings and an opportunity to choose the most suitable one for a particular plant.

Cash Flow

Study the cash flow in Table 11-1. It has been drawn up for a plant that will take two years to build at a total cost of $100 million. The plant's expected life is 15 years and depreciation of the $80 million fixed capital has been calculated over 10 years using the sum-of-years-digits method. The working capital is not depreciated (which is customary). Column A shows capital employed. Notice that the capital cost in the first year (year 0) is written inside

Table 11-1
A Typical Cash Flow Table
(see explanation in text)
$000

Year	Capital Income	Gross Sales	Total Operat-ing Costs	Depre-ciation SYD	Profit BFIT	Profit AFIT	Cash Flow	Dis-count Factor (R=0.2)	Dis-counted Cash Flow	Dis-count Factor (R=0.25)	Dis-counted Cash Flow	Dis-count Factor (R=0.235)	Dis-counted Cash Flow	Dis-count Factor (R=0.232)	Dis-counted Cash Flow
N	A	B	C	D	E=A+B−C−D	F=0.52 × E	G=F+D	$H=\frac{1}{(1+R)_n}$	I=G × H	$H=\frac{1}{(1+R)_N}$	I=G × H	$H=\frac{1}{(1+R)_N}$	I=G × H	$H=\frac{1}{(1+R)_N}$	I=G × H
0	(100,000)	—	—	—	(100,000)	(100,000)	(100,000)	1.000	(100,000)	1.000	(100,000)	1.000	(100,000)	1.000	(100,000)
1	—	25,000	10,000	14,545	455	237	14,782	0.833	12,313	0.800	11,826	0.810	11,973	0.812	12,003
2	—	46,000	15,000	13,091	17,909	9,313	22,404	0.094	15,548	0.640	14,339	0.656	14,697	0.659	14,764
3	—	70,000	20,000	11,636	38,364	14,749	26,385	0.579	15,277	0.512	13,509	0.531	14,010	0.535	14,116
4	—	70,000	20,000	10,182	39,818	20,705	30,887	0.482	14,888	0.410	12,664	0.430	13,281	0.434	13,405
5	—	70,000	20,000	8,727	41,273	21,462	30,189	0.402	12,136	0.328	9.902	0.348	10.506	0.352	10,627
6	—	70,000	20,000	7,273	42,727	22,218	29,491	0.335	9,879	0.262	7,727	0.282	8,316	0.286	8,434
7	—	70,000	20,000	5,818	44,182	22,975	28,793	0.279	8,033	0.210	6,047	0.228	6,565	0.232	6,680
8	—	70,000	20,000	4,364	45,636	23,731	28,095	0.233	6,546	0.168	4,720	0.185	5,198	0.188	5,282
9	—	70,000	20,000	2,909	47,091	24,487	27,396	0.194	5,315	0.134	3,671	0.150	4,109	0.153	4,192
10	—	70,000	20,000	1,455	48,545	25,243	26,698	0.162	4,325	0.107	2,857	0.121	3,230	0.124	3,311
11	—	70,000	20,000	—	50,000	26,000	26,000	0.134	3,484	0.086	2,236	0.098	2,548	0.101	2,626
12	—	70,000	20,000	—	50,000	26,000	26,000	0.112	2,912	0.069	1,794	0.079	2,054	0.082	2,132
13	—	47,000	16,000	—	31,000	16,120	16,120	0.093	1,499	0.055	887	0.064	1,032	0.066	1,064
14	—	30,000	12,000	—	18,000	9,360	9,360	0.078	730	0.044	411	0.052	487	0.054	505
15	20,000	25,000	10,000	—	35,000	18,200	18,200	0.065	1,183	0.035	640	0.042	764	0.044	801
				80,000			**260,800**	**Total DCF = 14,068**		**Total DCF = −6771**		**Total DCF = −1228**		**Total DCF = −59**	

Note: Of $100 million capital outlay, only $80 million is depreciable, the remainder being $20 million land cost and working capital, recouped in year 15.

parentheses. This is simply to honor an accounting convention which requires that capital expenditure (i.e. negative income) be shown in parentheses rather than with a negative sign. Notice, too, that there is no other entry in column A until year 15, when the land is sold and some working capital is regained.

Column B in the cash flow table contains the estimated plant operating costs. The figure in this column is the total of all running costs except depreciation. In other words, it contains the estimated total of fixed costs. (less depreciation), direct production costs, general plant overhead, administrative and office overhead, distribution costs and contingencies. Column C contains the estimated revenue from sales of the plant's product. It is perhaps the most difficult column to estimate since so many non-mathematical factors are involved, from federal policy to human fickleness. Column D contains the depreciation.

All the remaining columns are calculated from these first five. The next two (E and F) are called the P&L (Profit and Loss) entries. The first P&L column (E) is called "Profit BFIT", or profit before federal income tax. The value in this column is calculated by adding any capital income to the gross sales income and subtracting depreciation and total operating costs. Notice that the value for profit calculated in this way is not the true profit because the depreciation has never been spent: it is merely a book figure. You will see in just a moment how this is corrected in the Cash Flow column. By the way, in the first year or two of any project, the "Profit BFIT" will certainly not be a profit at all, but a loss, because there will be the expense of building, and, of course, during the plant's construction there will be no revenue. Next, federal income tax can be calculated at 48% (1978) and subtracted from the Profit BFIT to give Profit AFIT (after federal income tax) in column F.

Now to find out more about the cash that will be banked at the end of each operating year, we must look further than the value of Profit AFIT, which is not an indication since it includes the bogus expenditure of depreciation. Column G contains a value which is a true cash flow calculated by adding the depreciation back into the Profit AFIT. The term "cash flow" is appropriate since this figure is the amount expected to be in hand at the end of each year's operation. It is the amount of money that will be banked, distributed to stockholders, or used for whatever other purpose the company's board decides.

One way to work out a return on investment, then, is to add up the cash flows for each of the 15 years to get a total cash flow or income, and then work out how much more this is than the original investment. Unfortunately, the answer would not be particularly meaningful—suppose that instead of spending money on building the plant, it had been carefully banked with an organization giving, say, an interest rate of 8%. The company could then sit back and relax knowing that its money will grow at a steady 8% due to the efforts of others. However, if every company did that, there would be no production whatsoever, the country's economy would grind to a halt, and we would all be back to digging for roots in the forests. Industrial leaders decide that their companies can make more than the 8% by careful management and planning, and so they may consider putting their money into projects, such as building manufacturing complexes like chemical plants. If they do, they want to be sure that they get back more than the 8% they would have got from the bank because they would be taking a considerable financial risk as well as putting considerable effort into the venture.

Using Table 11-1 as an example, the total cash flow is $232.4 million: well over twice as much as the $100 million spent on the project. But if the $100 million had been banked at 8% in year 0, by year 15 it would have grown to $317,216,910—the cash flow would have been $217,216,910. This project, then, is a better use of the money than simply banking it at 8%.

There is another way of looking at the cash flows. They can all be adjusted back to the value they would have at year 0. For example, consider year 6 in Table 11-1. The cash flow is $26,436,000. To have that much money in year 6 by banking at 8%, the sum of $16,659,164 would have to be deposited at the bank in year 0. In other words, the *net present value* of the $26,436,000 from year 6 is $16,659,164 if banking is available at 8% interest. The means of carrying out this calculation is not complex. You know that compound interest works as follows:

$$\text{Cash in hand after } n \text{ years} = P(1+R)^n$$

where P is the principal deposited in year 0
R is the interest rate expressed as a decimal
n is the number of years the principal is on deposit.

For example, $100 banked at 5% (0.05 as a decimal) for 20 years becomes

$$100(1+0.05)^{20} = \$265.33$$

Therefore, working backwards,

$$P \text{ (at year 0)} = \frac{\text{Cash flow at year } n}{(1+R)^n}$$

Thus, for the example in Table 11-1,

$$\text{Present value of } \$29{,}491{,}000 = \frac{29{,}491{,}000}{(1+0.08)^6}$$

$$= \mathbf{\$18{,}584{,}000}$$

Looking at the cash flows this way offers a means of economic evaluation. We can ask the following question at the time of investment: If we were to deposit the investment at a bank, what rate of interest would give us the same financial gain as our proposed project? If the answer is 4%, then we would be much better off depositing our capital investment in a bank. On the other hand, if the answer is 25%, we would be foolish to settle for the much lower interest we would earn at a bank.

This is the basis for the method of economic evaluation called *Discounted Cash Flow*, or *DCF*. Having worked out the actual cash flows (as in column G in Table 11-1), those values can be discounted back to a present value (i.e. at year 0) by multiplying them by a *discount factor* calculated from an assumed interest rate:

$$\text{discount factor} = \frac{1}{(1 + R)^n}$$

When the discounted cash flows are added up, they will equal zero when the assumed interest rate would be the breakeven interest rate achieved at a bank.

The example in Table 11-1 will make this clear. First, I prepared the cash flow table up to column G as described above. Then I guessed an interest rate of 20%, worked out the discount factors shown in column H, and calculated the discounted cash flow shown in column I, which added up to $14,068,000: the project was worth that much more to the company than a bank would have paid out at 20%. I then set the interest rate (the rate of return on investment) at 25% and went through the procedure again to get a total of -$6,771,000 in the DCF column, meaning the company would be $6,771,000 worse off after 15 years by banking the money at 25%.

Although the $6,771,000 is small when dealing with hundreds of millions, I repeated the calculation to show you how you can "home in" on the solution. I guessed at 23.5% rate of return and you can see in Table 11-1 that I was still $1,128,000 away from the breakeven point. Very close! One more guess at 23.2% gave me a summed DCF column of -$59,000, which is negligible.. The conclusion to be drawn from this DCF calculation is that the proposed project breaks even with the earnings on a bank deposit if a bank will give a 23.2% interest rate. Highly unlikely! The company would very likely go ahead with this project and look forward to a 23.2% return on their investment after tax.

As you can see, the DCF calculation, though simple, is tedious to carry out by hand. Fortunately, most companies have a computer program which can be fed with all the economic data to carry out the iterative process just described and arrive at the DCF return.

The DCF method of economic evaluation is a good guide to profitability. If, for a project, the DCF return works out to be less than 8%, the project would be abandoned and a more attractive one sought. Indeed, unless the estimated DCF return worked out considerably higher than 8%, the company would still not take the risk, but would look around for a better investment. If the DCF return is around 15%, and there is little risk attached to the project (i.e., the plant is of a type with which the company is familiar) then this will be considered an adequate return on investment.

If, on the other hand, the project was considered a high risk, perhaps because it was a new process which had never been operated before or because the market for the product was considered fickle, then the DCF return would have to be up around 25% before the project will be considered. If the DCF return is found to be far greater than 25%, then the most likely conclusion is that a mistake has been made in the arithmetic, which should then be carefully checked. If the answer is still unusually high, the wise project manager will examine very critically all estimations of costs and sales revenues. If these prove to be the best that all concerned can make, then the project will certainly be launched.

After the Decision to Build

For the purposes of this book, I am going to assume that the entire design chain is carried out in the offices of the construction company. In other words, the client has contracted the construction company to develop a process to carry out a particular task and then design and construct the plant which will do the job.

The sequence begins with the appointment by both companies of project engineers who will be responsible to their respective companies for the design chain from start to finish. The project engineers will probably be chemical engineers and probably will have had many years of experience in some of the design offices in the design chain as well as experience with both start-up and plant operation. They will manage the project with the pride of professionals, approaching the work with the confidence of experience even though they will assume the responsibility of millions of dollars of capital and many people's lives and livelihoods. The job of a project engineer can be defined simply: He has an objective to reach within a given budget and before a given deadline.

Process Design

The project engineer appoints a chief process engineer who puts together a process design team to be responsible for all the chemical engineering aspects of the plant. Libraries are searched for literature pertinent to the particular process and the chemicals involved. Data are sought and collated. These include the physical properties data for all the substances in the feedstocks to the plant, the products from the plant, and any intermediate substances which can be expected to be made. In other words, physical properties data such as thermal conductivities, densities, viscosities, vapor pressure data, reactivities, solubilities and hosts of other kinds of information are acquired about every single species in the process. Just as cash is the lifeblood of industrial development, so physical properties information is the lifeblood of process design.

It is unfortunate that the data collected from different sources will contradict each other. One source will quote a value of thermal conductivity of some substance at a certain temperature and pressure condition as being a few percent different from a figure quoted for the same conditions in some technical article in a journal. Usually, the more conservative value will be used, but if the contradiction seems to be critical, the company may pay a laboratory to check the data experimentally.

The process designers, in the meantime, are putting together their ideas for the final flowsheet. A full heat and mass balance is carried out on the process. Knowing the flow rates of feed and product, and using the reaction and physical properties data that have been collected or measured, the flow rate, temperature, pressure and composition of every stream are calculated and carefully recorded. Similarly, the whereabouts of every Btu of energy is also calculated and recorded. All this information is put together on a schematic diagram called the process flowsheet, which shows the "connectedness" of the process. Figure 11-3 shows an example. Often, the flowsheet includes a table of all stream information beneath the diagram.

The job of developing the process flowsheet is called process design. It is carried out by chemical engineers and can take many man-years of effort to produce a flowsheet for a large plant. A recent innovation has shortened process design time and, by enhancing the skills of a competent designer, has given our industry the opportunity to produce more efficient designs. I am referring to the arrival of the computer and with it the comprehensive programming languages (such as FORTRAN) which allow design engineers to harness the computer's rapid calculating power and vast, accurate storage capability.

A number of companies and universities have written giant computer programs (one that I was involved with has over 250,000 FORTRAN instructions in it!), called *process simulators*, which work as follows. The process design engineer uses experience and knowledge to set down a preliminary flowsheet. The process units and the order in which they are connected are fed into the process simulator together with estimated values for process parameters such as pressures, temperatures, flow rates, and stream compositions. The simulator refers to a built-in library of process unit subroutines which describe each unit mathematically. Relevant subroutines are drawn from the library and strung together in the same order as the real units in the flowsheet. Similarly, the simulator has a databank of physical properties from which it draws data (heat capacities, densities, vapor-liquid-equilibrium data) for the flowsheet. Then, by calculating iteratively round and round the flowsheet, the simulator converges upon a steady-state heat and mass balance which is presented via a computer output device to the engineer using the program. After consideration of this first solution, perhaps changes will be made to the flowsheet and its operating parameters and the whole exercise repeated, again and again until a satisfactory design is reached. Often this activity is accompanied by a DCF evaluation of each design so that the choice of solution is an economic one.

Notice how the creative part of *computer-aided design* (the general title given to this kind of approach) is carried out by the engineer, whereas the tedious calculations and the looking-up of data are done by the computer.

Computer-aided design (now known universally as CAD) has become a welcome and accepted tool in the process design office and other offices further on in the design chain, as you will see.

The process flowsheet is handed to a number of design offices, which could be called the "engineering services offices."

The number and size of the reactors are decided using the criteria described in Chapter 7. The number, size and configuration of the distillation columns, heat exchangers, pumps, compressors, condensers, reboilers, furnaces, heaters and coolers are determined while considering their minimum cost of operation. Some design situations need special consideration. For instance, if there is a need for a large compressor in some part of the plant, the project engineer has to consider that, because of its continuous, high-speed running, there may be frequent maintenance problems during plant operation. To avoid excessive downtime, it would be better to have two compressors, each capable of handling the full flow of the material. Then, one could be on stand-by in case of breakdowns. On the other hand, maybe it would be preferable to have three compressors, each rated for half the total flow, so that two would always be working with one available on stand-by. The advantage with this arrangement is that three half-size compressors may be cheaper than two large ones. It could even work out that the best arrangement would be five quarter-size compressors—four in operation with one stand-by.

In this kind of situation the economics for each case would be worked out and compared. Not only does the cost of the compressors have to be included in the calculation, but so does the incredibly complex pipework and valve system necessary to isolate any one pump and carry out maintenance upon it without disrupting the plant's operation. Also, the future utility costs of operating the compressors will be taken into account. Similar decisions, often even more complicated than this, have to be taken for all the items of equipment on the site. Then, using the company's own operating experience, an estimate is made of the total weight of the steel and pipe required to build the plant. This is presented in the form of a preliminary order to the steel manufacturers so that they can plan to

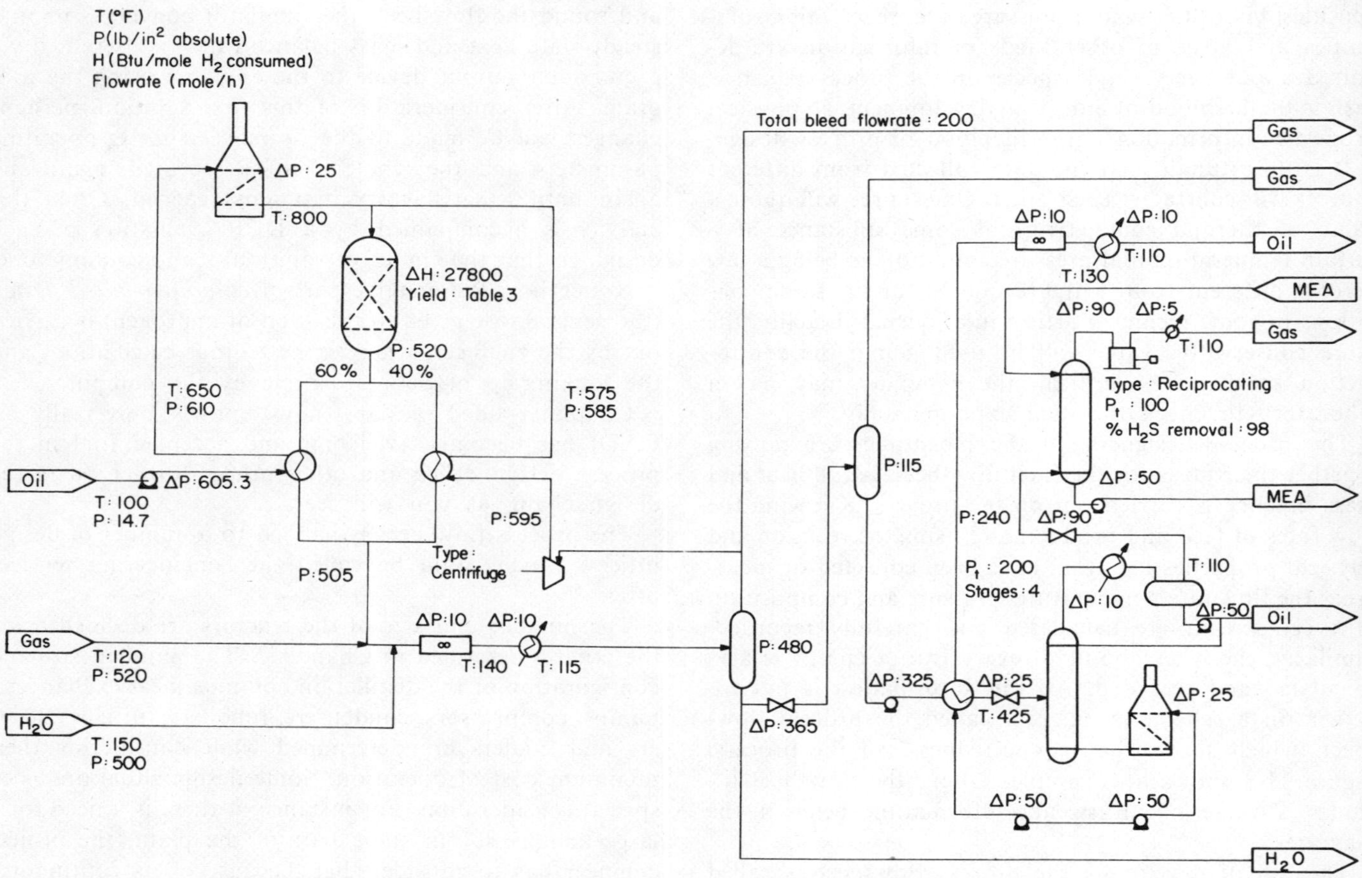

Figure 11-3. *A typical process flowsheet. Heat and material balance establishes material and thermal requirements. This hydrodesulfurization process removes sulfur from oil.*

be ready to roll the steel at the right time. Both the steel company and the construction company know that the figures in this preliminary order are approximate, but they are the best available at this stage.

The design evolves as control engineers work with the process engineers to decide upon the form of the instrumentation and automatic control for the plant. Remember that the more computer control that is built into the plant, the lower the labor costs will be because less men will be needed to run it. Calculations are carried out to find the required amounts of steam, cooling water, compressed air, electric power, refrigeration, and other utilities. These will be marked onto a new diagram called the Piping and Instrumentation Diagram, or the P&ID, which looks similar to the process flowsheet except that it contains considerably more information. Pipes, which are shown instead of process streams, are named for the first time. The name of the pipe is a simple code which includes information about the type of pipe, its size, material and its incidence number in the plant.

Naming a pipe requires knowledge of its diameter. For every pipe shown on the P&ID, a calculation must be carried out to decide on the best size for it. Consider, for example, a flow of 3,140 ft³/min of a gas. We would find that its velocity would be 576,000 ft/min (which is nearly nine times the speed of sound) in a 1-inch diameter pipe, 36,000 ft/min (about half the speed of sound) in a 4-inch diameter pipe, 4,000 ft/min in a 12-inch diameter pipe, and 1,000 ft/min in a 24-inch diameter pipe. Which one should be chosen? Tables of recommended fluid velocities would indicate that gas velocities around 4,000 ft/min are usually preferred. But, what about other pipe sizes in the region of 12 inches? The answer is found by an economics procedure as follows.

In Chapter 5 you saw how frictional pressure drop in a pipe can be calculated. You also saw how it depended upon the Reynolds number, which was in turn dependent upon the diameter. With that connection, it is a simple matter to work out an equation between the cost of pumping against the pressure drop and the diameter of the pipe. Figure 11-4 shows the relationship between them on a graph. Similarly, a mathematical relationship is derived between the capital costs (and therefore the loan repayments) and the diameter. This curve is plotted as shown on the same graph in Figure 11-4. Notice how the cost of pumping gets less as the diameter increases. This is what we would expect. It is easier to force a fluid through a large bore pipe than it is to force it through a narrow one. (Imagine drinking milk through a large straw and then through a narrow one.) As you would expect, the capital cost increases as the pipe diameter increases. The third line on the graph is obtained by adding the other two curves together. The example confirms a minimum cost with a 12-inch diameter pipe.

The pipe names are carefully recorded on the P&ID. Similarly, details of instrumentation, control, feedback, and so on, are determined and marked on the P&ID. For the first time, the sizes of the vessels are calculated and specified on the P&ID. An example P&ID is shown in Figure 11-5.

At this writing, there are no generally available CAD programs which help with the preparation of the P&ID. However, the success of process simulators and other CAD programs has spurred on a number of organizations to develop programs to help with P&IDs. It is interesting to consider their progress. Initially, developers were put off by the fact that the P&ID is essentially a graphical data presentation and computers were not able to draw. In the early 1970s, however, the computer manufacturers released a line of plotting devices which were both accurate and low cost. At about the same time, a number of drawing programs became available.

Thus the first barrier to CAD in the P&ID office had been overcome. The next problem was the sheer volume of data involved in a P&ID. FORTRAN programmers were used to dealing with tens of thousands of data items,

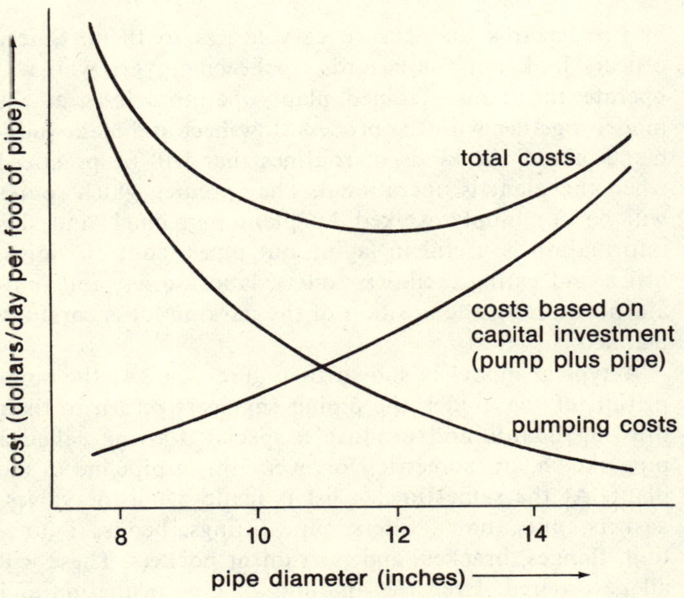

Figure 11-4. *Optimization of a pipe size.*

especially in design programs, but a P&ID office deals with many millions. Storage of this many numbers is trivial: each magnetic tape can hold well over a million numbers. The real problem is knowing *where* a number is stored and, thus, being able to find it when necessary.

A store of numbers held this way in a computer is called a *data base*. It is only in the late 1970s that the magnitude of the problem of using very large data bases in chemical plant design has been appreciated. Commercial computer programs for P&ID production will soon follow.

For every item of equipment in the plant, a specification sheet is made up. It contains information about the flow rate through the unit, the temperature and pressure, and other conditions of the input and output streams, all that is known about the physical properties of the substances in the streams, and any other relevant information. The specification sheets are sent to manufacturers who will be invited to bid on the design and fabrication of the particular item of equipment.

Here is another design activity for which the development of CAD programs to help the designer has proven simple. Heat exchangers, pressure vessels, tanks, towers and many other equipment items are designed to a rigorous code. The regimented nature of the code means that it is easy to write the instructions in a sequential manner and hence turn them into FORTRAN. Most equipment design programs not only work out the dimensions and geometry of the item but automatically produce an outline diagram as well.

The specification sheets and the equipment design data are also used to calculate the weight and types of steel that will be required to build the plant. The preliminary order that was given to a steel company can now be updated to a more accurate figure. When the subcontractors' bids are in and selected, a final and exact order will be placed with the steel company, which will have scheduled future production to include the steel for this project.

Around this time the project engineer has begun to consider the site and how the vessels will be laid out upon it. Perhaps a preliminary block model will be built using simple shapes to represent the vessels. The objective will not be to route the pipes inside the plant at this stage but merely to shuffle the plant items around until a working configuration is found which will minimize piping costs generally by siting the vessels as close as possible, particularly if they are connected by expensive, large-diameter piping.

It is essential that the preliminary block model be built as early in the design as possible in order that the position of nozzles is known on each vessel. For instance, it is preferable to bring the side stream pipes out of a distillation column in the direction that will take them to their destinations with minimum number of bends. The angular orientations on the column will be known if a preliminary block model has been built.

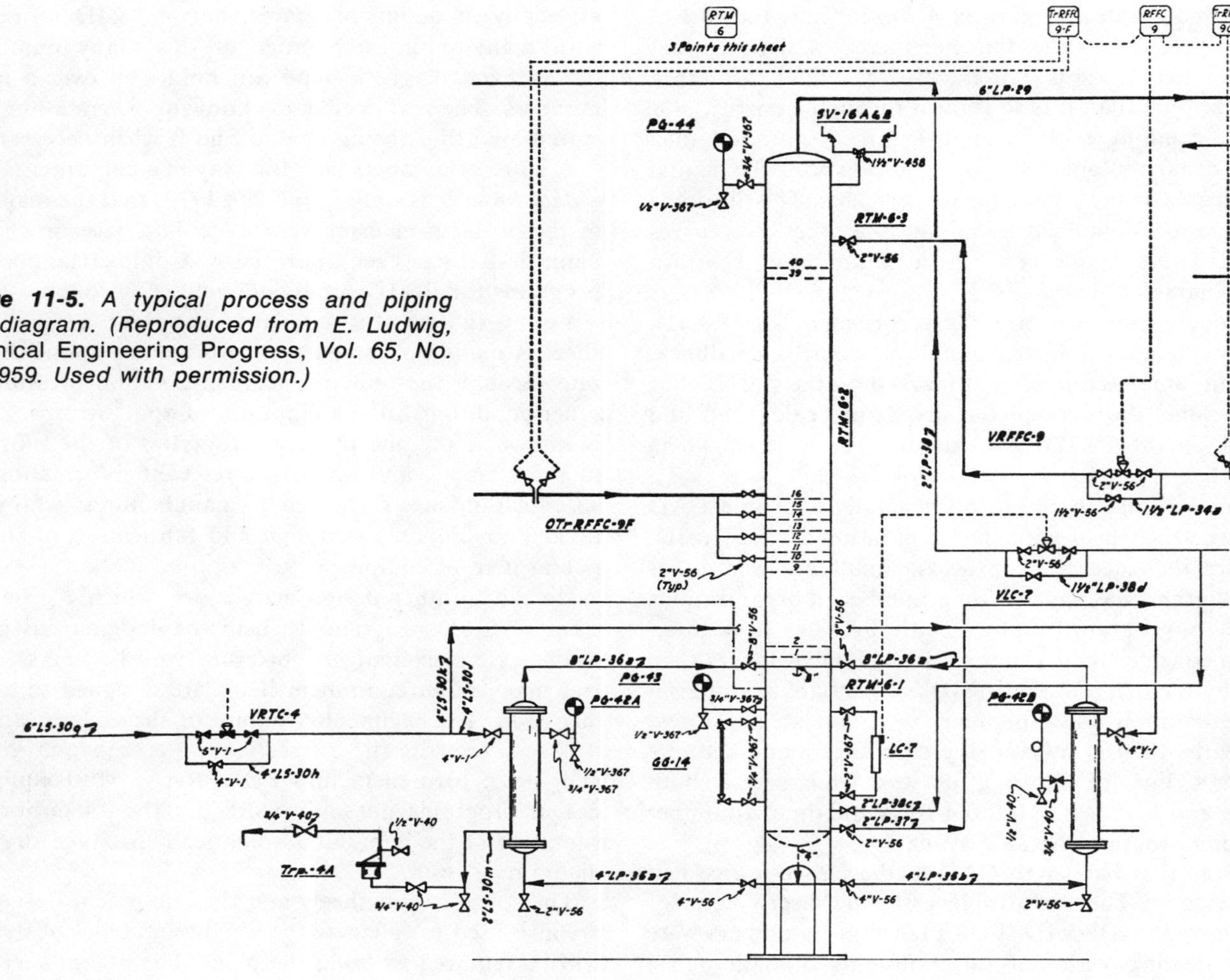

Figure 11-5. A typical process and piping flow diagram. (Reproduced from E. Ludwig, Chemical Engineering Progress, *Vol. 65, No. 10, 1959. Used with permission.)*

As details of the vessels come in from the subcontracting fabricators, small-scale models are made and used to build what is called the piping model which is crucial to the design of the plant. The scale is usually ¼ inch to 1 foot and there are many companies who sell model vessels, pipes, valves and other equipment to this scale.

The piping model is a design tool. The approximate positions decided on the preliminary block model are used to site the major items of equipment. Then, using scale-size pipe and considerable thought and time, the pipes are routed between the units and scale-size structures are added to support the vessels and carry platforms for human access, stairways, catwalks, and pipes. Concrete foundations are designed to hold the units. Electrical engineers visit the model and plan routes for their cables and signal wires around the plant. The construction workers who will actually build the plant examine the model to see whether they can swing units in overhead using cranes without damaging other units. Firefighting experts decide where to place their equipment and work out the location of fire hazards and ensure easy access to them. Safety officers look out for hazards to the employees who will operate the plant. Trained plant operators look at the model together with the process flowsheet and make judgments about the kinds of routines that will be practiced when the plant is operational. They predict which routes will be commonly walked by plant personnel, and this information is useful in laying out pipes, control rooms, office and eating facilities, toilets, laboratories, and first-aid posts. Even the position of the parking lot is carefully planned.

A typical model is shown in Figure 11-6. At the completion of the model, the piping engineers return to their drawing boards and produce a special drawing called a pipe sketch, or isometric, for every single pipeline in the plant. At the same time, a list is made of all the valves, gaskets, nuts, bolts, filters, pipe fittings, bends, T-junction, flanges, brackets and instrument pockets. These will all be ordered. Even the amounts of pipe insulation and paint which have to be bought will be jotted down, added

Figure 11-6. A simple block model plant layout. (Courtesy of Socony Mobil Oil Co., Inc.)

up, and ordered. When the final order for materials and equipment is made, all the relevant information about how closely it was estimated initially, and how the final figure was arrived at, is carefully logged. This information will perhaps enable a more accurate early estimation in the future.

The vessels, towers, heat exchangers, pumps, compressors, pipes, structures and other items of equipment are collectively called the *hardware* of the plant. Their design as individual items is carried out by CAD programs except in special cases where human ingenuity is required at all stages of the design. But the placement of the individual items together to make a total plant—the plant layout and design itself—is a different matter. Imagine the problem: Start with a process flowsheet, a P&ID, a sheaf of individual equipment drawings, and a few hundred acres of land. Now, build a plant. Where do we put what? As I explained earlier, the traditional approach is to build a piping model to scale, and use it as a thinking tool around which many different kinds of engineers can gather and discuss the layout.

The piping model, then, has been the key to plant layout. But since it is built by experienced and therefore highly-paid engineers, the model is expensive. It will cost about ½% of the total plant cost. The piping model of a $200 million refinery, for example, will cost about a million dollars. And so there has been a strong incentive to develop CAD techniques to help with the awesome tasks of plant layout and piping design. In the late 1970s, the first successful commercial CAD system for piping and plant design became available. Its name is PDMS (Plant Design and Management System) and I'm proud to tell you that I initiated its development in 1971 by describing the possibilities to a number of companies in order to put together a development consortium. Three years later, in 1974, I had two partners and enough funding to start work. PDMS was finished in 1977 and is now available worldwide.

The input data for PDMS are geometric and dimensional information about individual plant items and how they are to be connected. The program works out a three-dimensional route for all pipes, avoiding clashes between pipes and other plant equipment. PDMS tries to place pipes close together in order to minimize the cost of supporting them: it is considerably less expensive to make a suspension bracket to carry ten pipes rather than ten brackets each to hold one. The program also places all pipe fittings, valves, flanges and any other special equipment. It even shows the position of pipe welds. All the information is drawn into piping diagrams which can be inspected by the engineers who can then make whatever changes they think are necessary.

Other, similar programs will soon be available to compete with PDMS, and by the time you are a professional engineer, these new tools will be accepted and widely used in piping design offices. Of course, they will never replace the piping model which, as you saw earlier in this chapter, has many other uses. However, the information produced by PDMS-type programs will enable the model to be built more quickly and at less cost by junior engineers.

Construction and Start-Up

During the final phases of work on the piping model, site clearance will begin and, soon after, construction itself will commence. The skill of the project engineer is needed now more than ever. The project engineer will have carried out a "critical path analysis" for both the design and the construction phases of this project. The details of critical path analysis are beyond this book. However, you may be interested in its basis, which is to single out all those tasks that are in the main sequence of the design and construction of the plant. These are tasks which, if not carried out on schedule, will delay many other tasks.

An example is the laying of the concrete foundation for a distillation column which would be a dominant piece of equipment in the plant. Until the foundation is in position, no other construction task connected with that particular column can occur. Installation of the column, its piping, insulation, painting, and instrumentation all await the pouring and setting of the foundation, which then is said to be "on the critical path." On the other hand, the painting of the column holds back no other job and so it is not on the critical path. Having identified the tasks on the critical path, special attention can be given to them from the very start of the project.

Critical path analysis is another complex, tedious task that can require many hours of effort by highly paid engineers. Again, CAD techniques are useful, and most critical path analysis work is done by computer programs under the control of the project engineer.

Eventually, the plant is built and ready to be started up. First, it is tested for leaks by closing off all the inlets and outlets to the plant and pumping the entire plant—vessels, pipes and all—full of a gas under pressure. The pressure of the system is then watched, and if it falls, the leak must be found. It is usual to use a filler gas which can be traced very easily by mobile sensors. It is then a simple matter to probe with the sensors around gaskets, flanges, welds, valves, and other likely places until traces of the filler gas indicate the location of a leak.

After leak testing and the inevitable repairs, the plant is started up. This is a most difficult operation, and start-up engineers usually are the best people available, and, of course, the most highly paid. You have seen some of the difficulties of start-up in Chapters 6, 7, and 8. Reactors which rely on a recycle need many hours, even days, to settle down to their steady-state operating conditions. Similarly, distillation columns with refluxes have to settle down to their equilibrium states. In the meantime, the products from the plant are useless and must be stored in special start-up tanks whose contents will be emptied and fed back into the feed stream after the plant is fully operational.

At some point the two project engineers, one from the production company and one from the construction company, come together and agree that the plant does all that it was meant to do and that yields and flow rates are as specified. The final job is for the construction company to train the customer's own operational staff; then the plant is "handed over" to the customer, who assumes responsibility for it.

Summary

In this chapter you have seen something of the personalities of the chemical engineering design world. You will have noticed that the chemical plant is not a "product of technology" or a "product of science." It is the product of human minds which have gained knowledge, as yours will, in colleges and universities and have gone on to gain experience and know-how in their years of industrial practice. Men and women with experience and know-how are paid for their confidence in doing the job well, properly, and at the lowest cost. Professional engineers do not produce vague, awkward, and ambiguous writing full of misspellings and irritating grammatical errors. They do not rely on looking up formula after formula to push into a problem, hoping that one of them will eventually solve it. They are able to think a problem through and solve it by understanding it. If you learn now to write clearly and persuasively and develop the habit *now* of thinking into a problem in order to solve it, you will become a confident industrial leader yourself . . . more quickly than you can imagine at this point in your life.

Further Reading

Branan, C. *The Process Engineer's Pocket Handbook.* Houston: Gulf Publishing Co., 1976.

Evans, F.L. *Equipment Design Handbook for Refineries and Chemical Plants.* Houston: Gulf Publishing Co., 1979.

Ludwig, E.E. *Applied Process Design for Chemical and Petrochemical Plants.* Houston: Gulf Publishing Co., 1977.

Moore, S.M. *Plant Layout and Design.* New York: Macmillan, 1962.

Peters, M.S. and K.D. Timmerhaus. *Plant Design & Economics.* New York: McGraw-Hill, 1968.

Rase, H.F. *Piping Design for Process Plants.* New York: John Wiley & Sons, Inc., 1963.

Rase, H.F. and M.H. Barrow. *Project Engineering of Process Plants.* New York: John Wiley & Sons, Inc., 1957.

Reed, R. *Plant Location, Layout & Maintenance,* Homewood, Illinois: Irwin, 1967.

Appendix I

A Survey of Chemical Engineering

Table AI-1
Occurrences and Applications of Basic Chemical Engineering Principles

Principle	Application or Occurrence
Heat Transfer Conduction: transfer of heat through a body or between two bodies in physical contact without perceptible movement of the particles of the body(ies).	a) melting and solidifying metals, glass and polymers; b) heat treating metal products; c) firing pottery and bricks; d) transferring heat through pipes and tubes, e.g., heating, boiling, cooling and condensing fluids such as water and steam in power plants, hydrocarbons in oil refineries, etc.; e) losing heat from the exterior surfaces of pipes, furnaces, boilers, distillation towers, reactors, buildings, turbines and engines, etc.; f) baking breads and cakes and cooking other food products; g) making coke; h) smelting ores; i) measuring temperatures by thermometers; j) burning of solid fuels.
Heat Transfer Convection: transfer of heat in a fluid between regions of unequal density resulting from non-uniform heating.	Convection is always present when heat transfer to and from a fluid is involved. Since all of the above examples of conduction involve a fluid as well as a solid, they are also examples of convection. In addition, one might add: k) combustion involving two gases (e.g. burning of natural gas) or a gas and a liquid (e.g. burning of gasoline or fuel oil); l) heating or cooling of interiors of buildings; m) evaporation from liquid surfaces; n) meteorological phenomena, e.g. winds, cloud formations, rain and storms; o) oceanographic phenomena, e.g. currents and geothermal upwelling; p) draft in chimneys.
Heat Transfer Radiation: transfer of heat by absorption of radiant energy.	Radiation is always present when two bodies (fluids or solids) at different temperatures can "see" each other across a space. Therefore, any situation meeting this criterion involves radiation, even though it may also involve conduction and convection. Examples of radiation are found in: q) solar energy systems; r) heating of the earth by the sun; s) cooling of the earth at night; t) heat transfer from flames to boiler tubes, metal products, pottery, bricks, etc.; u) heat losses from all equipment and buildings which are hotter than their surroundings; v) heating up of space stations; w) coal particle and oil droplet combustion.

(*Continued*)

Table AI-1 (continued)
Occurrences and Applications of Basic Chemical Engineering Principles

Principle	Application or Occurrence
Mass Transfer	a) Evaporation—distillation (vapor rate), flashing, drying, sublimation, humidification, steam generation; b) condensation—distillation (reflux rate), cloud formation, dew formation, conversion of steam to water in reboilers, liquefaction of air and other gases; c) diffusion—leaching, liquid-liquid extraction, absorption, adsorption, osmosis, catalysis dissolution, crystallization, ion exchange.
Momentum Transfer	a) Flow of liquids and gases in pipes, pumps, compressors, distillation towers, absorbers, reactors, heat exchangers, boilers and other equipment containing moving fluids; b) measurement of viscosity; c) filtration; d) sedimentation and settling; e) bubbling; f) fluidization; g) spraying.
Conversion of Matter and Energy	Stoichiometry and mass and energy balances for furnaces, boilers, all kinds of reactors and process equipment involved in chemical plant design.

Table AI-2
Unit Operations and Their Industrial Applications

Unit Operation (and other names, if any)	Application
Size Reduction (crushing and grinding, comminution, milling, disintegration, attrition, pulverization, powdering)	Making pulverized fuel; preparing ores for smelting; milling cereal and other vegetable products; grinding clays for brick manufacture and kaolins for paper additives; manufacture of abrasives (sandpaper, carborundum stone, talc, soapstone); pulverizing cement, gypsum and lime; manufacture of furnace refractories; powdering pigments, dyes, coloring agents, cosmetics, soaps, polymers and chemicals.
Size Enlargement (flocculation, agglomeration, granulation, nodulizing, compacting, extrusion)	Making catalyst pellets; forming ceramics, cements and plasters; compacting and extruding plastics and rubber; briquetting ores; sintering metal powders; mixing and baking cereals; pelletizing and extruding feeds and fertilizers; flocculating coal slurries and particles in turbid waters.
Transport of Fluids (pumping and compression)	Pipelines; loading jetties; moving fluids about a chemical plant; fuel pumps in aircraft, automobiles and ships; fire-trucks; drying fans and blowers; air conditioning and refrigeration systems; water supply systems; pneumatic conveyors.
Heat Generation (combustion)	Boilers, furnaces, heaters, and stoves; all internal combustion engines; welding.

(Continued)

Table AI-2 (continued)
Unit Operations and Their Industrial Applications

Unit Operation (and other names, if any)	Application
Distillation (rectification, fractionation, stripping, stabilization)	Fractionation of crude oil; purification of natural gas; coke oven by-products recovery; refining of petroleum products; separation of petrochemical products; separation of liquid air into oxygen, nitrogen, inert gases; manufacture of liquor; purification of olefins prior to polymerization; dewaxing hydrocarbon fuels; desalination of seawater; separation of waxes; purification of drugs, dyes, sterols, hormones, vitamins, solvents, fatty acids, waxes, etc.; generally, separation of mixtures of components with different boiling points.
Absorption (gas absorption)	Recovery of heavier hydrocarbons from natural gas; removal of pollutants from gas streams; recovery of solvents; purification of CO_2 carburization of steel.
Solvent Extraction (liquid-liquid extraction—including leaching)	Separation of aromatic compounds from other hydrocarbons; production of penicillin and other drugs; extraction of soybean, peanut and other vegetable oils; finishing of lubricating oils; separation of uranium compounds from solutions containing other minerals; chromatography.
Drying	Dehumidification of air; drying of detergents, food products, chemicals and resins, clays, pharmaceuticals, catalysts, sewage sludge, etc.
Adsorption	Generally, separation of small amounts of materials from liquids and gases—e.g., desiccating, recovery of solvents, decolorizing of foods, oils, etc; deodorizing; filters in cigarettes.
Crystallization	Purification of chemicals and petrochemicals; zone refining; solar evaporation of seawater; production of semiconductors; metallurgy; production of refractories.
Sublimation	Freeze drying of food products, pharmaceuticals, blood and plasma specimens for electron microscopy and other heat-sensitive materials; purification of iodine and phthalic anhydrides.
Gaseous Diffusion	Separation of uranium isotopes; purification of gases.

(*Continued*)

Table AI-2 (continued)
Unit Operations and Their Industrial Applications

Unit Operation (and other names, if any)	Application
Dialysis and *Electrodialysis*	Recovery of mineral acids from aqueous solutions in metallurgical operations; recovery of colloidal organics in the manufacture of pharmaceuticals; recovery of caustic soda from a mixture with hemicellulose; desalting of brackish water.

Further Reading

Cook, G.A. *Survey of Modern Industrial Chemistry*. Ann Arbor: Ann Arbor Science, 1975.

Shreve, R.N. *Chemical Process Industries*, 4th Edition. New York: McGraw-Hill, 1977.

Appendix II

Assignments

Section A: Simple Computer Problems

1. Plan and write a program which: allows 50 values to be fed in; adds them up and prints out their sum; computes and prints out the sums of their squares, cubes, and square roots. Test the program with the first 50 telephone numbers in your local city directory (using seven digits, with a decimal point after the first three).

2. Plan and write a program to solve quadratic equations of the form

$$Ax^2 + Bx + C = 0$$

Test using $A = 7.31$; $B = -22.30$; $C = 14.36$.

3. Plan and write a program to find the mean of the squares of all odd numbers less than 1,000.

Section B: More Difficult Computer Problems

1. Develop a program to read in N values and find and print out the following:

a) the largest value;
b) the smallest value;
c) the average value;
d) the sum of their square roots;
e) their sum;
f) the sum of their squares;
g) the number of values which are odd.

2. Develop a program to read in N values, sort them into descending order, and print them out in rows each of which contains 10 values except, perhaps, the last row.

3. Develop a program to calculate curved grades from raw scores according to the following method:

a) read in the number of students, (N);
b) read in the raw scores of the class, (x_i);
c) arrange them in descending order;
d) find the median value, (x_m);
e) calculate, for each score, the difference between it and the median ($x_i - x_m$);
f) find the sum of the differences squared, $\Sigma[(x_i - x_m)^2]$;
g) find the Standard Deviation, $S = \Sigma[(x_i - x_m)^2]/N$;
h) determine the grade-letter breakpoints as follows:

A: $x_i \geqslant x_m + S$
B: $x_i \geqslant x_m$
C: $x_i \geqslant x_m - S$
D: $x_i \geqslant x_m - 2S$
F: $x_i < x_m - 2S$

i) print out the raw scores, the breakpoints for each grade letter, and the number of students who fall in each grade letter category.

4. Develop a program to calculate the temperature distribution in the section of furnace end-wall shown in Figure AII-1 using the method outlined in Figure AII-2.

5. A typical distillation column is shown in Figure AII-3. There are n stages. At stage 1 (the reboiler), V moles of liquid are evaporated and B moles of liquid are taken as bottoms product. The vapor rate is constant throughout the column. The liquid rate in the bottom of the column, L_B, is given by:

$$L_B = V + B$$

At stage f, F moles of liquid feed are introduced into the column. Above stage f, the liquid rate, L_T, is given by:

$$L_T = L_B - F = V + B - F$$

The vapor, V, from stage n is totally condensed. Some of it is taken as distillate product, D; the rest of it is returned to the column as reflux, R.

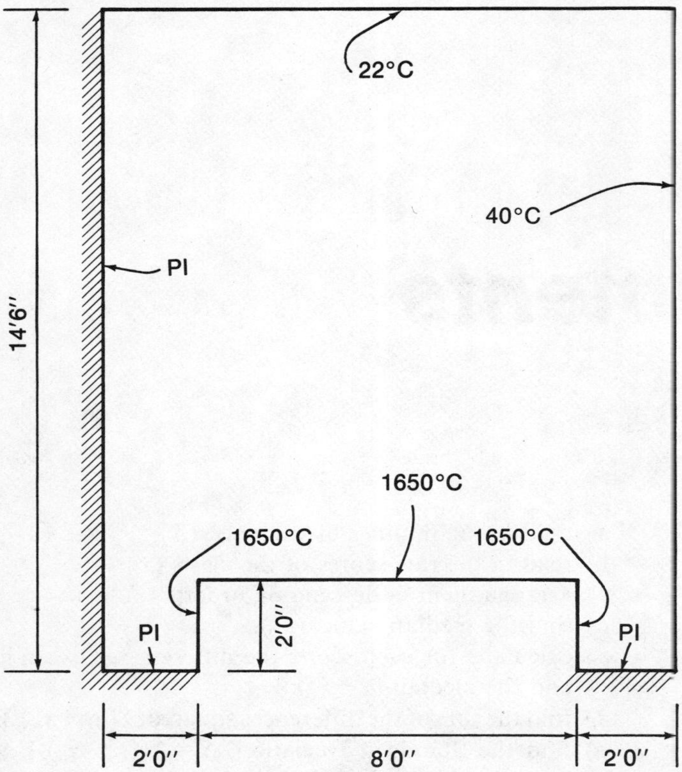

Key:

PI = this surface is assumed to be perfectly insulatec
1650°C = this surface is uniformly kept at 1650°C
40°C = this surface is uniformly kept at 40°C
22°C = this surface is uniformly kept at 22°C

Figure AII-1. *Diagram for Problem B-4.*

In a system containing two components (binary), the compositions of the liquid and vapor leaving stage i are given as the mole fractions of the more volatile component x_i and y_i, respectively. The composition of the bottoms product is x_l. The composition of the distillate product is y_n. The composition of the feed is XF. Ideally, the liquid composition of the feed stage, x_f, should be equal to the feed composition, XF. The feed is introduced at the stage where x_f is just greater than XF.

In a binary system, vapor/liquid equilibrium on each stage requires that

$$y_i = \frac{\alpha\, x_i}{1 + (\alpha - 1)\, x_i}$$

where α is the relative volatility of the two components.

If the bottoms composition is known, the liquid composition on each stage can be calculated from the following equations (based on conservation of mass):

For the bottom stage:

$$x_2 = \frac{V}{L_B}\, y_l + \frac{B}{L_B}\, x_l$$

For $1 < i < f$:

$$x_{i+1} = \frac{V}{L_B}\,(y_i - y_{i-1}) + x_i$$

For $i = f$:

$$x_{f+1} = \frac{V}{L_T}\,(y_f - y_{f-1}) + \frac{L_B}{L_T}\, x_f - \frac{F}{L_T} \cdot XF$$

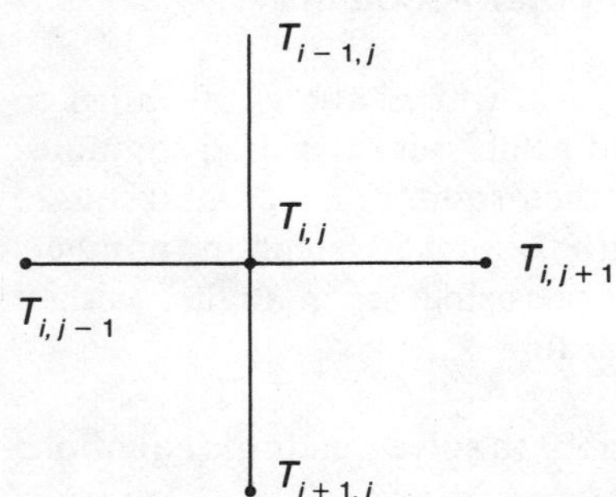

normal insulation

$$T_{i,j} = \frac{T_{i,j-1} + T_{i-1,j} + T_{i,j+1} + T_{i+1,j}}{4.0}$$

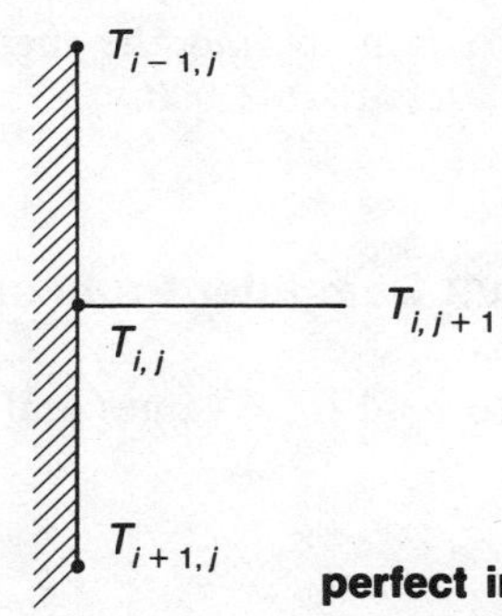

perfect insulation

$$T_{i,j} = \frac{T_{i-1,j} + T_{i+1,j} + 2\,T_{i,j+1}}{4.0}$$

1. Break down the area into small squares (suggest 6-inch squares)
2. Establish an array of temperatures at all node points
3. Guess values for unknown temperatures
4. Working from top left to bottom right (unknown temperatures only!), calculate a new temperature for each node point in terms of the surrounding temperatures
5. Repeat until two successive values are within 1%.

Figure AII-2. *Method of solution for Problem B-4.*

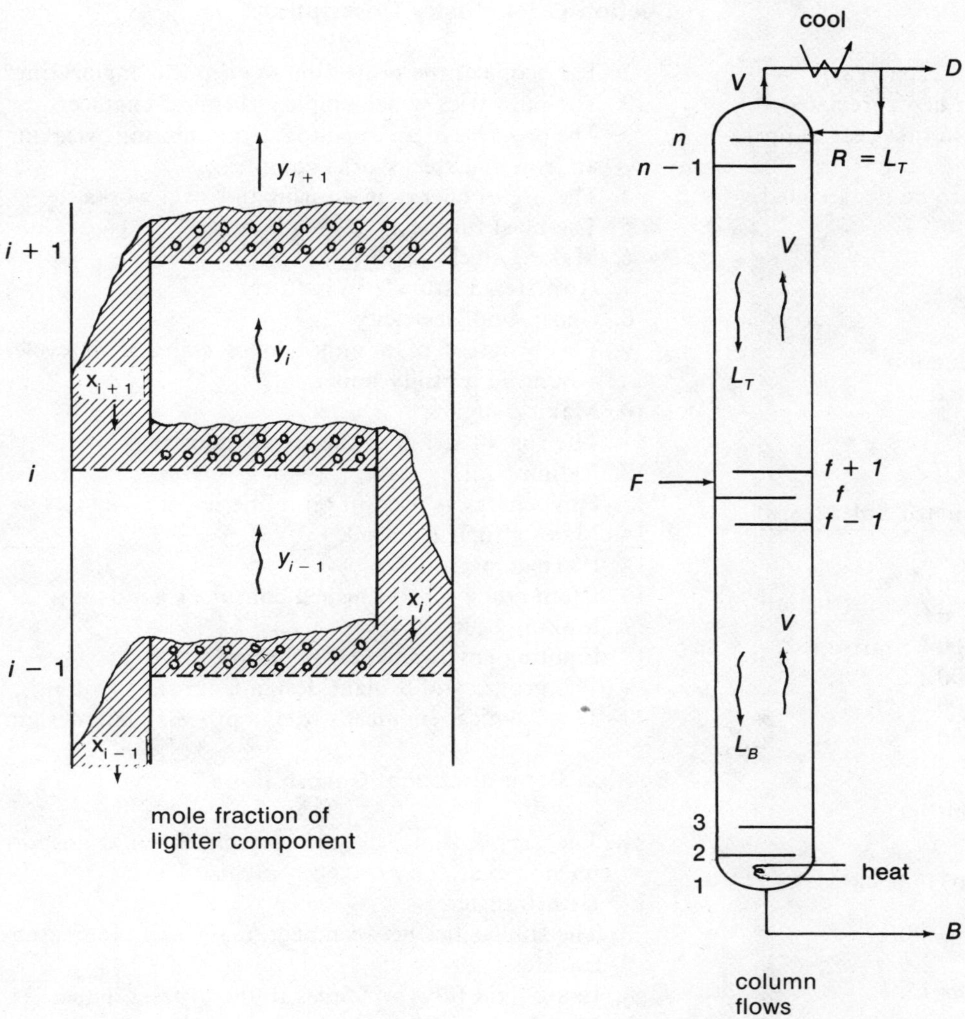

Figure AII-3. *Diagrams for Problem B-5.*

For $f < i < n$:

$$x_{i+l} = \frac{V}{L_T}(y_i - y_{i-l}) + x_i$$

The distillate, bottoms, and feed rates and compositions are related by the following equations:

$$B + D = F$$

$$x_l \cdot B + y_n \cdot D = XF \cdot F$$

If XF, F, x_l and y_n are given, you can calculate B and D.

Write a computer program that will accept α, F, XF, x_l, V, and y_n as input data and calculate the number of stages required by adding stages until y_i is just greater than y_n.

Output the input data, n, and gas and liquid compositions on each page.

With your program, solve the following cases.

Case	α	F	XF	x_l	y_n	V
1	1.2	100	0.5	0.01	0.99	800
2	1.2	100	0.5	0.10	0.95	800

6. Calculate the DCF Return for the 80,000 tons/year project outlined below:

Capital Employed

construction time:	2 years
capital cost:	$24,000,000 (half each year)
land cost:	$100,000 (not to be depreciated)
start-up cost:	$300,000 (spent in first year of operation)
working capital:	$1,000,000 (not to be depreciated)

Product Costs (excluding depreciation)

fixed costs: $1,700,000 per year
variable costs
raw material: $120/ton of production
all other costs: $28/ton of production

Sales Forecast at $370/Ton Market Price

Year of Operation	Estimated Sales (tons)	
1	0	Construction
2	0	
3	30,000	Start-up
4	40,000	
5	60,000	
6–14	80,000	

Depreciation: Assume 10-year straight line

Tax Rate: Federal Income Tax (FIT) is 48%.

Capital Recovery at Shutdown in Year 15:

land:	200% of purchase price
onsite and offsite facilities:	No value (scrap value = demolition costs)
working capital:	100% of value

Inflation: Assume there is *no* inflation

Schedule: Print out the cash flow table and DCF return for the above and do the same for these extra three cases:

Case #2—Use sum-of-years-digits depreciation
Case #3—Assume raw material cost increases by 10%
Case #4:—Assume market price drops by 10%

Section C: Industry Descriptions

1. The scope of the profession of chemical engineering.
2. The industries which employ chemical engineers.
3. The progress of an iron atom from entering to leaving an iron and steel works complex.
4. The use of energy in an iron and steel works.
5. The blast furnace.
6. Making steel.
7. How steam is made in industry.
8. Generating electricity.
9. The progress of a molecule of water as it cycles around in a utility boiler.
10. Making sugar.
11. The role of steam in the sugar industry.
12. Refining oil.
13. How energy is used in oil refineries.
14. Making fuels from oil.
15. Petroleum products.
16. From crude oil to finished consumer goods.
17. Making beer.
18. Fighting environmental pollution.
19. The progress of a plant design from start to finish.
20. The chemical engineer's role in process plant design.

Section D: Functional Descriptions

1. The application of the principles of transport phenomena.
2. Heat transfer.
3. The similarities between heat, mass, and momentum transfer.
4. Using fluid flow principles in the process industries.
5. Moving fluids.
6. The heat exchanger.
7. The use of heat and energy.
8. Distillation.
9. Reactors and their role in the process industries.
10. Why the petrochemical industry is sometimes called molecular architecture.
11. Boiling and evaporation.
12. Condensation applications.
13. Separation processes.
14. Filtration methods.
15. Taking suspended solids from liquids.
16. Economic evaluation of new chemical engineering projects.
17. The DCF method of project appraisal.
18. Exploiting basic scientific principles in chemical engineering.

Index